普通高等学校“十三五”规划教材

数值分析

Numerical Analysis

（第2版）

苏岐芳　主编

中国铁道出版社有限公司
CHINA RAILWAY PUBLISHING HOUSE CO., LTD.

内 容 简 介

本书介绍了科学计算中常用数值分析的基础理论及计算机实现方法。主要内容包括：误差分析、插值、函数逼近、数值积分和数值微分、非线性方程的数值解法、线性方程组的直接解法、线性方程组的迭代解法、常微分方程的数值解法及相应的上机实验内容等。各章都配有大量的习题及上机实验题目，并附有部分习题的参考答案及数学专业软件 Mathematica 和 Matlab 的简介。

本书采用中、英两种语言编写，适合作为数学、计算机和其他理工类各专业本科“数值分析（计算方法）”双语课程的教材或参考书，也可供从事科学计算的相关技术人员参考。

图书在版编目（CIP）数据

数值分析 = Numerical Analysis / 苏岐芳主编. —
2 版. —北京：中国铁道出版社，2017.2（2024.7 重印）
普通高等学校“十三五”规划教材
ISBN 978-7-113-22800-2

Ⅰ. ①数… Ⅱ. ①苏… Ⅲ. ①数值计算-高等学校-教材 Ⅳ. ①O241

中国版本图书馆 CIP 数据核字（2017）第 012289 号

书　　名：数值分析 Numerical Analysis
作　　者：苏岐芳

策　　划：李小军　　　　编辑部电话：（010）51873135
责任编辑：张文静　徐盼欣
封面设计：刘　颖
封面制作：白　雪
责任校对：张玉华
责任印制：樊启鹏

出版发行：中国铁道出版社有限公司（100054，北京市西城区右安门西街 8 号）
网　　址：https://www.tdpress.com/51eds/
印　　刷：北京铭成印刷有限公司
版　　次：2007 年 8 月第 1 版　2017 年 2 月第 2 版　2024 年 7 月第 3 次印刷
开　　本：710 mm×1 000 mm　1/16　印张：21.5　字数：431 千
书　　号：ISBN 978-7-113-22800-2
定　　价：39.80 元

第 2 版前言

本书第 1 版出版以来，得到了许多专家、同仁及读者的关心、支持和帮助，并提出了许多宝贵意见和建议。借再版之机，首先向关心本书的广大读者、专家、同行和本书的各位责任编辑表示由衷的谢意！

在修订中，为了更适合当前双语教学的需求，我们保留了原教材的系统和编写风格（理论部分以英文为主，软件实现部分以中文为主），注意吸收当前国内外教材改革中一些成功的经验，努力体现创新教学理念，以利于激发学生自主学习，提高实践应用能力，培养综合素质和创新能力。

本次再版修订的内容主要包括以下几方面:

1. 订正了语言文字表达方面的不足之处，力求用词规范，表达确切。

2. 剔除了个别内容重复和烦琐之处，使理论部分更好地体现“够用为度”的编写原则。

3. 恰当地处理有关定理的证明和有关例题的求解方法，使其更加通俗易懂。

4. 增补了多重积分、有理逼近、Padé 逼近等内容，进一步体现教材的先进性。

5. 结合增补内容，对习题配置作了进一步充实、完善。

6. 在实验部分，大量增加了算法的 Matlab 实现程序及相应的算例，以便于指导学生实践应用。

本书由浙江台州学院苏岐芳副教授主编，浙江台州学院郑学良教授、李希文副教授和应玮婷老师参与修订。具体写作分工为：第 1 章、第 2 章及附录由李希文修订；第 3 章由郑学良修订；第 4 章~第 8 章由苏岐芳修订；全书的计算机实验由应玮婷修订。

在本书修订过程中，浙江师范大学徐秀斌教授为本书提出了许多宝贵意见，在此表示衷心感谢！

编　者

2016 年 10 月

第 1 版前言

数值分析（计算方法），是研究数学问题数值解的构造性方法的一门科学。它既具有纯数学理论的抽象性和严谨性，更具有实用性和实验性的技术特征，是一门理论性和实践性都很强的学科。

随着计算机和计算方法的飞速发展，几乎所有学科都走向定量化和精确化，从而产生了一系列计算性的学科分支，如计算物理、计算化学、计算生物学、计算地质学、计算气象学和计算材料学等，计算数学中的数值计算方法则是解决“计算”问题的桥梁和工具。在科学研究和工程技术中都要用到各种计算方法。例如，在航天航空、地质勘探、汽车制造、桥梁设计、天气预报和汉字字样设计中都有计算方法的踪影。

在二十世纪七八十年代，大多数高校仅在数学系的计算数学专业和计算机系开设计算方法这门课程。现在，计算方法课程几乎已成为所有理工科学生的必修课程。在计算机上采用数值计算方法解决科学与工程计算中的实际问题已成为当今科学实验与理论研究的重要手段，利用计算机求解各类数学问题数值解的能力，已成为广大科技工作者、大学生和各类管理人员必备的能力。

本书是在作者多年来开设“数值分析”双语教学讲义的基础上，吸收国内外同类教材的精华，采用中英文双语编写而成。主要介绍了计算机中常用的、有效的各类数值问题的计算方法及相关数学理论。全书共分八章，包括误差分析、插值、函数逼近、数值积分和数值微分、非线性方程的数值解法、线性方程组的直接解法、线性方程组的迭代解法、常微分方程的数值解法等内容。每章由开篇、理论、应用和习题四个部分构成。

每章的开篇部分，给出中文内容提要和主要英语词汇解释，通过具体案例引入本章话题并提出本章的学习目标；理论部分以英文为主，重要的名词术语同时标注中文，从而**使读者在方便、有效地掌握知识的同时，渐进提高专业英语的阅读和应用能力。这是本书的第一个特点**。

每章的应用部分，采用算法分析、计算机示范程序、计算实例的结构安排，逐步引导读者将理论和方法转化为实际操作，并应用专业计算机软件实现更为复杂的数值计算，从而**培养读者利用计算机解决数学问题的能力**。这是本书的**第二个特点**。

每章还配备了大量的理论分析和计算机实验题目，书末附有部分习题参考答案及专业软件介绍，可以满足不同学习层次读者的学习需求；同时，还可以启发读者使用不同的计算机语言编写计算程序。**既体现了学习的自主性，又有利于培养读者的动手能力与创新能力**。这是本书的**第三个特点**。

本书由浙江台州学院苏岐芳主编，中欧国际工商学院苏岐英主审。各章编写情况如下：苏岐芳编写第1～5章，郑学良编写第6、7章，浙江海洋学院郝彦编写第8章，浙江台州学院李希文编写全书的计算机实验内容及附录。另外，浙江台州学院陈淑萍和浙江海洋学院朱玉辉对本书的编写也做了大量工作，在此一并致谢。

在本书的编写过程中，参阅了大量的文献，在此对这些作者表示衷心感谢！限于水平，书中错误和不足之处在所难免，恳请专家、教学同仁和广大读者批评指正，提出宝贵意见。

编　者

2007年2月

Contents

1 Error Analysis

提 要

本章主要介绍误差的来源、绝对误差和相对误差、误差的传播规律，以及数值计算中控制误差的一些原则等.

词 汇

accuracy	精度	integral	积分
absolute error	绝对误差	multiplication	乘法
algorithm	算法	polynomial	多项式
convergence	收敛性	round-off error	舍入误差
decimal	十进制（小数）	relative error	相对误差
discard	抛弃,舍弃	stability	稳定性
division	除法	series	级数
disturbance	扰动	significant digits	有效数字
elimination	消元	subtraction	减法
function	函数	truncation error	截断误差

1.1 Introduction

Although the computer is an ideal tool for performing complex numerical computations, casual or careless use of the output from a computer program can lead to highly undesirable **consequences** (推论). Indeed, one of the most common mistakes made by new users is to accept, almost as a matter of faith, the validity of numerical output produced by an operational computer program. A relatively begin **manifestation** (显示) of this phenomenon arises when one attributes more **precision** (精确) to a numerical output than the accuracy of the input data or the underlying mathematical model justifies.

In other instances, a value produced by an operational computer program can be totally meaningless in the sense that is not accurate to even one digit. This can result from an **accumulation** (累积) of round-off error due to the way a calculation is structured. As a simple illustration, consider the following **quadratic equation** (二次方程).

$$ax^2+bx+c=0 \tag{1.1.1}$$

Note that this equation is labeled (1.1.1) where the first digit identifies the chapter, the second digit identifies the section within in the chapter, and the last digit identifies the equation within in the section.

Suppose we want to find the two roots of (1.1.1). Using the well-known quadratic formula, the roots are

$$x_{1,2}=\frac{-b\pm\sqrt{b^2-4ac}}{2a} \tag{1.1.2}$$

To make the problem more specific, suppose the parameters are $a=1, b=-(10^4+10^{-4})$, and $c=1$. Carrying out the calculation in (1.1.2) yields the roots $x_1=10^4$ and $x_2=10^{-4}$. Interestingly enough, if we perform this simple calculation on a computer that has a precision of seven decimal digits (not uncommon), then the results are $x_1\approx 10^4$ and $x_2=0$. That is, the first root is easily obtained, but the error in the second root is 100 percent! This is a consequence of accumulated round-off error.

When you finish this chapter, you will know how the principal sources of error in numerical computations arise, including round-off error and formula truncation error. You will know how to control the **propagation** (传播) of errors. You will understand the significant digits of approximate number. You will achieve these overall goals by mastering the following chapter objectives.

Objectives

- Know how to define, calculate, and use the absolute and relative errors.
- Understand how round-off error and truncation error occur.
- Be able to analyze error propagation in basic arithmetic operations.
- Be able to specify bounds on the size of formula truncation error.
- Know how to choose proper algorithm for some mathematical model.

1.2 Sources of Errors（误差的来源）

Assessing (估计) the accuracy of the results of calculations is a paramount goal in numerical analysis. One distinguishes several kinds of errors which may limit this accuracy:

1. model errors（模型误差）

When we use computer to calculate a mathematical problem, we first set up a mathematical model which is **abstraction** (抽象化) and **simplification** (简单化) of the described practical problem, it is approximate. We call the error between mathematical model and practical problem **model error.**

Example 1.2.1

$$s(t)=\frac{1}{2}gt^2,\quad g\approx 9.812$$

2．Errors in the input data（输入数据误差）

Input data errors are beyond the control of the calculation. They may be due, for instance, to the inherent imperfections of physical measurements.

3．Round-off errors（舍入误差）

Round-off errors arise if one calculates with numbers whose representation is restricted to a finite number of digits, as is usually the case.

Rounding is an important concept in scientific computing. Consider a positive decimal number x of the form 0.□□□···□□□ with m digits to the right of the **decimal point** (小数点). One rounds x to n decimal places ($n<m$) in a manner that depends on the value of the (n+1)st digit. If this digit is a 0, 1, 2, 3, or 4, then the nth digit is not changed and all following digits are **discarded** (抛弃,舍弃). If it is a 5, 6, 7, 8 or 9, then the nth digit is increased by one unit and the remaining digits are discarded.

Here are some examples of seven-digit numbers being correctly rounded to four-digits

$$0.162\,5\leftarrow 0.162\,548\,9$$

$$1.000\quad\leftarrow 0.999\,960\,1$$

$$0.623\,3\leftarrow 0.623\,270\,9$$

If x is rounded so that $\overline{x}$ is the n-digit approximation to it, then

$$\left|x-\overline{x}\right|\leqslant\frac{1}{2}\times 10^{-n}$$

Another example, $\pi=3.141\,592\,6\cdots$, if we take $\pi\approx 3.14$, then the round-off error is $3.14-3.141\,592\,6\cdots=-0.001\,592\,6\cdots$.

4．Truncation errors（截断误差）

As for the fourth kind of the error, many methods will not yield the exact solution of the given problem, even if the calculations are carried out without rounding, but rather the solution of another simpler problem $\overline{P}$ which approximates P. For instance, the problem P of summing an infinite series, e.g.(例如),

$$\sin x=x-\frac{x^3}{3!}+\frac{x^5}{5!}-\frac{x^7}{7!}+\cdots$$

may be replaced by the simpler problem $\overline{P}$ of summing only up to a finite number of terms of the series.

$$\sin x\approx x-\frac{x^3}{3!}+\frac{x^5}{5!}-\frac{x^7}{7!}$$

The resulting approximation error is commonly called a **truncation error**.

Generally, $f(x)$ can be approximately replaced by **Taylor polynomial** (泰勒多项式)

$$P_n(x)=f(0)+\frac{f'(0)}{1!}x+\frac{f''(0)}{2!}x^2+\cdots+\frac{f^{(n)}(0)}{n!}x^n$$

the truncation error is

$$R_n(x)=f(x)-P_n(x)=\frac{f^{(n+1)}(\xi)}{(n+1)!}x^{n+1}$$

where ξ is between 0 and x .In this book,"ξ is between a and b" means that either $a<\xi<b$ or $b<\xi<a$.

1.3 Errors and Significant Digits（误差和有效数字）

Definition 1.3.1 If a real number x^* approximates to another number x, then the **absolute error (or error)** (绝对误差) is defined by

$$e=x^*-x$$

where e may be positive or negative number. If the absolute error is positive, the **approximate value** (近似值) is greater than the **exact value** (准确值), it is called **strong approximate value** . If the absolute error is negative, the approximate value is smaller than the exact value, it is called **weak approximate value** .

If there exists a positive number ε, such that

$$|e|=|x^*-x|\leqslant\varepsilon$$

Then ε is called the **limit of the (absolute) error** (绝对误差限). That is

$$x^*-\varepsilon\leqslant x\leqslant x^*+\varepsilon \quad \text{or} \quad x=x^*\pm\varepsilon$$

Example 1.3.1 The representation $x=0.310\,6\pm0.001\,4$ implies

$$0.310\,6-0.001\,4\leqslant x\leqslant 0.310\,6+0.001\,4$$

That is

$$0.309\,2\leqslant x\leqslant 0.312\,0$$

Example 1.3.2 We measure the length of an object with **mm-graduation ruler** (毫米单位刻度尺), then

$$|e|=|l^*-l|\leqslant\frac{1}{2}\text{ mm}$$

If the reading is $l^*=513\text{ mm}$, then $512.5\text{ mm}\leqslant l\leqslant 513.5\text{ mm}$.

Example 1.3.3 Discuss the calculation of e^{-x} for $0<x<1$ from the **series** (级数)

$$e^{-x}=1-\frac{x}{1!}+\frac{x^2}{2!}-\frac{1}{3!}x^3\cdots+\frac{(-x)^n}{n!}+\cdots$$

Solution Let

$$E(x)=1-\frac{x}{1!}+\frac{x^2}{2!}-\frac{1}{3!}x^3$$

Then

$$\left|e^{-x} - E(x)\right| \leqslant \frac{1}{24}$$

The limit of the error is $\frac{1}{24}$.

Definition 1.3.2 If a real number x^* approximates to another number x, then the **relative error** (相对误差) of x^* is

$$e_r = \frac{e}{x} = \frac{x^* - x}{x}$$

Frequently

$$e_r = \frac{e}{x^*} = \frac{x^* - x}{x^*}$$

If there exists a positive number ε_r, such that

$$|e_r| \leqslant \varepsilon_r$$

Then ε_r is called the **limit of relative error** (相对误差限).

Example 1.3.4 If $x = 10 \pm 1$, $y = 1000 \pm 5$

Then

$$x^* = 10,\ \varepsilon(x) = 1,\ y^* = 1\,000,\ \varepsilon(y) = 5$$

$$\varepsilon_r(x) = \frac{\varepsilon(x)}{|x^*|} = 10\%\ ,\quad \varepsilon_r(y) = \frac{\varepsilon(y)}{|y^*|} = 0.5\%$$

Example 1.3.5 $x = \pi = 3.141\,592\,65\cdots$.

If $x_3^* = 3.14$, then $\varepsilon_3^* \leqslant 0.002$; if $x_5^* = 3.141\,6$, then $\varepsilon_5^* \leqslant 0.000\,05$. We can see that the errors do not exceed the half unit of their end places . i.e.,

$$|\pi - 3.14| \leqslant \frac{1}{2} \times 10^{-2},\quad |\pi - 3.141\,6| \leqslant \frac{1}{2} \times 10^{-4}$$

Definition 1.3.3 If the limit of the error of approximate value x^* is a half unit of some place, and there are *n* digits from this place to the first nonzero digit of the front of x^*, then we say x^* has *n* **significant digits** (or **figures**).

x^* can be represented:

$$x^* = \pm 10^m \times (a_1 + a_2 \times 10^{-1} + \cdots + a_n \times 10^{-(n-1)}) \tag{1.3.1}$$

where $a_i \in \{0,1,2,\cdots,9\} (i = 1,\cdots,n)$, and $a_1 \neq 0$, m is an **integer** (整数), furthermore

$$|x - x^*| \leqslant \frac{1}{2} \times 10^{m-n+1} \tag{1.3.2}$$

e.g.,

$$31.415 = 10^1 \times (3 + 1 \times 10^{-1} + 4 \times 10^{-2} + 1 \times 10^{-3} + 5 \times 10^{-4})$$

where $m = 1, n = 5$, $|x - x^*| \leqslant \frac{1}{2} \times 10^{-3}$.

Example 1.3.6 Let $x^* = 3.14$ be the approximate value of π, then it has three significant digits. Let $x^* = 3.141\,6$ be the approximate value of π, then it has five significant digits.

Example 1.3.7 Write the approximate numbers with five significant digits of the following numbers by rounding-off

$$187.932\,5, \quad 0.037\,856, \quad 8.000\,0, \quad 2.718\,281\,8$$

Solution The results are respectively

$$187.93, \quad 0.037\,856, \quad 8.000\,0, \quad 2.718\,3$$

Example 1.3.8 Let $g \approx 9.80\ \mathrm{m/s^2}$, $g \approx 0.009\,80\ \mathrm{km/s^2}$. Then they have three significant digits, and

$$|g-9.80| \leqslant \frac{1}{2}\times 10^{-2}, \quad \text{where } m=0, n=3.$$

$$|g-0.009\,80| \leqslant \frac{1}{2}\times 10^{-5}, \quad \text{where } m=-3, n=3.$$

Their limits of the absolute errors are different,

$$\varepsilon_1 = \frac{1}{2}\times 10^{-2}\,\mathrm{m/s^2}, \quad \varepsilon_2 = \frac{1}{2}\times 10^{-5}\,\mathrm{km/s^2}$$

and their relative errors are equal,

$$\varepsilon_{\mathrm{r}} = 0.005/9.8 = 0.000\,005/0.009\,80$$

Remarks 1.3.1

(1) The number of significant digits is not relative to the place of the decimal point, e.g., both 3.14 and 314 have three significant digits.

(2) From $|x-x^*| \leqslant \frac{1}{2}\times 10^{m-n+1}$, we obtain that the more significant digits an approximate number has, the smaller limit of the absolute error it has.

(3) Zero at the end of an approximate number can not be **rounded down** (舍去).

Theorem 1.3.1 Let approximate number x^* be represented by

$$x^* = \pm 10^m \times (a_1 + a_2\times 10^{-1} + \cdots + a_n \times 10^{-(n-1)})$$

where $a_i (i=1,\cdots,n)$ is one of the digits from 0 to 9, $a_1 \neq 0$, m is an integer . If x^* has n significant digits , then its limit of the relative error satisfies

$$\varepsilon_{\mathrm{r}} \leqslant \frac{1}{2a_1}\times 10^{-(n-1)}$$

Contrarily, if the limit of the relative error of x^* satisfies

$$\varepsilon_{\mathrm{r}} \leqslant \frac{1}{2(a_1+1)}\times 10^{-(n-1)}$$

then x^* has n significant digits at least.

Proof Using (1.3.1) $x^* = \pm 10^m \times (a_1 + a_2\times 10^{-1} + \cdots + a_n \times 10^{-(n-1)})$

we obtain

$$a_1 \times 10^m \leqslant |x^*| < (a_1+1)\times 10^m$$

If x^* has n significant digits, then

$$\varepsilon_r = \frac{|x - x^*|}{|x^*|} \leqslant \frac{0.5 \times 10^{m-n+1}}{a_1 \times 10^m} = \frac{1}{2a_1} \times 10^{-n+1}$$

Contrarily, if

$$\varepsilon_r \leqslant \frac{1}{2(a_1+1)} \times 10^{-(n-1)}$$

from

$$a_1 \times 10^m < |x^*| < (a_1+1) \times 10^m$$

$$|x - x^*| = |x^*|\varepsilon_r < (a_1+1) \times 10^m \times \frac{1}{2(a_1+1)} \times 10^{-n+1} = 0.5 \times 10^{m-n+1}$$

Hence x^* has n significant digits at least. □

Example 1.3.9 By how many significant digits of the approximate value of $\sqrt{20}$ has such that its limit of the relative error is less than 0.1% ?

Solution Suppose it has n significant digits , From Theorem 1.3.1

$$\varepsilon_r \leqslant \frac{1}{2a_1} \times 10^{-n+1}$$

Let $n = 4$, then

$$\varepsilon_r \leqslant 0.125 \times 10^{-3} < 10^{-3} = 0.1\%$$

If we take four significant digits of the approximate value of $\sqrt{20}$, then its limit of the relative error is less than 0.1%. Now we can obtain $\sqrt{20} \approx 4.472$ from the **table of square roots** (平方根表).

Example 1.3.10 Determine the relative error if π is represented approximately by 3.141 6?

Solution 3.141 6 has five significant digits ,

$$n = 5, \quad a_1 = 3$$

therefore

$$|e_r| \leqslant \frac{1}{2 \times 3} \times 10^{-5+1} = \frac{1}{6} \times 10^{-4}$$

Example 1.3.11 By how many significant digits of the approximate value of $\sqrt{70}$ has such that its limit of the relative error is less than 0.1% ?

Solution Because

$$8 < \sqrt{70} < 9, \qquad a_1 = 8$$

Therefore

$$\frac{1}{2a_1} \times 10^{-n+1} = \frac{1}{2 \times 8} \times 10^{-n+1} < 0.1\%$$

to obtain $n = 3$. Now we have $\sqrt{70} \approx 8.37$ from the table of square roots.

Example 1.3.12 Measuring a length $x^* = 954\ \text{cm}$, determine its relative error.

Solution When

$$x^* = 954\ \text{cm}, \quad \varepsilon(x) = 0.5\ \text{cm}$$

Therefore

$$\varepsilon_r(954) = \frac{0.5}{954} \approx 0.000\,524\,1 < 0.052\,5\%$$

1.4 Error Propagation（误差的传播）

There are two approximate numbers x_1^* and x_2^*, their limits of error are respectively $\varepsilon(x_1^*)$ and $\varepsilon(x_2^*)$, then

$$\varepsilon(x_1^* \pm x_2^*) \leqslant \varepsilon(x_1^*) + \varepsilon(x_2^*)$$

$$\varepsilon(x_1^* x_2^*) \leqslant |x_1^*|\varepsilon(x_2^*) + |x_2^*|\varepsilon(x_1^*)$$

$$\varepsilon(x_1^* / x_2^*) \leqslant \frac{|x_1^*|\varepsilon(x_2^*) + |x_2^*|\varepsilon(x_1^*)}{|x_2^*|^2} \quad (x_2^* \neq 0)$$

Generally, the **functional value** (函数值) can yield error when its **independent variables** (自变量) have errors. Suppose that $f(x)$ is a function with one independent variable, x^* is an approximate value of x, and $f(x^*)$ is an approximate value of $f(x)$, whose limit of the error is denoted by $\varepsilon(f(x^*))$. By using **Taylor expansion** (泰勒展开式), we obtain

$$f(x) - f(x^*) = f'(x^*)(x - x^*) + \frac{f''(\xi)}{2}(x - x^*)^2$$

where ξ is between x and x^*,

$$|f(x) - f(x^*)| \leqslant |f'(x^*)|\varepsilon(x^*) + \frac{|f''(\xi)|}{2}\varepsilon^2(x^*)$$

We suppose that $|f''(x^*)| / |f'(x^*)|$ is small and the higher terms of $\varepsilon(x^*)$ are ignored, then $\varepsilon(f(x^*)) \approx |f'(x^*)|\varepsilon(x^*)$.

If f is a **function with several variables** (多元函数), e.g., $A = f(x_1, \cdots, x_n)$. The approximate values of $x_1, \cdots, x_n$ are respectively $x_1^*, \cdots, x_n^*$. The approximate value of A is $A^* = f(x_1^*, \cdots, x_n^*)$, then

$$e(A^*) = A^* - A = f(x_1^*, \cdots, x_n^*) - f(x_1, \cdots, x_n)$$

$$\approx \sum_{k=1}^{n} \left(\frac{\partial f(x_1^*, \cdots, x_n^*)}{\partial x_k} \right)(x_k^* - x_k) = \sum_{k=1}^{n} \left(\frac{\partial f}{\partial x_k} \right)^* e_k$$

$$\varepsilon(A^*) \approx \sum_{k=1}^{n} \left| \left(\frac{\partial f}{\partial x_k} \right)^* \right| \varepsilon(x_k)$$

$$\varepsilon_r(A^*) = \frac{\varepsilon(A^*)}{|A^*|} \approx \sum_{k=1}^{n} \left| \left(\frac{\partial f}{\partial x_k} \right)^* \right| \frac{\varepsilon(x_k)}{|A^*|}$$

Example 1.4.1 Measuring someplace, we obtain that its length is $l^* = 110\text{ m}$, its width is $d^* = 80\text{ m}$, and $|l - l^*| \leqslant 0.2\text{ m}$, $|d - d^*| \leqslant 0.1\text{ m}$, compute the limits of the absolute and relative errors of its area $s = ld$.

Solution Since

$$s = ld,\quad \frac{\partial s}{\partial l} = d,\quad \frac{\partial s}{\partial d} = l$$

$$\varepsilon(s^*) \approx \left|\left(\frac{\partial s}{\partial l}\right)^*\right| \varepsilon(l^*) + \left|\left(\frac{\partial s}{\partial d}\right)^*\right| \varepsilon(d^*)$$

where

$$\left(\frac{\partial s}{\partial l}\right)^* = d^* = 80\ \text{m},\quad \left(\frac{\partial s}{\partial d}\right)^* = l^* = 110\ \text{m}$$

and
$$\varepsilon(l^*) = 0.2\ \text{m},\quad \varepsilon(d^*) = 0.1\ \text{m}$$

Therefore

$$\varepsilon(s^*) \approx 80 \times 0.2 + 110 \times 0.1 = 27\ \text{m}^2$$

$$\varepsilon_{\text{r}}(s^*) = \frac{\varepsilon(s^*)}{|s^*|} = \frac{\varepsilon(s^*)}{l^* d^*} \approx \frac{27}{8\,800} \approx 0.31\%$$

Example 1.4.2 Let $I = (10 \pm 0.1)\ \text{A}$, $V = (220 \pm 2)\ \text{V}$, solve R and estimate its absolute and relative errors.

Solution Since $R = \dfrac{V}{I},\ R^* = \dfrac{V^*}{I^*} = \dfrac{220}{10} = 22\ \Omega$

$$e(R^*) \approx \left(\frac{\partial R}{\partial V}\right)^* e(V^*) + \left(\frac{\partial R}{\partial I}\right)^* e(I^*) = \frac{1}{I^*} e(V^*) - \frac{V^*}{(I^*)^2} e(I^*)$$

and

$$V^* = 220\ \text{V},\ |e(V^*)| \leqslant 2\ \text{V};\quad I^* = 10\ \text{A},\ |e(I^*)| \leqslant 0.1\ \text{A}$$

Therefore

$$|e(R^*)| \leqslant \left|\frac{1}{I^*}\right| \cdot |e(V^*)| + \left|\frac{V^*}{(I^*)^2}\right| \cdot |e(I^*)| \leqslant \frac{1}{10} \times 2 + \frac{220}{(10)^2} \times 0.1 = 0.42\ \Omega$$

$$|e_{\text{r}}(R^*)| = \left|\frac{e(R^*)}{R^*}\right| \leqslant \frac{0.42}{22} \approx 0.019\ 1 = 1.91\%$$

1.5 Qualitative Analysis and Control of Errors （定性分析和误差控制）

1.5.1 Ill-condition Problem and Condition Number（病态问题与条件数）

When we compute a functional value $f(x)$, if x has disturbance (error) $\Delta x = x - x^*$, which relative error is $\dfrac{\Delta x}{x}$, then the relative error of function $f(x)$ is $\dfrac{f(x) - f(x^*)}{f(x)}$, the ratio of the relative errors

$$\left|\frac{f(x)-f(x^*)}{f(x)}\right| \Big/ \left|\frac{\Delta x}{x}\right| \approx \left|\frac{xf'(x)}{f(x)}\right| = C_p$$

is called **condition number of computing the functional value.** If the condition number C_p is very big, then it gives rise to very big relative error of the functional value, this problem is called **ill-conditioned problem.**

For example, if $f(x)=x^n$, then $C_p = \left|\frac{xf'(x)}{f(x)}\right| = \left|\frac{x \cdot nx^{n-1}}{x^n}\right| = n$, it shows that the relative error enlarge n times. If $n=10$, then $f(1)=1, f(1.02)\approx 1.24$. Let $x=1$, $x^*=1.02$, the relative error of independent variable is 2% , and the relative error of the functional value is 24 %. Now this problem can be regarded as ill-conditioned problem.

If the condition number $C_p \geqslant 10$, then it is regarded as ill-conditioned.

1.5.2 The Stability of Algorithm（算法的稳定性）

Example 1.5.1 Compute integral

$$E_n = \int_0^1 x^n e^{x-1} dx \quad (n=1,2,\cdots)$$

Solution 1 From the definition it is easy to see that

$$E_1 > E_2 > \cdots > E_{n-1} > E_n > \cdots > 0$$

To obtain a recursion, integrate by parts to get

$$E_n = x^n e^{x-1}\Big|_0^1 - n\int_0^1 x^{n-1} e^{x-1} dx = 1 - nE_{n-1}$$

The first member of the family is

$$E_1 = 1 - \int_0^1 e^{x-1} dx = e^{-1}$$

and from it we can easily compute any E_n. If this is done in single precision on a PC, it is found the results in Table 1-1.

A little analysis helps us understand what is happening. Suppose we had started with $E_1^* = E_1 + \varepsilon$ and made no arithmetic errors when evaluating the recurrence. Then

$$E_2^* = 1 - 2E_1^* = 1 - 2E_1 - 2\varepsilon = E_2 - 2!\varepsilon$$

$$E_3^* = 1 - 3E_2^* = 1 - 3E_2 + 6\varepsilon = E_3 + 3!\varepsilon$$

……

$$E_n^* = E_n \pm n!\varepsilon$$

A small change in the first value E_1 grows very rapidly in the later E_n. The effect is worse in a relative sense because the desired quantities E_n decrease as n increases.

Solution 2 For this example there is a way to get a stable algorithm. If we could find an approximation E_n^* to E_n for some n, we could evaluate the recursion in reverse order

$$E_{n-1} = \frac{1-E_n}{n}$$

to approximate $E_{n-1}, E_{n-2}, \cdots, E_1$. Studying the stability of this recursion as before, if

$E_n^* = E_n + \delta$, then

$$E_{n-1}^* = \frac{1-E_n^*}{n} = \frac{1-E_n}{n} - \frac{\delta}{n} = E_{n-1} - \frac{\delta}{n}$$
$$E_{n-2}^* = E_{n-2} + \frac{\delta}{n(n-1)}$$
$$\cdots\cdots$$
$$E_1^* = E_1 \pm \frac{\delta}{n!}.$$

The recursion is so cheap and the error damps out so quickly that we can start with a poor approximation E_n^* for some large n and get accurate answers in expensively for the E_n that really interest us. Notice that recurring in this direction, the E_n increase, making the relative errors damps out even faster. The inequality

$$0 < E_n = \int_0^1 x^n e^{x-1} dx < \int_0^1 x^n dx = \frac{1}{n+1} \to 0 \quad (n \to \infty)$$

shows how to easily get an approximation to E_n with an error that we can bound. For example, if we take $n = 20$, the crude approximation $E_{20}^* = 0$ has an absolute error less than $1/21$ in magnitude. The magnitude of the absolute error in E_{19}^* is then less than $1/(20 \times 21) = 0.002\,4,\ldots,$ and that in E_{15}^* is less than 4×10^{-8}. The approximations to $E_{14}, \cdots, E_1$ will be even more accurate. See Table 1-2.

Table 1-1 Some results of integral

k	E_k
1	0.367 879
2	0.264 242
3	0.207 274
4	0.170 904
5	0.145 480
6	0.127 120
7	0.216 0
8	−0.728 0
⋮	⋮

Table 1-2 Some rseults of integral

k	E_k
20	0.000 000 0
19	0.050 000 0
18	0.050 000 0
17	0.052 777 8
16	0.055 719 0
15	0.059 017 6
14	0.062 732 2
13	0.066 947 7
12	0.071 773 3
11	0.077 352 3
10	0.083 877 1
9	0.091 612 3

For an algorithm, if the input data have errors, and the round-off errors do not increase during computing, then this algorithm is called **numerical stable** (数值稳定的). Otherwise, this algorithm is called **unstable**(不稳定的).

1.5.3 The Control of Errors

1. Avoiding the divisor (除数) too small

Since

$$e\left(\frac{x_1}{x_2}\right) \approx \left(\frac{\partial f}{\partial x_1}\right)^* e(x_1^*) + \left(\frac{\partial f}{\partial x_2}\right)^* e(x_2^*) = \frac{1}{x_2^*} e(x_1^*) - \frac{x_1^*}{(x_2^*)^2} e(x_2^*)$$

$$= \frac{x_1^*}{x_2^*}\left(e_r(x_1^*) - e_r(x_2^*)\right)$$

Therefore, it is necessary to avoid the divisor too small .

2. Avoiding the subtraction of two nearly equal numbers

$$e_r(x_1 - x_2) \approx \frac{x_1^*}{x_1^* - x_2^*} e_r(x_1) - \frac{x_2^*}{x_1^* - x_2^*} e_r(x_2)$$

$$|e_r(x_1 - x_2)| \leqslant \left|\frac{x_1^*}{x_1^* - x_2^*}\right| |e_r(x_1)| + \left|\frac{x_2^*}{x_1^* - x_2^*}\right| |e_r(x_2)|$$

Therefore, it is necessary to avoid the subtraction of two nearly equal numbers.

Example 1.5.2 Finding the smallest positive root of the following equation $x^2 - 16x + 1 = 0$.

Solution The quadratic formula states that the roots of $ax^2 + bx + c = 0$, when $a \neq 0$, are

$$x_1 = \frac{-b + \sqrt{b^2 - 4ac}}{2a} \quad \text{and} \quad x_2 = \frac{-b - \sqrt{b^2 - 4ac}}{2a}$$

This formula is applied to the equation $x^2 - 16x + 1 = 0$, whose roots are approximately

$$x_1 = 8 + \sqrt{63} \quad \text{and} \quad x_2 = 8 - \sqrt{63} \approx 8 - 7.94 = 0.06 = x_2^*$$

x_2^* has only one significant digit.

If $x_2 = 8 - \sqrt{63} \approx 8 - 7.937\,25 \approx 0.062\,7$, then it has 3 significant digits.

Example 1.5.3 Calculate $A = 10^7(1 - \cos 2^\circ)$, by using mathematical table with 4 units .

Solution 1 Since $\cos 2^\circ \approx 0.999\,4$

Therefore

$$A = 10^7(1 - \cos 2^\circ) \approx 10^7(1 - 0.999\,4) = 6 \times 10^3$$

It has only one significant digit.

Solution 2

$$A = 10^7(1 - \cos 2^\circ) = 2 \times 10^7 \sin^2 1^\circ$$

$$\approx 2 \times 10^7 \times 0.017\,5^2 = 6.13 \times 10^3$$

It has 3 significant digits. $(\sin 1^\circ \approx 0.017\,5)$

We can avoid loss of significant digits by changing the computational formula. If x_1 is nearly to x_2, then

$$\lg x_1 - \lg x_2 = \lg \frac{x_1}{x_2}$$

$$\sqrt{x+1}-\sqrt{x}=\frac{1}{\sqrt{x+1}+\sqrt{x}} \quad \text{(Let } x \text{ be very big)}$$

In general, if $f(x)\approx f(x^*)$, then

$$f(x)-f(x^*)=f'(x^*)(x-x^*)+\frac{f''(x^*)}{2}(x-x^*)^2+\cdots$$

Example 1.5.4 For the smaller positive value of x, calculate e^x-1.

Solution Since

$$e^x=1+x+\frac{x^2}{2!}+\frac{x^3}{3!}+\cdots$$

Therefore

$$e^x-1=x\left(1+\frac{x}{2!}+\frac{x^2}{3!}+\cdots\right)$$

3. Avoiding big number eats small number

Example 1.5.5 Solve the equation $x^2-(10^9+1)x+10^9=0$.

Solution The exact solution is $x_1=10^9, x_2=1$. If we use the quadratic formulas of Example 1.5.3 to operate on computer with 8 wordlength and decimal system, then

$$\begin{aligned}-b&=10^9+1=0.1\times10^{10}+0.000\,000\,000\,1\times10^{10}\\&\approx0.1\times10^{10}=10^9\end{aligned}$$

Hence

$$\sqrt{b^2-4ac}=\sqrt{\left(-(10^9+1)\right)^2-4\times10^9}\approx10^9$$

$$x_1=\frac{-b+\sqrt{b^2-4ac}}{2a}\approx\frac{10^9+10^9}{2}=10^9$$

$$x_2=\frac{-b-\sqrt{b^2-4ac}}{2a}\approx\frac{10^9-10^9}{2}=0$$

x_1 is reliable, we calculate x_2 by using the formula

$$x_1x_2=\frac{c}{a},\quad x_2=\frac{c}{a}\cdot\frac{1}{x_1}=\frac{10^9}{1\times10^9}=1$$

Remark 1.5.1

For a quadratic equation, we first solve the bigger root of absolute value by using the quadratic formula and then solve the second root by using the relations between the roots and coefficients.

4. Reduce (缩减) the number of operations

Example 1.5.6 Calculate $P(x)=a_nx^n+a_{n-1}x^{n-1}+\cdots+a_1x+a_0$.

Solution 1 Calculate $a_ix^i\,(i=0,1,2,\cdots,n)$ and then add all terms. The number of multiplications it needs is $1+2+\cdots+(n-1)+n=\frac{n(n+1)}{2}$.

If $n=10$, then the number of additions is 10, the number of multiplications is 55.

Solution 2 Method of Qinjiushao (秦九韶算法)

$$P(x)=((\cdots((a_n x+a_{n-1})x+a_{n-2})x+\cdots+a_2)x+a_1)x+a_0$$

It needs n multiplications.

If $n=10$, then the number of additions is 10, the number of multiplications is 10.

1.6 Computer Experiments

1.6.1 Functions Needed in the Experiments by Mathematica

1．变量替换（见表 1-3）

表 1-3 变量替换函数

函数	意义
expr/.x->value	用value 替代表达式expr中的x
expr/.{x->xval,y->yval}	分别用xval、yval 替代expr中的 x,y

2．方程求根（见表 1-4）

表 1-4 方程求根函数

函数	意义
Solve[lhs=rhs,vars]	给出方程的解集
NSolve[lhs=rhs,vars]	直接给出方程的数值解集
Roots[lhs=rhs,vars]	求表达式的根
FindRoot[lhs=rhs,{x,x0}]	求初值x=x0时，方程的解值

3．近似计算

在 Mathematica 中，对数值运算的结果，系统总是以尽可能精确的形式输出.使用 N 函数，可以输出所期望的近似结果. N 函数的使用有三种格式（见表 1-5）.

表 1-5 近似计算函数

格式	意义
<expr>//N	计算表达式expr的机器近似值，通常输出6位有效数字
N[<expr>]	计算表达式expr的机器近似值，通常输出6位有效数字
N[<expr>, n]	计算表达式expr的值，输出n位有效数字

1.6.2 Experiments by Mathematica

1．近似计算

Example 1.6.1 Solve approximate value to $\sqrt{3}$.

```
In[1]:=Sqrt[3.0]//N
Out[1]=1.73205
In[2]:=N[Sqrt[3.0]]
```

```
Out[2]=1.73205
In[3]:=N[Sqrt[3.0], 21]
Out[3]=1.73205080756887729353
In[4]:=N[Sqrt[3.0], 50]
Out[4]=1.7320508075688772935274463415058723669428052538104
```

说明：

（1）当输入参数为整数时，系统认为是精确的，输出结果也会是精确的；若输入的内容中含小数，则系统输出的计算结果也会是近似的；

（2）系统可以对输出结果进行化简；

（3）若对系统常数不要求输出近似结果，则输出其本来符号表示准确值.

Example 1.6.2 Evaluate $f(x)=1+3x-6x^2+x^3$ at $x=3.8$, using three-digit rounding arithmetic and estimate the relative error.

Table 1-6 gives the intermediate results in the calculations.

Table 1-6 Intermediate results

results	x	x^2	x^3	$3x$	$6x^3$
Exact	3.8	14.44	54.872	11.4	86.64
Three-digit(rounding)	3.8	14.4	54.8	11.4	86.6

Exact: $f(3.8)=1+11.4-86.64+54.872=-19.368$

Three-digit(rounding): $f(3.8)=1+11.4-86.6+54.8=-19.4$

The relative error for the three-digit rounding method is

$$\left|\frac{-19.368+19.4}{-19.368}\right|\approx 0.001\,65$$

Using mathematica:

```
In[1]:=1+3x-6x^2+x^3/.x->3.8
Out[1]=-19.368
In[2]:=Abs[(-19.368+19.4)/(-19.368)]
Out[2]=0.00165221
```

2. 解方程(组)

Example 1.6.3 Find the root of equation $x^2-5x+5=0$.

```
In[1]:= Solve[x^2-5x+5==0,x]
```

Out[1]={{x->$\frac{1}{2}$(5-$\sqrt{5}$)},{x->$\frac{1}{2}$(5+$\sqrt{5}$)}}

下面是其数值解

```
In[2]:=N[%]
Out[2]={{x 1.38197},{x 3.61803}}
```

用 FindRoot 函数可求解超越方程(方程组).

Example 1.6.4 Find the root of the following system near $(x_0,y_0)=(1,0)$.

$$\begin{cases}\sin x = x - y \\ \cos y = x + y\end{cases}$$

Using Mathematica：

```
In[3]:= FindRoot[{Sin[x]==x-y,Cos[y]==x+y},{x,1},{y,0}]
Out[3]= {x->0.883401,y->0.1105}
```

1.6.3 Functions Needed in the Experiments by Matlab

实验中需要用到的 Matlab 函数见表 1-7.

表 1-7 方程求根函数

函　数	意　义	函　数	意　义
solve('eq', 'var')	给出方程eq关于var为未知数的解	roots(P)	给出多项式的根，P为系数（降幂）向量
solve('eq1', 'eq2',⋯,'eqn')	给出方程的数值解集	linsolve(A,b)	求线性方程组$Ax=b$的解

1.6.4 Experiments by Matlab

1．解方程

（1）求解方程$x^2-x-6=0$.

```
>>clear
>>solve('x^2-x-6=0')
ans =
     -2
      3
```

或者

```
>> P=[1 -1 -6];
>> x=roots(P)
x =
     3
    -2
```

（2）求解方程$\cos(2x)-\sin(x)=1$.

```
>>clear
>>s=solve('cos(2*x)+sin(x)=1')
s =
         0
        pi
    1/6*pi
    5/6*pi
```

2．解方程组

（1）解方程组$\begin{cases}x^2+xy+y=3 \\ x^2-4x+3=0\end{cases}$.

```
>>clear
>>[x,y]=solve('x^2+x*y+y=3','x^2-4*x+3=0')
x =
     1
```

```
        3
y =
        1
     -3/2
```

（2）解线性方程组$\begin{cases} 5x_1 + 4x_3 + 2x_4 = 3 \\ x_1 - x_2 + 2x_3 + x_4 = 1 \\ 4x_1 + x_2 + 2x_3 = 1 \\ x_1 + x_2 + x_3 + x_4 = 0 \end{cases}$.

```
>>clear
>>A=[5 0 4 2;1 -1 2 1;4 1 2 0;1 1 1 1];
>>b=[3 1 1 0]';
>>x=linsolve(A,b)
x =
     1
    -1
    -1
     1
```

Exercises 1

Questions

1. Let $x>0$, the relative error be ε, solve the error of $\ln x$.

2. Let the relative error of x be 1%, compute the relative error of x^n.

3. The approximate numbers of the following are obtained by round-off , determine their significant digits.

$$x_1^* = 2.1021, \quad x_2^* = 0.085, \quad x_3^* = 685.2, \quad x_4^* = 98.44300$$

4. Solve the limits of the errors of the following approximate numbers where $x_1^*, x_2^*, x_3^*, x_4^*$ are given by problem 3.

(1) $x_1^* + x_2^* + x_3^*$;　　(2) $x_1^* x_2^* x_3^*$;　　(3) x_1^* / x_3^*.

5. Computing the volume of a sphere, such that the limit of the relative error is 1% , show the limit of the relative error of the radius of the sphere.

6. How to compute $\int_M^{M+1} \frac{1}{1+x^2} \mathrm{d}x$ if M sufficiently big?

7. Let $Y_0 = 28$, by the **recurrence formula** (递推公式)

$$Y_n = Y_{n-1} - \frac{1}{100}\sqrt{783} \qquad (n = 1, 2, \cdots)$$

show the error of Y_{100} if $\sqrt{783} \approx 27.982$ (5 significant digits).

8. Find the roots of the equation $x^2 - 26x - 1 = 0$, such that they have four significant digits ($\sqrt{680} \approx 26.0768$).

9. Given the edge of a square is about 100 cm, by how to measure it such that the error of its area is less than 1 cm^2.

10. Suggest ways to avoid loss of significance in these calculations.

(1) $\sqrt{x^2+1}-x$;　　(2) $\ln x-\ln y$;

(3) $\sin x-\tan x$;　　(4) $(1-x)/(1+x)-1/(3x+1)$.

11. Given the recurrence formula by

$$y_n=10y_{n-1}-1 \qquad (n=1,2,\cdots)$$

Let $y_0=\sqrt{2}\approx 1.41$ (3 significant digits), solve the error of y_{10}, is this computational process stable ?

12. Compute $f=(\sqrt{2}-1)^6$, take $\sqrt{2}\approx 1.4$, which is the best result by using the following equations?

(1) $(3-2\sqrt{2})^3$;　(2) $99-70\sqrt{2}$;　(3) $\dfrac{1}{(\sqrt{2}+1)^6}$;　(4) $\dfrac{1}{(3+2\sqrt{2})^3}$.

Computer Questions

1. Find the approximations to 7^{49} with 8 and 80 significant digits respectively.

2. Compute the absolute error and relative error in approximations of x by x^*.

(1) $x=\pi,\quad x^*=22/7$;　　(2) $x=\pi,\quad x^*=3.14$;

(3) $x=\mathrm{e},\quad x^*=2.718$;　　(4) $x=\sqrt{2},\quad x^*=1.414$.

3. Use three-digit rounding arithmetic to perform the following calculations.

(1) $(121-0.327)-119$;　　(2) $(121-119)-0.327$;

(3) $\left(\dfrac{13}{14}-\dfrac{6}{7}\right)/(2\mathrm{e}-5.4)$;　　(4) $-10\pi+6\mathrm{e}-\dfrac{3}{62}$.

4. Let

$$f(x)=\frac{x\cos x-\sin x}{x-\sin x}$$

(1) Use four -digit rounding arithmetic to evaluate $f(0.1)$.

(2) The actual value is $f(0.1)=-1.998\,999\,98$. Find the relative error for the value obtained in part (1).

5. Use four-digit rounding arithmetic and the quadratic formulas of Example 1.5.2 to find the most accurate approximations to the roots of the following quadratic equations. Compute the absolute errors and relative errors.

(1) $1.002x^2-11.01x+0.012\,65=0$;

(2) $\dfrac{1}{3}x^2+\dfrac{123}{4}x-\dfrac{1}{6}=0$.

2 Interpolating

提　要

插值是广泛应用于理论研究和实际工程的一种重要数值方法．一方面，大量实际问题中的函数关系是用表格法给出的，如观测或实验而得到的函数数据表.从提供的部分离散点去进行理论分析和计算都很不方便，甚至是不可能的，因此，需要寻找与已知函数值相符而形式简单的插值函数．另一方面，函数表达式虽已给定，但其形式复杂，计算不便，因此，也需要寻找一个计算简单的函数去近似原函数．本章主要介绍常用的拉格朗日 (Lagrange) 插值、牛顿 (Newton) 插值、埃尔米特 (Hermite) 插值及样条插值.

词　汇

representation	表达式	interpolation interval	插值区间
polynomial interpolation.	多项式插值	piecewise interpolation	分段插值
spline interpolation	样条插值	parabolic interpolation	抛物线插值
linear interpolation	线性插值	quadratic interpolation	二次插值
cubic interpolation	三次插值	geometric significance	几何意义
convergence	收敛	estimate	估计
base (cardinal) function	基函数	node	节点
uniqueness	唯一性	remainder	余项
step width (step length)	步长	divided-difference	均差
difference quotient	差商	difference	差分
linear combination	线性组合	forward difference	向前差分
backward difference	向后差分	center difference	中心差分
invariant operator	不变算子	shift operator	移位算子
polygonal function	折线函数	ill-posed	病态
diverge	发散	multiple root	重根
analytic function	解析函数	equidistant	等距的
derive	求导数	abscissa	横坐标
acceleration	加速度	integrate	积分
ordinate	纵坐标	latitude	纬度

2.1 Introduction

There are many instances in engineering where a relationship between two variables can be characterized experimentally, but no **underlying theory** (基本理论) is available to accurately represent it mathematically. Curve fitting is a natural way to obtain a working mathematical description in these cases.

Gravitational Acceleration（重力加速度）

The **law of universal gravitation** (万有引力定律) says that the force of attraction between two masses m_1 and m_2 is proportional to the product of the masses divided by the square of the distance r between them.

$$F = \frac{Gm_1 m_2}{r^2}$$

Here,G=6.673×10^{-11} N • m^2/kg^2 is the constant of gravitation. Suppose m_1 represents the mass of the earth and m_2 represents the mass of an object on the surface of the earth. In this case, the force on the object caused by the gravitational attraction of the earth reduces to

$$F = m_2 g$$

The new parameter $g=Gm_1/r^2$ is the acceleration due to gravity measured at the surface of the earth. Since the earth is not perfectly spherical, the acceleration due to gravity is not a constant, but instead varies with **latitude** (纬度). Measurements of the mean value of g at seal level at various latitudes are listed in Table 2-1.

Suppose a scientific experiment that requires a highly accurate value for acceleration due to gravity is performed on ship. In this case, it is appropriate to fit a curve f to the data in Table 2-1 and then use $g=f(x)$ where x is the latitude. An example of a curve fit to the data in Table 2-1 using Lagrange interpolation is shown in Figure 2-1.

Table 2-1 Variation in g with latitude at sea level

Latitude/(deg)	$g/(m/s^2)$
0	9.780 39
10	9.781 95
20	9.786 41
30	9.793 29
40	9.801 71
50	9.810 71
60	9.819 18
70	9.826 08
80	9.830 59
90	9.832 17

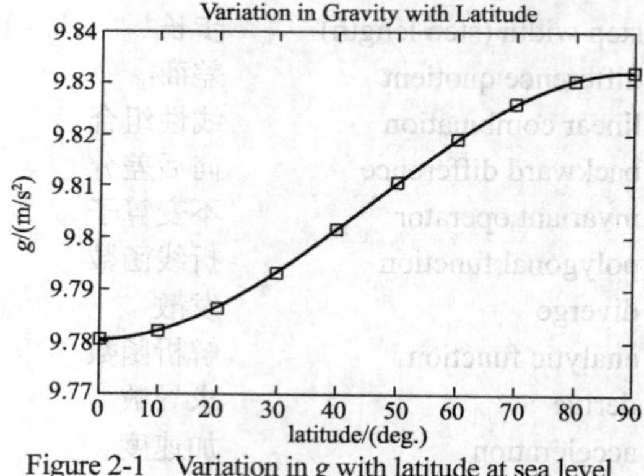

Figure 2-1 Variation in g with latitude at sea level

When you finish this chapter, you will understand what an interpolating polynomial is and how to efficiently evaluate it. You will know how to express an interpolating polynomial directly in terms of the data points using Newton's difference formula. You will understand why higher-order polynomials are not good candidates for curve fitting and how this problem can be corrected by using cubic splines. These overall goals will be achieved by mastering the following chapter objectives.

Objectives

- Know how to construct and use Lagrange interpolating polynomials to perform interpolation.
- Know how to construct the interpolating polynomial using Newton's formula.
- Know how to express piecewise-linear interpolation.
- Know how to construct cubic spline interpolation.
- Understand the divided-difference and difference.

2.2 Basic Concepts

Let $y=f(x)$ be defined on $[a,b]$, and given n+1 values $y_0,y_1,\cdots,y_n$ at the given points $a\leqslant x_0<x_1<\cdots<x_n\leqslant b$, if there exists a simple function $p(x)$, such that

$$p(x_i)=y_i \quad (i=0,1,\cdots,n)$$

holds, then we call $p(x)$ an **interpolation function** (插值函数) of $f(x)$, $x_0,x_1,\cdots,x_n$ **interpolation nodes** (插值节点), and $[a,b]$ **interpolation interval** (插值区间), this method is called **interpolation method**. If $p(x)$ is a polynomial whose degree does not exceed n, i.e.,

$$p(x)=a_0+a_1x+\cdots+a_nx^n \tag{2.2.1}$$

where $a_0,a_1,\cdots,a_n$ are real numbers, we call $p(x)$ an **interpolation polynomial** (插值多项式), the method is called **polynomial interpolation**. If $p(x)$ is a piecewise polynomial, then it is called **piecewise interpolation**(分段插值). If $p(x)$ is a **trigonometric polynomial** (三角多项式), then it is called **trigonometric interpolation** (三角插值).

The geometric significance of interpolation (see Figure 2-2)

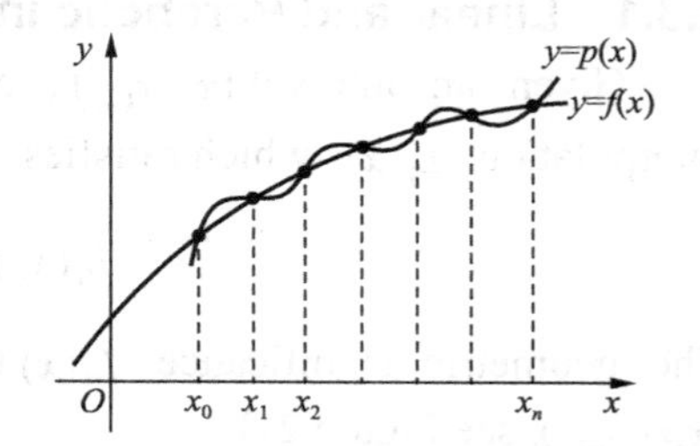

Figure 2-2 Interpolation polynomial

Polynomials have attracted the attention of mathematicians for centuries because of their many beautiful properties. For numerical purposes they have the advantage that their computation reduces to additions and multiplications only. Therefore, it is quite natural to use polynomials for the approximation of more complicated functions. We begin with the case of polynomials, which is the oldest and simplest.

Theorem 2.2.1 If $x_0,x_1,\cdots,x_n$ are distinct real numbers, then for arbitrary values

$y_0, y_1, \cdots, y_n$, there is a unique polynomial p_n of degree at most n such that

$$p_n(x_i) = y_i \quad (i = 0,1,\cdots,n) \tag{2.2.2}$$

Proof From polynomial (2.2.1) and conditions (2.2.2), we have $n+1$ equations with $n+1$ **unknown coefficients**(未知系数) $a_0, a_1, \cdots, a_n$,

$$\begin{cases} a_0 + a_1 x_0 + \cdots + a_n x_0^n = y_0 \\ a_0 + a_1 x_1 + \cdots + a_n x_1^n = y_1 \\ \cdots\cdots \\ a_0 + a_1 x_n + \cdots + a_n x_n^n = y_n \end{cases} \tag{2.2.3}$$

the matrix of coefficients

$$\begin{pmatrix} 1 & x_0 & \cdots & x_0^n \\ 1 & x_1 & \cdots & x_1^n \\ \vdots & \vdots & & \vdots \\ 1 & x_n & \cdots & x_n^n \end{pmatrix}$$

is a **Vandermonde** matrix. Since $x_0, x_1, \cdots, x_n$ are distinct, therefore

$$\det A = \prod_{\substack{i,j=0 \\ i>j}}^{n-1} (x_i - x_j) \neq 0$$

The system of equations (2.2.3) has a unique solution. □

2.3 Lagrange Interpolation

We now present an **alternative form** (一般形式) for the interpolating polynomial associated with a table of data points (x_i, y_i) for $0 \leqslant i \leqslant n$. It is important to understand that there is one and only one interpolating polynomial of degree $\leqslant n$ associated with the data (assuming, of course, that the $n+1$ abscissas x_i are distinct). However, the possibility certainly exists for expressing this polynomial in different forms and for arriving at it by different algorithms.

2.3.1 Linear and Parabolic Interpolation（线性插值与抛物线插值）

Given an interval (x_k, x_{k+1}), and $y_k = f(x_k)$, $y_{k+1} = f(x_{k+1})$. We seek a linear interpolation $L_1(x)$ which satisfies

$$L_1(x_k) = y_k, \quad L_1(x_{k+1}) = y_{k+1} \tag{2.3.1}$$

The geometric significance $L_1(x)$ is the unique straight line passing through (x_k, y_k), (x_{k+1}, y_{k+1}), see Figure 2-3.

$$L_1(x) = y_k + \frac{y_{k+1} - y_k}{x_{k+1} - x_k}(x - x_k) \qquad \text{(point slope form)}$$

or

$$L_1(x) = \frac{x - x_{k+1}}{x_k - x_{k+1}} y_k + \frac{x - x_k}{x_{k+1} - x_k} y_{k+1} \qquad \text{(two-point form)}$$

The two-point form shows that $L_1(x)$ consists of the linear combination of two linear functions, i.e.,

$$l_k(x)=\frac{x-x_{k+1}}{x_k-x_{k+1}}$$

$$l_{k+1}(x)=\frac{x-x_k}{x_{k+1}-x_k} \tag{2.3.2}$$

Its coefficients are respectively y_k and y_{k+1}, i.e.,

$$L_1(x)=y_k l_k(x)+y_{k+1}l_{k+1}(x) \tag{2.3.3}$$

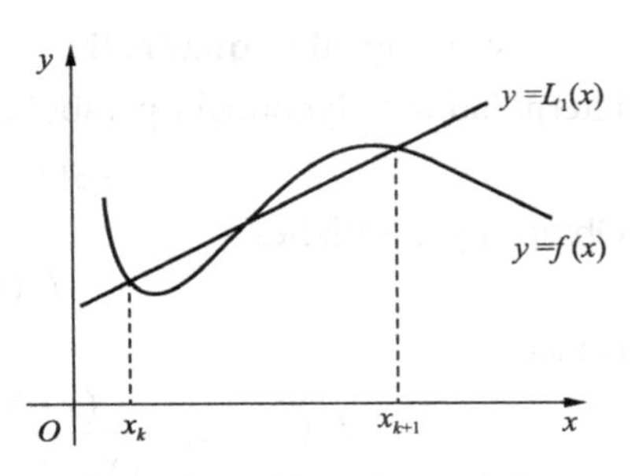

Figure 2-3 Linear interpolation

and

$$l_k(x_k)=1, \quad l_k(x_{k+1})=0, \quad l_{k+1}(x_k)=0, \quad l_{k+1}(x_{k+1})=1$$

We call $l_k(x)$ and $l_{k+1}(x)$ **base (cardinal) functions of linear interpolation** (线性插值基函数).

Given interpolation nodes x_{k-1}, x_k and x_{k+1}, we can seek an interpolation polynomial $L_2(x)$ with degree 2 which satisfies

$$L_2(x_j)=y_j \quad (j=k-1,k,k+1)$$

As we all know, $L_2(x)$ is a parabola passing through (x_{k-1},y_{k-1}), (x_k, y_k) and (x_{k+1},y_{k+1}), see Figure 2-4, we construct the base functions of interpolation $l_{k-1}(x)$, $l_k(x)$,$l_{k+1}(x)$, which satisfy

$$\left.\begin{aligned} &l_{k-1}(x_{k-1})=1, \quad l_{k-1}(x_j)=0 \quad (j=k,k+1) \\ &l_k(x_k)=1, \quad l_k(x_j)=0 \quad (j=k-1,k+1) \\ &l_{k+1}(x_{k+1})=1, \quad l_{k+1}(x_j)=0 \quad (j=k-1,k) \end{aligned}\right\} \tag{2.3.4}$$

These base functions can be determined easily, e.g., for $l_{k-1}(x)$. Since it has two zeros x_k and x_{k+1}, therefore it can be represented in the form

$$l_{k-1}(x)=A(x-x_k)(x-x_{k+1})$$

where A is **undetermined coefficient** (待定系数). From $l_{k-1}(x_{k-1})=1$, we can obtain that

$$A=\frac{1}{(x_{k-1}-x_k)(x_{k-1}-x_{k+1})}$$

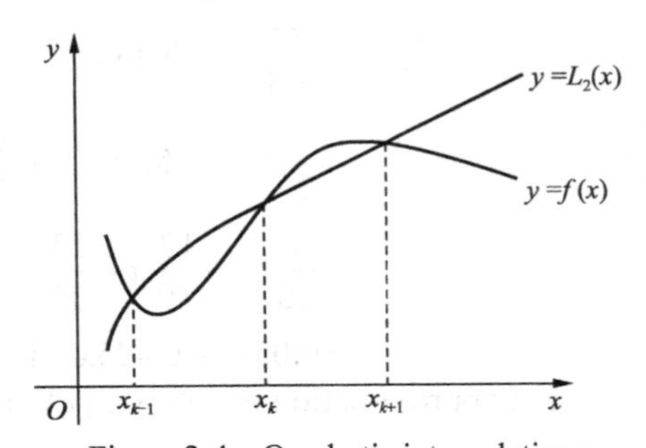

Figure 2-4 Quadratic interpolation

Hence

$$l_{k-1}(x)=\frac{(x-x_k)(x-x_{k+1})}{(x_{k-1}-x_k)(x_{k-1}-x_{k+1})}$$

Analogously

$$l_k(x)=\frac{(x-x_{k-1})(x-x_{k+1})}{(x_k-x_{k-1})(x_k-x_{k+1})}$$

$$l_{k+1}(x)=\frac{(x-x_{k-1})(x-x_k)}{(x_{k+1}-x_{k-1})(x_{k+1}-x_k)}$$

By using the **quadratic** (二次) base functions we can obtain the quadratic interpolation polynomial (parabolic interpolation polynomial) immediately

$$L_2(x) = y_{k-1}l_{k-1}(x) + y_k l_k(x) + y_{k+1}l_{k+1}(x) \tag{2.3.5}$$

Obviously, it satisfies

$$L_2(x_j) = y_j \quad (j = k-1,\ k,\ k+1)$$

Hence

$$L_2(x) = y_{k-1}\frac{(x-x_k)(x-x_{k+1})}{(x_{k-1}-x_k)(x_{k-1}-x_{k+1})} + y_k\frac{(x-x_{k-1})(x-x_{k+1})}{(x_k-x_{k-1})(x_k-x_{k+1})} + y_{k+1}\frac{(x-x_{k-1})(x-x_k)}{(x_{k+1}-x_{k-1})(x_{k+1}-x_k)}$$

Example 2.3.1 Using the numbers (or nodes) $x_0=2$, $x_1=2.5$ and $x_2=4$ to find the second interpolating polynomial for $f(x)=1/x$.

Solution It requires that we determine the coefficient polynomials $l_0(x)$, $l_1(x)$, and $l_2(x)$. In nested form they are

$$l_0(x) = \frac{(x-2.5)(x-4)}{(2-2.5)(2-4)} = x^2 - 6.5x + 10$$

$$l_1(x) = \frac{(x-2)(x-4)}{(2.5-2)(2.5-4)} = -\frac{4}{3}x^2 + 8x - \frac{32}{3}$$

and

$$l_2(x) = \frac{(x-2)(x-2.5)}{(4-2)(4-2.5)} = \frac{1}{3}x^2 - \frac{3}{2}x + \frac{5}{3}$$

Since $f(x_0) = f(2) = 1/2$, $f(x_1) = f(2.5) = 2/5$, and $f(x_2) = f(4) = 1/4$, we have

$$\begin{aligned} L_2(x) &= \sum_{k=0}^{2} f(x_k) l_k(x) \\ &= \frac{1}{2}(x^2 - 6.5x + 10) + \frac{2}{5}\left(-\frac{4}{3}x^2 + 8x - \frac{32}{3}\right) + \frac{1}{4}\left(\frac{1}{3}x^2 - \frac{3}{2}x + \frac{5}{3}\right) \\ &= \frac{1}{20}x^2 - \frac{17}{40}x + \frac{23}{20} \\ &= 0.05x^2 - 0.425x + 1.15 \end{aligned}$$

An approximation to $f(3) = 1/3$ is $f(3) \approx L_2(3) = 0.325$.

2.3.2 Lagrange Interpolation Polynomial

In this section we will discuss the interpolation polynomial $L_n(x)$ of degree n for $n+1$ nodes, which satisfies

$$L_n(x_j) = y_j \quad (j = 0, 1, \cdots, n) \tag{2.3.6}$$

Definition 2.3.1 If the polynomials $l_0(x), l_1(x), \cdots, l_n(x)$ of degree n satisfy the following conditions on $n+1$ nodes

$$l_j(x_k) = \begin{cases} 1 & k = j \\ 0 & k \neq j \end{cases} \quad (j, k = 0, 1, \cdots, n) \tag{2.3.7}$$

Then these polynomials $l_0(x), l_1(x), \cdots, l_n(x)$ are known as **base functions of interpolation of degree *n* (Lagrange factors)** on nodes $x_0, x_1, \cdots, x_n$.

We have discussed the cases of $n=1$ and $n=2$. Analogously, let

$$l_k(x) = A_k(x-x_0)(x-x_1)\cdots(x-x_{k-1})(x-x_{k+1})\cdots(x-x_n)$$

Since $l_k(x_k)=1$, then

$$A_k = \frac{1}{(x_k-x_0)\cdots(x_k-x_{k-1})(x_k-x_{k+1})\cdots(x_k-x_n)}$$

Therefore

$$l_k(x) = \frac{(x-x_0)\cdots(x-x_{k-1})(x-x_{k+1})\cdots(x-x_n)}{(x_k-x_0)\cdots(x_k-x_{k-1})(x_k-x_{k+1})\cdots(x_k-x_n)} \tag{2.3.8}$$

We have the *n*th Lagrange interpolating polynomial

$$\begin{aligned} L_n(x) &= y_0 l_0(x) + y_1 l_1(x) + \cdots + y_n l_n(x) \\ &= \sum_{k=0}^{n} y_k \frac{(x-x_0)\cdots(x-x_{k-1})(x-x_{k+1})\cdots(x-x_n)}{(x_k-x_0)\cdots(x_k-x_{k-1})(x_k-x_{k+1})\cdots(x_k-x_n)} \end{aligned} \tag{2.3.9}$$

We denote

$$\omega_{n+1}(x) = (x-x_0)(x-x_1)\cdots(x-x_n) \tag{2.3.10}$$

Then

$$\omega'_{n+1}(x_k) = (x_k-x_0)\cdots(x_k-x_{k-1})(x_k-x_{k+1})\cdots(x_k-x_n)$$

The Formula (2.3.9) can be rewritten in the form

$$L_n(x) = \sum_{k=0}^{n} y_k \frac{\omega_{n+1}(x)}{(x-x_k)\omega'_{n+1}(x_k)} \tag{2.3.11}$$

In what follows, we denote by P_n **the set of all real or complex polynomials whose degrees do not exceed** n (次数不超过 n 的实或复多项式的集合).

2.3.3 Interpolation Remainder and Error Estimate（插值余项与误差估计）

If $L_n(x)$ is an approximation of $f(x)$ on $[a,b]$, then the truncation error is $R_n(x) = f(x) - L_n(x)$, it is also called **remainder of interpolation polynomial**.

Theorem 2.3.1 Let $f(x)$ be $(n+1)$**-times continuously differentiable** ($n+1$次连续可微的) on [*a*, *b*]. Then the remainder $R_n(x) = f(x) - L_n(x)$ for polynomial interpolation with $(n+1)$ distinct points $x_0, \cdots, x_n \in [a,b]$ can be represented in the form

$$R_n(x) = \frac{f^{(n+1)}(\xi)}{(n+1)!}\prod_{j=0}^{n}(x-x_j), \quad x \in [a,b] \tag{2.3.12}$$

for some $\xi \in [a,b]$ depending on x.

Proof Since (2.3.12) is trivially satisfied if *x* coincides with one of the interpolation points $x_0, \cdots, x_n$, we need be concerned only with the case where *x* does not coincide with one of the interpolation points. We define

$$\omega_{n+1}(x)=\prod_{j=0}^{n}(x-x_j)$$

and, keeping x fixed, consider $g:[a,b]\to\mathbf{R}$ given by

$$g(y)=f(y)-L_n(y)-\omega_{n+1}(y)\frac{f(x)-L_n(x)}{\omega_{n+1}(x)},\quad y\in[a,b] \tag{2.3.13}$$

By the assumption on f, the function g is also $(n+1)$ -times continuously differentiable. Obviously, g has at least $n+2$ zeros, namely x and $x_0,\cdots,x_n$. Then, by Rolle's theorem the derivative g' has at least $n+1$ zeros. Repeating the argument, by induction we deduce that the derivative $g^{(n+1)}$ has at least one zero in (a,b), which we denote by ξ. For this zero we have that

$$0=f^{(n+1)}(\xi)-(n+1)!\frac{R_n(x)}{\omega_{n+1}(x)}$$

and from this we obtain (2.3.12). □

The limit of error

$$|R_n(x)|\leqslant\frac{M_{n+1}}{(n+1)!}|\omega_{n+1}(x)| \tag{2.3.14}$$

where $M_{n+1}=\max\limits_{a\leqslant x\leqslant b}\left|f^{(n+1)}(x)\right|$.

Using Theorem 2.3.1, if $f(x)=x^k, k=0,1,\cdots,n,$ then

$$\sum_{j=0}^{n}x_j^k l_j(x)=x^k \tag{2.3.15}$$

and

$$\sum_{j=0}^{n}l_j(x)=1 \tag{2.3.16}$$

In fact, let $f(x)=x^k$, when $k=0,1,\cdots,n,\quad f^{(n+1)}(x)=0.$

Construct Lagrange interpolation

$$L_n(x)=\sum_{j=0}^{n}x_j^k l_j(x)$$

Then $$R_n(x)=f(x)-L_n(x)=\frac{f^{(n+1)}(\xi)}{(n+1)!}\omega_{n+1}(x)=0$$

where $\xi\in[a,b]$.

Therefore $$\sum_{j=0}^{n}x_j^k l_j(x)\equiv x^k\quad(k=0,1,\cdots,n)$$

Especially, if $k=0$, we have $\sum\limits_{j=0}^{n}l_j(x)=1$.

Example 2.3.2 The linear interpolation is given by

$$L_1(x)=\frac{1}{h}\big(f(x_0)(x_1-x)+f(x_1)(x-x_0)\big)$$

with the **step width**（步长）$h=x_1-x_0$. For the polynomial

$$\omega_2(x)=(x-x_0)(x-x_1)$$

We have

$$\max_{x\in[x_0,x_1]}|\omega_2(x)|=\frac{h^2}{4}$$

Therefore, by Theorem 2.3.1, the error occurring in linear interpolation of a twice continuously differentiable function f can be estimated by

$$R_1(x)=\frac{1}{2}f''(\xi)(x-x_0)(x-x_1),\quad \xi\in(x_0,x_1) \tag{2.3.17}$$

$$|R_1(x)|\leqslant\frac{1}{2}|f''(\xi)||\omega_2(x)|\leqslant\frac{h^2}{8}\max_{x\in[x_0,x_1]}|f''(x)|$$

Let $n=2$, the remainder of parabolic interpolation is

$$R_2(x)=\frac{1}{6}f'''(\xi)(x-x_0)(x-x_1)(x-x_2),\quad \xi\in[x_0,x_2] \tag{2.3.18}$$

Example 2.3.3 Let $f(x)=\sin x$ and let $x_0,\cdots,x_n\in[0,\pi]$ be $n+1$ distinct points. Since

$$\left|f^{(n+1)}(x)\right|\leqslant 1,\quad x\in(0,\pi)$$

and

$$|\omega_{n+1}(x)|\leqslant\pi^{n+1},\quad x\in(0,\pi)$$

By using Theorem 2.3.1, we have the estimate

$$|R_n(x)|\leqslant\frac{\pi^{n+1}}{(n+1)!},\quad x\in(0,\pi)$$

Hence the sequence $\{L_n(x)\}$ converges to the interpolated function $f(x)$ **uniformly** (一致地) on $[0,\pi]$ as $n\to\infty$.

Example 2.3.4 Given $\sqrt{100}=10$, $\sqrt{121}=11$, $\sqrt{144}=12$, We have computed approximate value of $\sqrt{115}$ with linear and parabolic interpolation respectively, please estimate their errors.

Solution $f(x)=\sqrt{x}$. For linear interpolation, since $115\in(100,121)$, therefore the interpolation nodes are

$$x_0=100,\quad x_1=121,\quad y_0=10,\quad y_1=11$$

$$R_1(x)=\frac{f''(\xi)}{2!}\omega_2(x)=-\frac{1}{8}\xi^{-3/2}(x-x_0)(x-x_1),\quad \xi\in(x_0,x_1)$$

$$|R_1(115)|\leqslant\frac{1}{8}\times|(115-100)(115-121)|\times\max_{\xi\in[100,121]}\xi^{-3/2}$$

$$=\frac{1}{8}\times 15\times 6\times 10^{-3}=0.011\,25$$

For parabolic interpolation

$$R_2(x)=\frac{f'''(\xi)}{3!}\omega_3(x)=\frac{1}{16}\xi^{-5/2}(x-x_0)(x-x_1)(x-x_2),\quad \xi\in(x_0,x_1)$$

$$|R_2(115)|\leqslant\frac{1}{16}\times|(115-100)(115-121)(115-144)|\times 10^{-5}<0.001\,7$$

Example 2.3.5 Given

$$\sin 0.32 = 0.314\,567, \quad \sin 0.34 = 0.333\,487, \quad \sin 0.36 = 0.352\,274$$

Calculate $\sin 0.336\,7$ with linear and parabolic interpolation respectively, then estimate their truncation errors.

Solution $x_0 = 0.32,\ x_1 = 0.34,\ x_2 = 0.36$

$$y_0 = 0.314\,567,\ y_1 = 0.333\,487,\ y_2 = 0.352\,274$$

Using linear interpolation with points x_0, x_1,

$$\sin 0.336\,7 \approx L_1(0.336\,7) = y_0 + \frac{y_1 - y_0}{x_1 - x_0}(0.336\,7 - x_0)$$

$$= 0.314\,567 + \frac{0.018\,92}{0.02} \times 0.016\,7 = 0.330\,365$$

Its limit of the truncation error is

$$|R_1(x)| \leqslant \frac{M_2}{2}|(x - x_0)(x - x_1)|$$

Where $M_2 = \max\limits_{x_0 \leqslant x \leqslant x_1} |f''(x)|$.

Since

$$f(x) = \sin x, \quad f''(x) = -\sin x$$

Therefore

$$M_2 = \max_{x_0 \leqslant x \leqslant x_1} |f''(x)| = \sin x_1 \leqslant 0.333\,5$$

Hence

$$|R_1(0.336\,7)| = |\sin 0.336\,7 - L_1(0.336\,7)|$$

$$\leqslant \frac{1}{2} \times 0.333\,5 \times 0.016\,7 \times 0.003\,3 \leqslant 0.92 \times 10^{-5}$$

Using parabolic interpolation

$$\sin 0.336\,7 \approx L_2(0.336\,7)$$

$$= y_0 \frac{(x - x_1)(x - x_2)}{(x_0 - x_1)(x_0 - x_2)} + y_1 \frac{(x - x_0)(x - x_2)}{(x_1 - x_0)(x_1 - x_2)} + y_2 \frac{(x - x_0)(x - x_1)}{(x_2 - x_0)(x_2 - x_1)}$$

$$= 0.314\,567 \times \frac{0.768\,9 \times 10^{-4}}{0.000\,8} + 0.333\,487 \times \frac{3.89 \times 10^{-4}}{0.000\,4} + 0.352\,274 \times \frac{-0.5511 \times 10^{-4}}{0.000\,8}$$

$$= 0.330\,374$$

Its limit of the truncation error is

$$|R_2(x)| \leqslant \frac{M_3}{6}|(x-x_0)(x-x_1)(x-x_2)|$$

where $M_3 = \max\limits_{x_0 \leqslant x \leqslant x_2} |f'''(x)| = \cos x_0 < 0.828$.

Hence

$$\begin{aligned}|R_2(0.336\,7)| &= |\sin 0.336\,7 - L_2(0.336\,7)| \\ &\leqslant \frac{1}{6}\times 0.828 \times 0.016\,7 \times 0.033 \times 0.023\,3 \\ &< 0.178\times 10^{-6}\end{aligned}$$

2.4 Divided-differences and Newton Interpolation (均差与牛顿插值)

In the preceding section, we discussed the problem of interpolating a function by a polynomial. We return to that problem now. Let f be a function whose values are known or computable at a set of points (nodes) $x_0, x_1, \cdots, x_n$. We assume in this section that these points are distinct, but they need not be ordered on the real line. We know that there exists a unique polynomial P_n of degree at most n that interpolates f at the n+1 nodes:

$$p_n(x_j) = f(x_j) = f_j \quad (j = 0,1,\cdots,n)$$

Of course, the polynomial P_n can be constructed as a linear combination of the basic polynomials $1, x, x^2, \cdots, x^n$. We prefer to use a basis appropriate to the Newton form of the interpolating polynomial:

$$\begin{aligned}&q_0(x) = 1\\ &q_1(x) = (x-x_0)\\ &q_2(x) = (x-x_0)(x-x_1)\\ &q_3(x) = (x-x_0)(x-x_1)(x-x_2)\\ &\cdots\cdots\\ &q_n(x) = (x-x_0)(x-x_1)(x-x_2)\cdots(x-x_{n-1})\end{aligned}$$

These lead to the Newton form

$$\begin{aligned}N_n(x) &= a_0 + a_1(x-x_0) + a_2(x-x_0)(x-x_1) + \cdots + \\ &\quad a_n(x-x_0)\cdots(x-x_{n-1}) \\ &= \sum_{j=0}^{n} a_j q_j(x)\end{aligned} \tag{2.4.1}$$

Where $a_0, a_1, \cdots, a_n$ are undetermined coefficients that can be determined by the following conditions

$$N_n(x_i) = f(x_i) = f_i \quad (i = 0,1,\cdots,n)$$

When $x = x_0$, we have $N_n(x_0) = a_0 = f_0$.

When $x = x_1$, we have $N_n(x_1) = a_0 + a_1(x_1 - x_0) = f_1$.

Inferring that (推出)

$$a_1 = \frac{f_1 - f_0}{x_1 - x_0}$$

When $x = x_2$, we have

$$N_n(x_2) = a_0 + a_1(x_2 - x_0) + a_2(x_2 - x_0)(x_2 - x_1) = f_2$$

Inferring that

$$a_2 = \frac{\dfrac{f_2 - f_0}{x_2 - x_0} - \dfrac{f_1 - f_0}{x_1 - x_0}}{x_2 - x_1}$$

Analogously, we can obtain $a_3, \cdots, a_n$.

Definition 2.4.1 Representation

$$f[x_0, x_1] = \frac{f(x_1) - f(x_0)}{x_1 - x_0}$$

is called **first divided-difference** (一阶均差) of function $f(x)$ with points x_0, x_1, and

$$f[x_0, x_1, x_2] = \frac{f[x_1, x_2] - f[x_0, x_1]}{x_2 - x_0}$$

is called **second divided-difference** of function $f(x)$.

In general,

$$f[x_0, x_1, \cdots, x_n] = \frac{f[x_1, x_2, \cdots, x_n] - f[x_0, x_1, \cdots, x_{n-1}]}{x_n - x_0} \tag{2.4.2}$$

is called ***n*th divided-difference** or **difference quotient** (差商) of function $f(x)$, or **divided-difference of order** *n*. The **zeroth divided-difference** of the function *f* with respect to x_i, denoted $f[x_i]$, is simply the value of f at x_i:

$$f[x_i] = f(x_i) \tag{2.4.3}$$

Obviously, $f[x_0]$ is the coefficient of x^0 in the polynomial of degree 0 interpolating f at x_0 and $a_0 = f[x_0] = f(x_0)$. The quantity $f[x_0, x_1]$ is the coefficient of *x* in the polynomial of degree at most 1 interpolating f at x_0 and x_1, and $a_1 = f[x_0, x_1]$. Analogously, $f[x_0, x_1, \cdots, x_n]$ is the coefficient of x^n in the polynomial of degree at most *n* interpolating *f* at $x_0, x_1, \cdots, x_n$ and $a_n = f[x_0, x_1, \cdots, x_n]$. Hence

$$\begin{aligned} N_n(x) = f(x_0) &+ f[x_0, x_1](x - x_0) + \\ &f[x_0, x_1, x_2](x - x_0)(x - x_1) + \cdots + \\ &f[x_0, x_1, \cdots, x_n](x - x_0)(x - x_1)\cdots(x - x_{n-1}) \end{aligned} \tag{2.4.4}$$

We call $N_n(x)$ **Newton interpolation polynomial.**

The properties of divided-differences

Theorem 2.4.1 The divided-difference is a **symmetric function** (对称函数) of its arguments. Thus, if $(z_0, z_1, \cdots, z_n)$ is a **permutation** (排列) of $(x_0, x_1, \cdots, x_n)$, then

$$f[z_0, z_1, \cdots, z_n] = f[x_0, x_1, \cdots, x_n] \tag{2.4.5}$$

Proof The divided-difference on the left side of Equation (2.4.5) is the coefficient of x^n in the polynomial of degree at most n that interpolates f at the points $(z_0, z_1, \cdots, z_n)$. The divided-difference on the right is the coefficient of x^n in the polynomial of degree at most n that interpolates f at the points $(x_0, x_1, \cdots, x_n)$. These two polynomials are, of course, the same. □

Theorem 2.4.2 Let p be the polynomial of degree at most n that interpolates a function f at a set of n+1 distinct nodes $x_0, x_1, \cdots, x_n$. If t is a point different from the nodes, then

$$f(t) - p(t) = f[x_0, x_1, \cdots, x_n, t]\prod_{j=0}^{n}(t - x_j) \tag{2.4.6}$$

Proof First, let q be the polynomial of degree at most n+1 that interpolates f at the nodes $x_0, x_1, \cdots, x_n, t$. We know that q is obtained from p by adding one term. In fact,

$$q(x) = p(x) + f[x_0, x_1, \cdots, x_n, t]\prod_{j=0}^{n}(x - x_j)$$

Since $q(t) = f(t)$, we obtain at once (by letting $x = t$)

$$f(t) = p(t) + f[x_0, x_1, \cdots, x_n, t]\prod_{j=0}^{n}(t - x_j)$$ □

Theorem 2.4.3 Suppose that $f \in C^n[a,b]$ and $x_0, x_1, \cdots, x_n$ are distinct numbers in $[a,b]$, then a number ξ exists in (a,b) with

$$f[x_0, x_1, \cdots, x_n] = \frac{f^{(n)}(\xi)}{n!} \tag{2.4.7}$$

Proof Let

$$g(x) = f(x) - p_n(x)$$

Since $f(x_i) = p_n(x_i)$, for each $i = 0, 1, \cdots, n$, the function g has $n+1$ distinct zeros in $[a,b]$. The generalized Rolle's theorem implies that a number ξ in (a,b) exists with $g^{(n)}(\xi) = 0$, so

$$0 = f^{(n)}(\xi) - p_n^{(n)}(\xi)$$

Since $p_n(x)$ is a polynomial of degree n whose leading coefficient is $f[x_0, x_1, \cdots, x_n]$

$$p_n^{(n)}(x) = n!\, f[x_0, x_1, \cdots, x_n]$$

for all values of x. As a consequence,

$$f[x_0, x_1, \cdots, x_n] = \frac{f^{(n)}(\xi)}{n!}$$ □

Theorem 2.4.4 The kth divided-difference can be represented by **linear combination** (线性组合) of the values $f(x_0), \cdots, f(x_k)$. i.e.,

$$f[x_0, x_1, \cdots, x_k] = \sum_{j=0}^{k} \frac{f(x_j)}{(x_j - x_0)\cdots(x_j - x_{j-1})(x_j - x_{j+1})\cdots(x_j - x_k)}$$

It can be proved by **induction** (归纳法). An algorithm for computing a divided difference table can be very efficient and is recommended as the best means for producing an interpolating polynomial. The divided difference table is shown here:(see Table 2-2)

Table 2-2 The table of divided-differences (The divided-difference scheme)

x_k	$f(x_k)$	First Divided-difference	Second Divided-difference	Third Divided-difference	Fourth Divided-difference
x_0	$f(x_0)$				
x_1	$f(x_1)$	$f[x_0,x_1]$			
x_2	$f(x_2)$	$f[x_1,x_2]$	$f[x_0,x_1,x_2]$		
x_3	$f(x_3)$	$f[x_2,x_3]$	$f[x_1,x_2,x_3]$	$f[x_0,x_1,x_2,x_3]$	
x_4	$f(x_4)$	$f[x_3,x_4]$	$f[x_2,x_3,x_4]$	$f[x_1,x_2,x_3,x_4]$	$f[x_0,x_1,x_2,x_3,x_4]$
⋮	⋮	⋮	⋮	⋮	⋮

Example 2.4.1 For the points $x_0=0,x_1=1,x_2=3,x_3=4$ and the values $y_0=0,y_1=2,y_2=8,y_3=9$, the table of the divided-differences is given by Table 2-3.

The interpolation polynomial is given by

$$N_3(x)=2x+\frac{1}{3}x(x-1)-\frac{1}{4}x(x-1)(x-3)$$

Example 2.4.2 Given $\sqrt{100}=10,\ \sqrt{121}=11,\ \sqrt{144}=12.$ Solve approximate value of $\sqrt{115}$ with linear and parabolic interpolation respectively.

Solution Construct the table of divided-differences, see Table 2-4.

Table 2-3 The divided-differences

x_k	$f(x_k)$	order 1	order 2	order 3
0	0			
		2		
1	2		1/3	
		3		−1/4
3	8		−2/3	
		1		
4	9			

Table 2-4 The divided-differences

x	$\sqrt{x}$	order 1	order 2
100	10		
		0.047 619	
121	11	0.043 478	−0.000 094
144	12		

$$f(x_0)=10$$
$$f[x_0,x_1]=0.047\ 619$$
$$f[x_0,x_1,x_2]=-0.000\ 094$$

Hence, by using linear interpolation we have

$$\sqrt{115}\approx N_1(115)=10+0.047\ 619\times(115-100)=10.714\ 3$$

By using parabolic interpolation we have

$$\sqrt{115}\approx N_2(115)=N_1(115)+(-0.000\ 094)\times(115-100)\times(115-121)$$
$$=10.722\ 8$$

2.5 Differences and Newton Difference Formulae
（差分与牛顿差分公式）

2.5.1 Differences

Given the values $f_k = f(x_k)$ of function $y = f(x)$ with **equally spaced nodes** (等距点) $x_k = x_0 + kh \ (k = 0,1,\cdots,n)$, where h is a constant, and called **step length (step size, step width)** (步长).

Definition 2.5.1 Notations

$$\Delta f_k = f_{k+1} - f_k \tag{2.5.1}$$

$$\nabla f_k = f_k - f_{k-1} \tag{2.5.2}$$

$$\delta f_k = f\left(x_k + \frac{h}{2}\right) - f\left(x_k - \frac{h}{2}\right) = f_{k+\frac{1}{2}} - f_{k-\frac{1}{2}} \tag{2.5.3}$$

are called **first forward-difference** (一阶向前差分), **first backward-difference** (一阶向后差分) and **first center-difference** (一阶中心差分) of f evaluated at x_k with step size h respectively.

Signs Δ, ∇, δ are called **forward-difference operator** (向前差分算子), **backward-difference operator** (向后差分算子) and **center-difference operator** (中心差分算子) respectively.

Notation

$$\Delta^2 f_k = \Delta f_{k+1} - \Delta f_k = f_{k+2} - 2f_{k+1} + f_k$$

is called **second difference.**

Analogously, we have m**th difference:**

$$\Delta^m f_k = \Delta^{m-1} f_{k+1} - \Delta^{m-1} f_k$$

$$\nabla^m f_k = \nabla^{m-1} f_k - \nabla^{m-1} f_{k-1}$$

$$\delta f_{k+\frac{1}{2}} = f_{k+1} - f_k$$

$$\delta f_{k-\frac{1}{2}} = f_k - f_{k-1}$$

$$\delta^2 f_k = \delta f_{k+\frac{1}{2}} - \delta f_{k-\frac{1}{2}}$$

invariant operator (不变算子) I: $\quad If_k = f_k$

shift operator (移位算子) E: $\quad Ef_k = f_{k+1}$

Since

$$\Delta f_k = f_{k+1} - f_k = Ef_k - If_k = (E - I)f_k$$

Therefore

$$\Delta = E - I$$

Furthermore

$$\nabla = I - E^{-1} \qquad (E^{-1} f_k = f_{k-1})$$

$$\delta = E^{\frac{1}{2}} - E^{-\frac{1}{2}} \qquad \left(E^{\frac{1}{2}} f_k = f_{k+\frac{1}{2}}\right)$$

Property 1 Each difference can be represented by functional values, e.g.,

$$\begin{aligned}\Delta^n f_k &= (E - I)^n f_k = \sum_{j=0}^{n} (-1)^j \binom{n}{j} E^{n-j} f_k \\ &= \sum_{j=0}^{n} (-1)^j \binom{n}{j} f_{n+k-j}\end{aligned} \tag{2.5.4}$$

$$\begin{aligned}\nabla^n f_k &= (I - E^{-1})^n f_k = \sum_{j=0}^{n} (-1)^{n-j} \binom{n}{j} E^{j-n} f_k \\ &= \sum_{j=0}^{n} (-1)^{n-j} \binom{n}{j} f_{k+j-n}\end{aligned} \tag{2.5.5}$$

where $\binom{n}{j} = \dfrac{n(n-1)\cdots(n-j+1)}{j!}$.

Property 2 Functional value can be represented by differences, e.g.,

Since
$$f_{n+k} = E^n f_k = (I + \Delta)^n f_k = \left(\sum_{j=0}^{n} \binom{n}{j} \Delta^j\right) f_k$$

Therefore
$$f_{n+k} = \sum_{j=0}^{n} \binom{n}{j} \Delta^j f_k$$

Property 3 Divided-difference has relationship with difference, e.g.,

$$f[x_k, x_{k+1}] = \frac{f_{k+1} - f_k}{x_{k+1} - x_k} = \frac{\Delta f_k}{h}$$

$$f[x_k, x_{k+1}, x_{k+2}] = \frac{f[x_{k+1}, x_{k+2}] - f[x_k, x_{k+1}]}{x_{k+2} - x_k} = \frac{1}{2h^2} \Delta^2 f_k$$

Generally

$$f[x_k, \cdots, x_{k+m}] = \frac{1}{m!} \frac{1}{h^m} \Delta^m f_k \tag{2.5.6}$$

Analogously , we have

$$f[x_k, x_{k-1}, \cdots, x_{k-m}] = \frac{1}{m!} \frac{1}{h^m} \nabla^m f_k \tag{2.5.7}$$

Using (2.5.6) and (2.4.7), we can obtain

$$\Delta^n f_k = h^n f^{(n)}(\xi), \quad \xi \in (x_k, x_{k+n})$$

See Table 2-5.

Table 2-5 The table of differences

f_k	$\Delta\ (\nabla)$	$\Delta^2(\nabla^2)$	$\Delta^3(\nabla^3)$	$\Delta^4(\nabla^4)$	$\cdots$
f_0					
	$\Delta f_0\,(\nabla f_1)$				
f_1		$\Delta^2 f_0\,(\nabla^2 f_2)$			
	$\Delta f_1\,(\nabla f_2)$		$\Delta^3 f_0\,(\nabla^3 f_3)$		
f_2		$\Delta^2 f_1\,(\nabla^2 f_3)$		$\Delta^4 f_0\,(\nabla^4 f_4)$	$\vdots$
	$\Delta f_2\,(\nabla f_3)$		$\Delta^3 f_1\,(\nabla^3 f_4)$	$\vdots$	
f_3		$\Delta^2 f_2\,(\nabla^2 f_4)$	$\vdots$		
	$\Delta f_3\,(\nabla f_4)$	$\vdots$			
f_4	$\vdots$				
$\vdots$					

2.5.2 Newton Difference Formulae

Given points $x_k = x_0 + kh \ (k = 0,1,\cdots,n)$, we want to compute the functional value near x_0 of $f(x)$. Let

$$x = x_0 + th \ \ (0 \leqslant t \leqslant 1), \quad x_j = x_0 + jh \quad (j = 0,\cdots,n)$$

then

$$\omega_{n+1}(x) = \prod_{j=0}^{n}(x - x_j) = t(t-1)\cdots(t-n)h^{n+1}$$

and **Newton forward-difference formula** (牛顿向前差分公式) is of the form

$$N_n(x_0 + th) = f_0 + t\Delta f_0 + \frac{t(t-1)}{2!}\Delta^2 f_0 + \cdots + \frac{t(t-1)\cdots(t-n+1)}{n!}\Delta^n f_0$$

Its remainder is

$$R_n(x) = \frac{t(t-1)\cdots(t-n)}{(n+1)!}h^{n+1}f^{(n+1)}(\xi), \quad \xi \in (x_0, x_n)$$

If we want to compute the functional value near x_n of $f(x)$, using Newton's interpolation formula, the order of points should be $x_n, x_{n-1}, \cdots, x_0$, then

$$\begin{aligned} N_n(x) = f(x_n) + f[x_n, x_{n-1}](x - x_n) + \\ f[x_n, x_{n-1}, x_{n-2}](x - x_n)(x - x_{n-1}) + \cdots + \\ f[x_n, x_{n-1}, \cdots, x_0](x - x_n)\cdots(x - x_1) \end{aligned}$$

Let $x = x_n + th\ (-1 \leqslant t \leqslant 0)$, then we have **Newton backward-difference formula** (牛顿向后差分公式)

$$\begin{aligned} N_n(x_n + th) = f_n + t\nabla f_n + \frac{t(t+1)}{2!}\nabla^2 f_n + \cdots + \\ \frac{t(t+1)\cdots(t+n-1)}{n!}\nabla^n f_n \end{aligned}$$

Its remainder is

$$R_n(x) = \frac{t(t+1)\cdots(t+n)}{(n+1)!}h^{n+1}f^{(n+1)}(\xi), \quad \xi \in (x_0, x_n)$$

Example 2.5.1 Given the table of $\sin x$, calculate sin(0.12) and sin(0.58), and estimate their truncation errors respectively (see Table 2-6).

Table 2-6 The differences

x	$\sin x$	Δy	$\Delta^2 y$	$\Delta^3 y$
0.1	0.099 83			
		0.098 84		
0.2	0.198 67		−0.001 99	
		0.096 85		−0.000 96
0.3	0.295 52		−0.002 95	
		0.093 90		−0.000 94
0.4	0.389 42		−0.003 89	
		0.090 01		−0.000 91
0.5	0.479 43		−0.004 80	
		0.085 21		
0.6	0.564 64			

Solution (1) Since 0.12 is between 0.1 and 0.2, let $x_0 = 0.1$, then

$$t = \frac{x - x_0}{h} = \frac{0.12 - 0.1}{0.1} = 0.2$$

Using linear interpolation

$$N_1(x_0 + th) = y_0 + t\Delta y_0$$

we have

$$\sin(0.12) \approx N_1(0.12) = 0.099\,83 + 0.2 \times 0.098\,84 = 0.119\,60$$

Using quadratic interpolation, we have

$$\begin{aligned}\sin(0.12) &\approx N_2(0.12)\\ &= 0.099\,83 + 0.2 \times 0.098\,84 + \frac{0.2 \times (0.2-1)}{2} \times (-0.001\,99)\\ &= N_1(0.12) + 0.000\,16 = 0.119\,76\end{aligned}$$

Using cubic interpolation, we have

$$\begin{aligned}\sin(0.12) &\approx N_3(0.12)\\ &= N_2(0.12) + \frac{0.2 \times (0.2-1) \times (0.2-2)}{6} \times (-0.000\,96) = 0.119\,71\end{aligned}$$

$$\begin{aligned}|R_3(0.12)| &\leqslant \left|\frac{0.2 \times (0.2-1) \times (0.2-2) \times (0.2-3)}{24}\right| \times (0.1)^4 \times \sin(0.4)\\ &< 0.000\,002\end{aligned}$$

(2) Using Newton's backward interpolation formula, we have

$$t = \frac{x - x_5}{h} = \frac{0.58 - 0.6}{0.1} = -0.2$$

$$\begin{aligned}\sin(0.58) &\approx N_3(0.58)\\ &= 0.564\,64 + (-0.2) \times 0.085\,21 + \frac{(-0.2) \times (-0.2+1)}{2} \times (-0.004\,80) +\\ &\quad \frac{(-0.2) \times (-0.2+1) \times (-0.2+2)}{6} \times (-0.000\,91)\\ &= 0.548\,02\end{aligned}$$

$$\begin{aligned}|R_3(0.58)| &\leqslant \left|\frac{-0.2 \times (-0.2+1) \times (-0.2+2) \times (-0.2+3)}{24}\right| \times (0.1)^4 \times \sin(0.6)\\ &< 0.000\,002\end{aligned}$$

Example 2.5.2 Given Table 2-7.

(1) Compute approximate $f(0.7)$ with **cubic** (三次) interpolation polynomial.

(2) Compute approximate $f(0.95)$ with **quadratic** (二次) interpolation polynomial.

(3) How many significant digits do we get to compute $f(x)(0.2\leqslant x\leqslant 1.2)$ with quadratic interpolation polynomial, given $\max\limits_{0.2\leqslant x\leqslant 1.2}|f'''(x)|\leqslant 600$, not consider round-off error?

Table 2-7 The experimental data

x	0.2	0.4	0.6	0.8	1.0	1.2
$f(x)$	21	25	23	20	21	24

Solution see Table 2-8.

Table 2-8 The differences

x	y	Δy	$\Delta^2 y$	$\Delta^3 y$	$\Delta^4 y$	$\Delta^5 y$
0.2	21					
		4				
0.4	25		−6			
		−2		5		
0.6	23		-1		0	
		−3		5		−7
0.8	20		4		−7	
		1		−2		
1.0	21		2			
		3				
1.2	24					

(1) $x_1=0.4, x_2=0.6, x_3=0.8, x_4=1.0$

$$N_3(x_1+th)=y_1+t\Delta y_1+\frac{t(t-1)}{2!}\Delta^2 y_1+\frac{t(t-1)(t-2)}{3!}\Delta^3 y_1$$

$$N_3(0.4+0.2t)=25-2t-\frac{t(t-1)}{2}+\frac{5t(t-1)(t-2)}{6}$$

Since $\quad 0.4+0.2t=0.7,\ t=1.5$

Therefore $\quad f(0.7)\approx N_3(0.7)=21.312\,5$

(2) $x_3=0.8, x_4=1.0, x_5=1.2$

$$N_2(x_3+th)=y_3+t\Delta y_3+\frac{t(t-1)}{2!}\Delta^2 y_3$$

$$N_2(0.8+0.2t)=20+t+t(t-1)$$

Since $\quad 0.8+0.2t=0.95,\quad t=0.75$

Therefore $\quad f(0.95)\approx N_2(0.95)=20.562\,5$

(3) $R_2(x_0+th)=\dfrac{f'''(\xi)}{3!}h^3t(t-1)(t-2),\quad 0.2<\xi<1.2,\ 0\leqslant t\leqslant 2$

$$|R_2(x_0+th)|=\left|\frac{f'''(\xi)}{3!}h^3t(t-1)(t-2)\right|$$

$$\leqslant\frac{600}{3!}\times 0.2^3\max_{0\leqslant t\leqslant 2}|t(t-1)(t-2)|$$

$$=\frac{600}{3!}\times 0.008\times 0.384\,900=0.307\,92<0.5$$

It has two significant digits.

2.6 Hermite Interpolation（埃尔米特插值）

The term **Hermite interpolation** refers to the interpolation of a function and some of its derivatives at a set of nodes.

Consider the distinct points $a = x_0, x_1, x_2, \cdots, x_n = b$

and

$$y_j = f(x_j),\quad m_j = f'(x_j)\quad (j = 0,1,\cdots,n)$$

find an interpolation polynomial $H(x)$ which satisfies the following interpolation conditions:

$$H(x_j) = y_j, H'(x_j) = m_j \quad (j = 0,1,\cdots,n) \tag{2.6.1}$$

Here are $2n+2$ conditions which can determine a unique polynomial $H_{2n+1}(x) = H(x)$ and its degree is less than or equal to $2n+1$:

$$H_{2n+1}(x) = a_0 + a_1 x + \cdots + a_{2n+1} x^{2n+1}$$

It is very complex to determine $2n+2$ coefficients by using conditions (2.6.1), therefore, we use the method of the base functions. We first solve the base functions of interpolation $\alpha_j(x)$ and $\beta_j(x)$ $(j = 0,1,\cdots,n)$, each base function is a polynomial of degree $2n+1$ and satisfies :

$$\left.\begin{aligned} &\alpha_j(x_k) = \delta_{jk} = \begin{cases} 0 & k \neq j \\ 1 & k = j \end{cases} \quad \alpha_j'(x_k) = 0 \\ &\beta_j(x_k) = 0,\ \beta_j'(x_k) = \delta_{jk} \quad (j,k = 0,1,\cdots,n) \end{aligned}\right\} \tag{2.6.2}$$

then the interpolation polynomial $H(x) = H_{2n+1}(x)$ can be written as follows

$$H_{2n+1}(x) = \sum_{j=0}^{n} [y_j \alpha_j(x) + m_j \beta_j(x)] \tag{2.6.3}$$

By (2.6.2), it is obvious that

$$H_{2n+1}(x_k) = y_k, H_{2n+1}'(x_k) = m_k \quad (k = 0,1,\cdots,n)$$

Let

$$\alpha_j(x) = (ax + b) l_j^2(x)$$

where $l_j(x)$ is a base function of Lagrange interpolation.

By using (2.6.2), we obtain

$$\alpha_j(x_j) = (ax_j + b) l_j^2(x_j) = 1$$

$$\alpha_j'(x_j) = l_j(x_j)[a l_j(x_j) + 2(ax_j + b) l_j'(x_j)] = 0$$

By arranging, we obtain

$$\begin{cases} ax_j + b = 1 \\ a + 2l_j'(x_j) = 0 \end{cases}$$

That is

$$a = -2l_j'(x_j),\quad b = 1 + 2x_j l_j'(x_j)$$

Since

$$l_j(x) = \frac{(x-x_0)\cdots(x-x_{j-1})(x-x_{j+1})\cdots(x-x_n)}{(x_j-x_0)\cdots(x_j-x_{j-1})(x_j-x_{j+1})\cdots(x_j-x_n)}$$

Take the logarithm and then compute the derivative of the two sides of the equation, we obtain that

$$l_j'(x_j) = \sum_{\substack{k=0 \\ k\neq j}}^{n} \frac{1}{x_j - x_k}$$

Therefore

$$\alpha_j(x) = \left(1 - 2(x-x_j)\sum_{\substack{k=0 \\ k\neq j}}^{n} \frac{1}{x_j - x_k}\right) l_j^2(x) \tag{2.6.4}$$

Analogously

$$\beta_j(x) = (x - x_j) l_j^2(x) \tag{2.6.5}$$

We can prove that the interpolation polynomial satisfying (2.6.1) is unique with the proof by contradiction.

In fact, we assume that both $H_{2n+1}(x)$ and $\overline{H}_{2n+1}(x)$ satisfy the condition (2.6.1) ,then $\varphi(x) = H_{2n+1}(x) - \overline{H}_{2n+1}(x)$ has 2-multiple root at each point x_k, that is, $\varphi(x)$ has altogether $2n+2$ roots. Since the degree of $\varphi(x)$ does not exceed $2n+1$, hence $\varphi(x) \equiv 0$, the uniqueness has been proved.

The remainder of Hermite interpolation is

$$R(x) = f(x) - H_{2n+1}(x) = \frac{f^{(2n+2)}(\xi)}{(2n+2)!}\omega_{n+1}^2(x) \tag{2.6.6}$$

where $\xi \in (a,b)$ and depending on x.

Especially, let $n=1$, the interpolation polynomial $H_3(x)$ with points x_k and x_{k+1}, which satisfies :

$$\left.\begin{aligned} H_3(x_k) = y_k \quad H_3(x_{k+1}) = y_{k+1} \\ H_3'(x_k) = m_k \quad H_3'(x_{k+1}) = m_{k+1} \end{aligned}\right\} \tag{2.6.7}$$

The base functions of interpolation are $\alpha_k(x), \alpha_{k+1}(x), \beta_k(x), \beta_{k+1}(x)$, which satisfy:

$$\alpha_k(x_k) = 1, \alpha_k(x_{k+1}) = 0, \quad \alpha_k'(x_k) = \alpha_k'(x_{k+1}) = 0$$

$$\alpha_{k+1}(x_k) = 0, \alpha_{k+1}(x_{k+1}) = 1, \quad \alpha_{k+1}'(x_k) = \alpha_{k+1}'(x_{k+1}) = 0$$

$$\beta_k(x_k) = \beta_k(x_{k+1}) = 0, \quad \beta_k'(x_k) = 1, \beta_k'(x_{k+1}) = 0$$

$$\beta_{k+1}(x_k) = \beta_{k+1}(x_{k+1}) = 0, \quad \beta_{k+1}'(x_k) = 0, \beta_{k+1}'(x_{k+1}) = 1$$

By using (2.6.4) and (2.6.5), we obtain :

$$\left.\begin{aligned} \alpha_k(x) &= \left(1 + 2\frac{x - x_k}{x_{k+1} - x_k}\right)\left(\frac{x - x_{k+1}}{x_k - x_{k+1}}\right)^2 \\ \alpha_{k+1}(x) &= \left(1 + 2\frac{x - x_{k+1}}{x_k - x_{k+1}}\right)\left(\frac{x - x_k}{x_{k+1} - x_k}\right)^2 \end{aligned}\right\} \tag{2.6.8}$$

$$\left.\begin{aligned}\beta_k(x) &= (x-x_k)\left(\frac{x-x_{k+1}}{x_k-x_{k+1}}\right)^2 \\ \beta_{k+1}(x) &= (x-x_{k+1})\left(\frac{x-x_k}{x_{k+1}-x_k}\right)^2\end{aligned}\right\} \tag{2.6.9}$$

Finally

$$H_3(x) = y_k\alpha_k(x) + y_{k+1}\alpha_{k+1}(x) + m_k\beta_k(x) + m_{k+1}\beta_{k+1}(x) \tag{2.6.10}$$

Using (2.6.6), we obtain the remainder:

$$R_3(x) = \frac{1}{4!}f^{(4)}(\xi)(x-x_k)^2(x-x_{k+1})^2, \quad \xi \in (x_k, x_{k+1})$$

Example 2.6.1 Solve an interpolation polynomial and its remainder, such that

$$P(x_j) = f(x_j) \ (j = 0,1,2), \quad P'(x_1) = f'(x_1)$$

Solution Since this polynomial cross the points $(x_0, f(x_0))$, $(x_1, f(x_1))$, $(x_2, f(x_2))$, therefore its form is of

$$P(x) = f(x_0) + f[x_0,x_1](x-x_0) + f[x_0,x_1,x_2](x-x_0)(x-x_1) + A(x-x_0)(x-x_1)(x-x_2)$$

where A is an undetermined constant, which can be determined by the condition $P'(x_1) = f'(x_1)$. By computing, we obtain:

$$A = \frac{f'(x_1) - f[x_0,x_1] - (x_1-x_0)f[x_0,x_1,x_2]}{(x_1-x_0)(x_1-x_2)}$$

Let

$$R(x) = f(x) - P(x) = k(x)(x-x_0)(x-x_1)^2(x-x_2)$$

where $k(x)$ is an undetermined function, we structure

$$\varphi(t) = f(t) - P(t) - k(x)(t-x_0)(t-x_1)^2(t-x_2)$$

Obviously, $\varphi(x_j) = 0 (j = 0,\ 1,\ 2)$ and $\varphi'(x_1) = 0$, $\varphi(x) = 0$, so $\varphi(t)$ has 5 zeros in (a,b). Using Rolle's theorem repeatedly, we obtain that $\varphi^{(4)}(t)$ has one point ξ at least in (a,b), therefore

$$\varphi^{(4)}(\xi) = f^{(4)}(\xi) - 4!k(x) = 0$$

Hence

$$k(x) = \frac{1}{4!}f^{(4)}(\xi)$$

$$R(x) = \frac{1}{4!}f^{(4)}(\xi)(x-x_0)(x-x_1)^2(x-x_2) \tag{2.6.11}$$

where ξ is in the range determined by x_0, x_1, x_2 and x.

Example 2.6.2 Solve an interpolation polynomial which satisfies the conditions

$$P(0) = P'(0) = 0, \quad P(1) = 1, \quad P(2) = 2$$

Solution Let

$$P(x)=x^2(ax+b)$$

Then from the given conditions, we have

$$P(x)=-\frac{1}{2}x^3+\frac{3}{2}x^2$$

What happen in Hermite interpolation when there is only one node? In this case, we require a polynomial p of degree k, say , for which

$$p^{(j)}(x_0)=c_{0j}\quad (0\leqslant j\leqslant k)$$

The solution is the Taylor polynomial

$$p(x)=c_{00}+c_{01}(x-x_0)+\frac{c_{02}}{2!}(x-x_0)^2+\cdots+\frac{c_{0k}}{k!}(x-x_0)^k\,.$$

Now let us explain how the Newton divided difference method can be extended to solve Hermite interpolation problems. We begin with a simple case in which a quadratic polynomial p is sought, taking prescribed values:

$$p(x_0)=c_{00},\ p'(x_0)=c_{01},\ p(x_1)=c_{10} \tag{2.6.12}$$

We write the divided difference table in this form:

x_0	c_{00}	c_{01}	?
x_0	c_{00}	?	
x_1	c_{10}		

The question marks stand for entries that are not yet computed. Observe that x_0 appears twice in the argument column since two conditions are being imposed on p at x_0. Note further that the prescribed value of $p'(x_0)$ has been placed in the column of first-order divided differences. This is in accordance with the equation

$$\lim_{x\to x_0} f[x_0,x]=\lim_{x\to x_0}\frac{f(x)-f(x_0)}{x-x_0}=f'(x_0)$$

This equation justifies our defining

$$f[x_0,x]\equiv f'(x_0)$$

The remaining entries in the divided difference table can be computed in the usual way. The difficulty to be expected when the nodes are repeated occurs only at the entry c_{01}, and the value of c_{01} has been supplied by the data so it does not need to be computed. The entries denoted by question marks are then computed in the way:

$$p[x_0,x_1]=\frac{p(x_1)-p(x_0)}{x_1-x_0}=\frac{c_{10}-c_{00}}{x_1-x_0}$$

and

$$p[x_0,x_0,x_1]=\frac{p[x_0,x_1]-p[x_0,x_0]}{x_1-x_0}=\frac{c_{10}-c_{00}}{(x_1-x_0)^2}-\frac{c_{01}}{x_1-x_0}$$

The interpolating polynomial is written in the usual way:

$$p(x)=p(x_0)+p[x_0,x_0](x-x_0)+p[x_0,x_0,x_1](x-x_0)^2$$

We can obtain the interpolation polynomial

$$p(x)=f(x_0)+f'(x_0)(x-x_0)+f[x_0,x_0,x_1](x-x_0)^2+$$
$$f[x_0,x_0,x_1,x_1](x-x_0)^2(x-x_1)$$

directly from the following divided difference table

x_0	$f(x_0)$	$f'(x_0)$	$f[x_0,x_0,x_1]$	$f[x_0,x_0,x_1,x_1]$
x_0	$f(x_0)$	$f[x_0,x_1]$	$f[x_0,x_1,x_1]$	
x_1	$f(x_0)$	$f'(x_1)$		
x_1	$f(x_0)$			

From

$$f[x_0,x_1,\cdots,x_k]=\frac{f^{(k)}(\xi)}{k!}$$

here it must be assumed that $f^{(k)}$ exists and is continuous in the smallest interval containing the nodes $x_0,x_1,\cdots,x_k$. The point ξ will lie in the same interval. If the length of that interval shrinks to 0, we obtain in the limit

$$f[x_0,x_0,\cdots,x_0]=\frac{f^{(k)}(x_0)}{k!} \tag{2.6.13}$$

Notice that when $k\geqslant 2$, we must be careful to include the factor $\frac{1}{k!}$.

Example 2.6.3 Use the extended Newton divided difference algorithm to determine a polynomial that takes these values:

$$p(1)=2,\ p'(1)=3,\ p(2)=6,\ p'(2)=7,\ p''(2)=4$$

Solution We put the data in the divided difference array as follows, using question marks to signify that quantities are to be computed. The final result is on the right

1	2	3	?	?	?
1	2	?	?	?	
2	6	7	4		
2	6				
2	6				

1	2	3	1	2	−1
1	2	4	3	1	
2	6	7	4		
2	6				
2	6				

Notice that in the third row, a second difference of 4 is inserted in accordance with Formula (2.6.13) in the case $k=2$. When the array is completed, the numbers in the top row (excluding the node) are the coefficients in the interpolating polynomial:

$$p(x)=2+3(x-1)+(x-1)^2+2(x-1)^2(x-2)-(x-1)^2(x-2)^2$$

2.7 Piecewise Low Degree Interpolation（分段低次插值）

2.7.1 Ill-posed Properties of High Degree Interpolation

See Figure 2-5. For function y=1/(1+25x^2), $P_5(x)$ and $P_{10}(x)$ are interpolation polynomials of degree 5 and 10 respectively.

As we have seen in our considerations of the convergence of interpolation polynomials, increasing the number of interpolation points , i.e., increasing the degree of the polynomials, does not always lead to an improvement in the approximation. The piecewise interpolation of low degree and the spline interpolation (样条插值) that we will study in this section **remedy** (弥补) this deficiency of interpolation by high-degree polynomials.

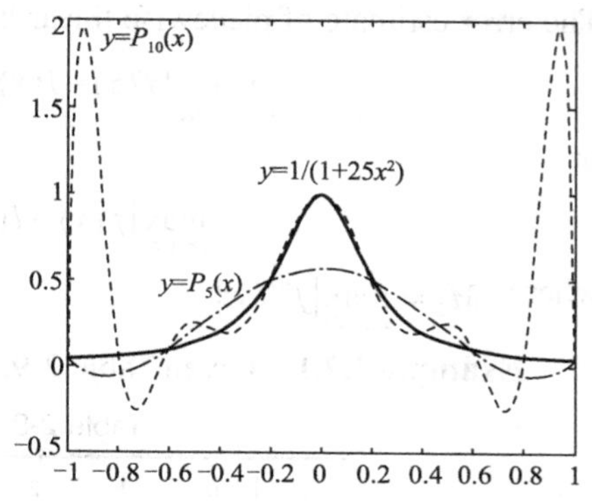

Figure 2-5 Runge's phenomenon

2.7.2 Piecewise Linear Interpolation

Given the functional values $f_0, f_1, \cdots, f_n$ at points $a = x_0 < x_1 < \cdots < x_n = b$. Denote $h_k = x_{k+1} - x_k, h = \max\limits_k h_k$ solve a **polygonal function** (折线函数) $I(x)$ which satisfies :

(1) $I(x) \in \mathbf{C}[a,b]$.

(2) $I(x_k) = f_k \quad (k = 0,1,\cdots,n)$.

(3) $I(x)$ is a linear function on each subinterval $[x_k, x_{k+1}]$. We call $I(x)$ **piecewise linear interpolation** (see Figure 2-6).

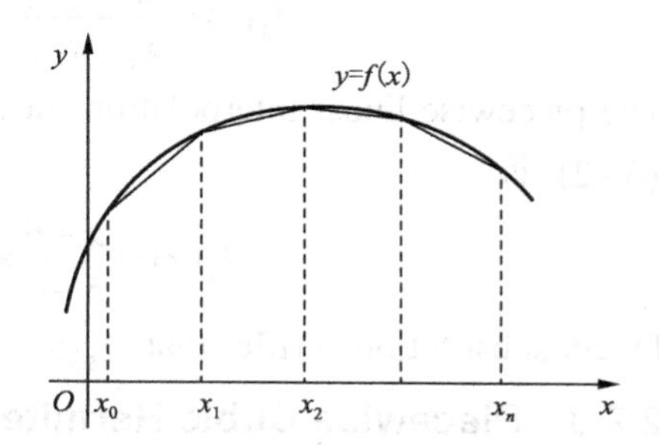

Figure 2-6 Piecewise linear interpolation

From the definition of piecewise linear interpolation, it is clearly that $I(x)$ can be represented on each subinterval $[x_k, x_{k+1}]$ as follows:

$$I(x) = \frac{x - x_{k+1}}{x_k - x_{k+1}} f_k + \frac{x - x_k}{x_{k+1} - x_k} f_{k+1} \quad (x_k \leqslant x \leqslant x_{k+1}) \tag{2.7.1}$$

If we use base functions of interpolation, then on the whole interval $[a,b]$, $I(x)$ can be represented by

$$I(x) = \sum_{j=0}^{n} f_j l_j(x) \tag{2.7.2}$$

where $l_j(x)$ satisfies the condition

$$l_j(x_k) = \delta_{jk} \quad (j,\ k = 0,1,\cdots,n)$$

and $l_j(x)$ is of the form

$$l_j(x)= \begin{cases} \dfrac{x - x_{j-1}}{x_j - x_{j-1}} & x_{j-1} \leqslant x \leqslant x_j \quad (j = 0 \text{ is omited}) \\ \dfrac{x - x_{j+1}}{x_j - x_{j+1}} & x_j \leqslant x \leqslant x_{j+1} \quad (j = n \text{ is omited}) \\ 0 & x \in [a,b], x \notin [x_{j-1}, x_{j+1}] \end{cases} \tag{2.7.3}$$

The error estimate of piecewise linear interpolation is given by

$$\max_{x_k \leqslant x \leqslant x_{k+1}} |f(x)-I(x)| \leqslant \frac{M_2}{2} \max_{x_k \leqslant x \leqslant x_{k+1}} |(x-x_k)(x-x_{k+1})|$$

or

$$\max_{a \leqslant x \leqslant b} |f(x)-I(x)| \quad \frac{M_2}{8} h^2 \to 0 \quad (h \to 0) \tag{2.7.4}$$

where $M_2 = \max_{a \leqslant x \leqslant b} |f''(x)|$.

Example 2.7.1 Given Table 2-9.

Table 2-9 The experimental data

x	0	1	2	3	4	5	6	7	8
$f(x)$	6	0	2	–1	3	4	–2	5	4

Find a piecewise linear fit.

Solution In this case, $x_k = k,\ k = 0,\cdots,8$,

$$I_k(x) = \frac{x-x_{k+1}}{x_k - x_{k+1}} \cdot f_k + \frac{x-x_k}{x_{k+1}-x_k} \cdot f_{k+1}, \quad k = 0,\cdots,7$$

The piecewise linear interpolation, say, between the points $(x_5, y_5) = (5,4)$ and $(x_6, y_6) = (6,-2)$, is

$$I_5(x) = \frac{x-6}{5-6} \times 4 + \frac{x-5}{6-5} \times (-2) = -6x+34$$

Direct substitution verifies that $I_5(5) = 4$ and $I_5(6) = -2$ as required.

2.7.3 Piecewise Cubic Hermite Interpolation（分段三次埃尔米特插值）

The derivative of piecewise linear interpolation $I_h(x)$ is discontinuous, if given $f_k' = m_k (k = 0,1,\cdots,n)$, then we can structure a piecewise interpolation function $I_h(x)$ who satisfies:

(1) $I_h(x) \in \mathbf{C}^1[a,b]$. ($\mathbf{C}^1[a,b]$ is the set of functions whose derivative are continuous on interval $[a,b]$)

(2) $I_h(x_k) = f_k, I_h'(x_k) = f_k' \quad (k = 0,1,\cdots,n)$.

(3) $I_h(x)$ is a cubic polynomial on each subinterval $[x_k, x_{k+1}]$.

From interpolation polynomial (2.6.10), on interval $[x_k, x_{k+1}]$, $I_h(x)$ can be represented:

$$I_h(x) = \left(\frac{x-x_{k+1}}{x_k - x_{k+1}}\right)^2 \left(1+2\frac{x-x_k}{x_{k+1}-x_k}\right) f_k + \left(\frac{x-x_k}{x_{k+1}-x_k}\right)^2 \left(1+2\frac{x-x_{k+1}}{x_k - x_{k+1}}\right) f_{k+1} + \left(\frac{x-x_{k+1}}{x_k - x_{k+1}}\right)^2 (x-x_k) f_k' + \left(\frac{x-x_k}{x_{k+1}-x_k}\right)^2 (x-x_{k+1}) f_{k+1}' \tag{2.7.5}$$

If we define a group of piecewise cubic interpolation base functions $\alpha_j(x)$ and $\beta_j(x)$ $(j=0,1,\cdots,n)$, then

$$I_h(x)=\sum_{j=0}^{n}[f_j\alpha_j(x)+f_j'\beta_j(x)] \tag{2.7.6}$$

where
$$\alpha_j(x)=\begin{cases}\left(\dfrac{x-x_{j-1}}{x_j-x_{j-1}}\right)^2\left(1+2\dfrac{x-x_j}{x_{j-1}-x_j}\right) & x_{j-1}\leqslant x\leqslant x_j \ (j=0 \text{ is omited})\\ \left(\dfrac{x-x_{j+1}}{x_j-x_{j+1}}\right)^2\left(1+2\dfrac{x-x_j}{x_{j+1}-x_j}\right) & x_j\leqslant x\leqslant x_{j+1} \ (j=n \text{ is omited})\\ 0 & \text{others}\end{cases} \tag{2.7.7}$$

$$\beta_j(x)=\begin{cases}\left(\dfrac{x-x_{j-1}}{x_j-x_{j-1}}\right)^2(x-x_j) & x_{j-1}\leqslant x\leqslant x_j \ (j=0 \text{ is omited})\\ \left(\dfrac{x-x_{j+1}}{x_j-x_{j+1}}\right)^2(x-x_j) & x_j\leqslant x\leqslant x_{j+1} \ (j=n \text{ others})\\ 0 & \text{others}\end{cases} \tag{2.7.8}$$

$$I_h(x)=f_k\alpha_k(x)+f_{k+1}\alpha_{k+1}(x)+f_k'\ \beta_k(x)+f_{k+1}'\beta_{k+1}(x) \quad (x_k\leqslant x\leqslant x_{k+1}) \tag{2.7.9}$$

Theorem 2.7.1 Let $f\in \mathbf{C}^1[a,b]$, then $I_h(x)$ uniformly converges to $f(x)$ if $h\to 0$.

2.8 Cubic Spline Interpolation（三次样条插值）

2.8.1 Definition of Cubic Spline

Definition 2.8.1 For given Table 2-10.

$(a=x_0<x_1<\cdots<x_n=b)$ If function $S(x)$ satisfies:

(1) $S(x)$ is a polynomial of degree at most 3 on each subinterval

$$[x_j,x_{j+1}] \quad (j=0,1,\cdots,n-1)$$

(2) $S(x),S'(x),S''(x)$ are continuous on $[a, b]$.

(3) $S(x_j)=y_j \quad (j=0,1,\cdots,n)$.

Then $S(x)$ is called **cubic spline interpolation of function** $f(x)$ **with points** $x_0,x_1,\cdots,x_n$ (see Figure 2-7).

Let $S_j(x)$ be the representation of $S(x)$ in $[x_j,x_{j+1}]$ $(j=0,1,\cdots,n-1)$, then

$$S_j(x)=a_{j0}+a_{j1}x+a_{j2}x^2+a_{j3}x^3, \qquad x\in[x_j,x_{j+1}]$$

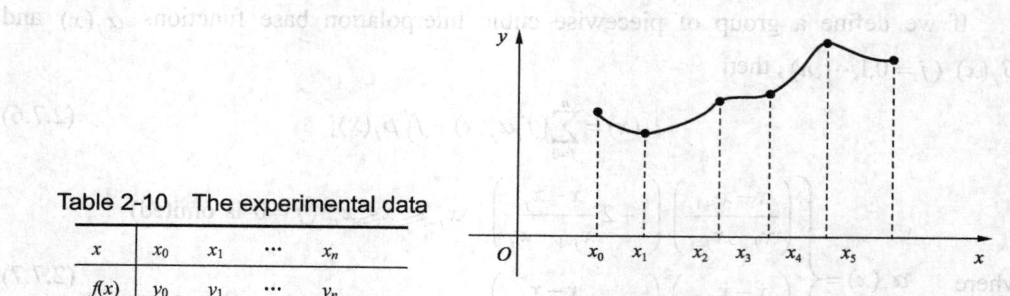

Table 2-10 The experimental data

x	x_0	x_1	$\cdots$	x_n
$f(x)$	y_0	y_1	$\cdots$	y_n

Figure 2-7 Cubic spline interpolation

There are $4n$ undetermined coefficients, from the condition (2) and (3) in the definition, we have $4n-2$ equations :

$$\begin{cases} S(x_i-0)=S(x_i+0) & (i=1,2,\cdots,n-1) \\ S'(x_i-0)=S'(x_i+0) & (i=1,2,\cdots,n-1) \\ S''(x_i-0)=S''(x_i+0) & (i=1,2,\cdots,n-1) \\ S(x_i)=y_i & (i=0,1,\cdots,n) \end{cases} \tag{2.8.1}$$

In order to determine $4n$ undetermined coefficients, it also needs two conditions. For this reason, we give the **boundary conditions** (边界条件):

(a)
$$S'(x_0)=f_0', \quad S'(x_n)=f_n' \tag{2.8.2}$$
(b)
$$S''(x_0)=f_0'', \quad S''(x_n)=f_n'' \tag{2.8.3}$$
Especially,

$$S''(x_0)=S''(x_n)=0$$

It is called **free or natural boundary**.

When the free boundary conditions occur, the spline is called a **natural cubic spline**.

(c) If $f(x)$ is a **periodic function** (周期函数) with period x_n-x_0, then it requires that $S(x)$ is a periodic function, now, the boundary conditions are $S(x_0+0)=S(x_n-0)$, $S'(x_0+0)=S'(x_n-0)$, $S''(x_0+0)=S''(x_n-0)$ and $y_0=y_n$.

2.8.2 The Construction of Cubic Spline

Let

$$S''(x_j)=M_j \quad (j=0,1,\cdots,n) \tag{2.8.4}$$

On subinterval $[x_j, x_{j+1}]$, $S(x)$ is a polynomial of degree $\leqslant 3$, therefore, its second derivative $S''(x)$ is a linear function which can be represented (where $h_j=x_{j+1}-x_j$)

$$S''(x)=M_j\frac{x_{j+1}-x}{h_j}+M_{j+1}\frac{x-x_j}{h_j} \tag{2.8.5}$$

Integrate (积分) this equation for two times and determine the integral constants by using $S(x_j)=y_j$ and $S(x_{j+1})=y_{j+1}$, we obtain

$$S(x)=M_j\frac{(x_{j+1}-x)^3}{6h_j}+M_{j+1}\frac{(x-x_j)^3}{6h_j}+\left(y_j-\frac{M_jh_j^2}{6}\right)\frac{x_{j+1}-x}{h_j}+\left(y_{j+1}-\frac{M_{j+1}h_j^2}{6}\right)\frac{x-x_j}{h_j} \tag{2.8.6}$$

where $M_j, j=0,1,\cdots,n,$ are unknown. In order to determine $M_j, j=0,1,\cdots,n,$ let's differentiate $S(x)$ and obtain

$$S'(x)=-M_j\frac{(x_{j+1}-x)^2}{2h_j}+M_{j+1}\frac{(x-x_j)^2}{2h_j}+\frac{y_{j+1}-y_j}{h_j}-\frac{M_{j+1}-M_j}{6}h_j \tag{2.8.7}$$

Hence, we obtain

$$S'(x_j+0)=-\frac{h_j}{3}M_j-\frac{h_j}{6}M_{j+1}+\frac{y_{j+1}-y_j}{h_j}$$

$$S'(x_j-0)=\frac{h_{j-1}}{6}M_{j-1}+\frac{h_{j-1}}{3}M_j+\frac{y_j-y_{j-1}}{h_{j-1}}$$

Using $S'(x_j+0)=S'(x_j-0)$, we obtain

$$\mu_jM_{j-1}+2M_j+\lambda_jM_{j+1}=d_j \quad (j=1,\ 2,\cdots,n-1) \tag{2.8.8}$$

Where

$$\mu_j=\frac{h_{j-1}}{h_{j-1}+h_j},\ \lambda_j=\frac{h_j}{h_{j-1}+h_j},\quad j=1,\cdots,n \tag{2.8.9}$$

$$d_j=6\frac{f[x_j,x_{j+1}]-f[x_{j-1},x_j]}{h_{j-1}+h_j}=6f[x_{j-1},x_j,x_{j+1}] \tag{2.8.10}$$

For the first boundary condition, we can deduce the equations:

$$\left.\begin{aligned} 2M_0+M_1&=\frac{6}{h_0}(f[x_0,x_1]-f_0') \\ M_{n-1}+2M_n&=\frac{6}{h_{n-1}}(f_n'-f[x_{n-1},x_n]) \end{aligned}\right\} \tag{2.8.11}$$

Let

$$\lambda_0=1,\ \ d_0=\frac{6}{h_0}(f[x_0,x_1]-f_0')$$

$$\mu_n=1,\ \ d_n=\frac{6}{h_{n-1}}(f_n'-f[x_{n-1},x_n])$$

then (2.8.8) and (2.8.11) can be represented in the form of matrix

$$\begin{pmatrix} 2 & \lambda_0 & & & \\ \mu_1 & 2 & \lambda_1 & & \\ & \ddots & \ddots & \ddots & \\ & & \mu_{n-1} & 2 & \lambda_{n-1} \\ & & & \mu_n & 2 \end{pmatrix} \begin{pmatrix} M_0 \\ M_1 \\ \vdots \\ M_{n-1} \\ M_n \end{pmatrix} = \begin{pmatrix} d_0 \\ d_1 \\ \vdots \\ d_{n-1} \\ d_n \end{pmatrix} \tag{2.8.12}$$

For the second boundary condition we can obtain equations of end points:

$$M_0 = f_0'' \ , \quad M_n = f_n''$$

that is

$$\begin{cases} 2M_0 = 2f_0'' \\ 2M_n = 2f_n'' \end{cases} \tag{2.8.13}$$

If let $\lambda_0 = \mu_n = 0, \ d_0 = 2f_0'', \ d_n = 2f_n''$

Then (2.8.8) and (2.8.13) can be written in the form of (2.8.12).

For the third boundary condition (c), we can obtain

$$M_0 = M_n, \qquad \lambda_n M_1 + \mu_n M_{n-1} + 2M_n = d_n \tag{2.8.14}$$

Where

$$\lambda_n = \frac{h_0}{h_{n-1} + h_0}, \quad \mu_n = 1 - \lambda_n = \frac{h_{n-1}}{h_{n-1} + h_0}$$

$$d_n = 6\frac{f[x_0, x_1] - f[x_{n-1}, x_n]}{h_0 + h_{n-1}}$$

Then (2.8.8) and (2.8.14) can be written in the form

$$\begin{pmatrix} 2 & \lambda_1 & & & \mu_1 \\ \mu_2 & 2 & \lambda_2 & & \\ & \ddots & & \ddots & \\ & & \mu_{n-1} & 2 & \lambda_{n-1} \\ \lambda_n & & & \mu_n & 2 \end{pmatrix} \begin{pmatrix} M_1 \\ M_2 \\ \vdots \\ M_{n-1} \\ M_n \end{pmatrix} = \begin{pmatrix} d_1 \\ d_2 \\ \vdots \\ d_{n-1} \\ d_n \end{pmatrix} \tag{2.8.15}$$

Example 2.8.1 Given Table 2-11.

Table 2-11 The experimental data

x	−1.5	0	1	2
y	0.125	−1	1	9

Solve a cubic spline function $S(x)$ on the interval$[-1.5,2]$ which satisfies

$$S'(-1.5) = 0.75, \quad S'(2) = 14$$

Solution This is the first case of boundary conditions

$$h_0 = x_1 - x_0 = 1.5, \ h_1 = x_2 - x_1 = 1, \ h_2 = x_3 - x_2 = 1$$

$$f[x_0, x_1] = -0.75, \ f[x_1, x_2] = 2, \ f[x_2, x_3] = 8$$

$$\mu_1 = \frac{h_0}{h_0 + h_1} = \frac{1.5}{1.5 + 1} = 0.6, \ \lambda_0 = 1$$

$$\mu_2 = \frac{h_1}{h_1 + h_2} = \frac{1}{1 + 1} = 0.5, \ \lambda_1 = 1 - \mu_1 = 0.4$$

$$\mu_3 = 1, \quad \lambda_2 = 1 - \mu_2 = 0.5$$

$$d_0 = \frac{6}{h_0}(f[x_0,x_1] - y_0') = -6$$

$$d_3 = \frac{6}{h_2}(f_3' - f[x_2,x_3]) = 36$$

$$d_1 = 6f[x_0,x_1,x_2] = \frac{6}{1.5+1}(f[x_1,x_2] - f[x_0,x_1])$$

$$= \frac{6}{2.5}\left(\frac{1-(-1)}{1-0} - \frac{-1-0.125}{0-(-1.5)}\right) = 6.6$$

$$d_2 = 6f[x_1,x_2,x_3] = \frac{6}{1+1}(f[x_2,x_3] - f[x_1,x_2])$$

$$= 3\left(\frac{9-1}{2-1} - \frac{1-(-1)}{1-0}\right) = 18$$

The system of equations in the form of matrix

$$\begin{pmatrix} 2 & 1 & 0 & 0 \\ 0.6 & 2 & 0.4 & 0 \\ 0 & 0.5 & 2 & 0.5 \\ 0 & 0 & 1 & 2 \end{pmatrix}\begin{pmatrix} M_0 \\ M_1 \\ M_2 \\ M_3 \end{pmatrix} = \begin{pmatrix} -6 \\ 6.6 \\ 18 \\ 36 \end{pmatrix}$$

Solve the system, we obtain

$$M_0 = -5,\quad M_1 = 4,\quad M_2 = 4,\quad M_3 = 16$$

Hence, from (2.8.6), we have

$$S_1(x) = x^3 + 2x^2 - 1,\quad x \in [-1.5, 0]$$

$$S_2(x) = 2x^2 - 1,\quad x \in [0, 1]$$

$$S_3(x) = 2x^3 - 4x^2 + 6x - 3,\quad x \in [1, 2]$$

The cubic spline function

$$S(x) = \begin{cases} x^3 + 2x^2 - 1 & -1.5 \leqslant x < 0 \\ 2x^2 - 1 & 0 \leqslant x < 1 \\ 2x^3 - 4x^2 + 6x - 3 & 1 \leqslant x \leqslant 2 \end{cases}$$

2.9 Computer Experiments

2.9.1 Functions Needed in the Experiments by Mathematica

1．Product——计算连乘积函数

格式 1：`Product [f,{i,m,n}]`

功能：i 从 m 到 n 按自然数递增计算表达式 f 的连乘积 $\prod_{i=m}^{n} f$．

格式 2：`Product [f,{i,m,n,step}]`

功能：i 从 m 到 n 按步长 step 递增计算表达式 f 的连乘积 $\prod_{i=m}^{n} f$.

2. Sum——求和函数

格式 1：`Sum [f,{i,m,n}]`

功能：i 从 m 到 n 按自然数递增计算表达式 f 的和 $\sum_{i=m}^{n} f$.

格式 2：`Sum [f,{i,m,n,step}]`

功能：i 从 m 到 n 按步长 step 递增计算表达式 f 的和 $\sum_{i=m}^{n} f$.

格式 3：`Sum [f,{i,m1,m2},{j,n1,n2},…]`

功能：多重求和 $\sum_{i=m1}^{m2}\sum_{j=n1}^{n2}\cdots f$.

3. Simplify——多项式化简函数

格式：`Simplify [exp]`

功能：将多项式 exp 化为最简形式.

4. Length——计算元素个数函数

格式：`Length[exp]`

功能：计算表达式 exp 中元素的个数.

5. Table——建表函数

格式 1：`Table[f(i),{i,n}]`

功能：生成一个以关于 i 的表达式为元素的表，i 从 1 变化到 n.

格式 2：`Table[f(i),{i,imin,imax,h}]`

功能：生成一个一维表，i 从 $i$min 到 $i$max，增量为 h.若 h 为 1 可省略.

2.9.2 Experiments by Mathematica

1. Lagrange 插值

（1）算法.

① 输入 n+1 个插值点 (已知)：(x_i,y_i), i=0,1,2,⋯,n;

② 计算插值基函数;

③ 得出 n 次 Lagrange 插值多项式.

（2）程序清单.

```
(*Lagrange Interpolation*)
Clear[Ln,xi,yi];
xi=Input["xi="]
yi=Input["yi="]
n=Length[xi]-1;
la=Sum[yi[[i]]*(Product[(x-xi[[j]])/(xi[[i]]-xi[[j]]),{j,1,i-1}]*
   Product[(x-xi[[j]])/(xi[[i]]-xi[[j]]), {j,i+1,n+1}]),{i,1,n+1}]
Ln[x_]=Simplify[la]
```

（3）变量说明.

xi：保存插值基点$\{x_0, x_1, \cdots, x_n\}$；

yi：保存对应函数值$\{y_0, y_1, \cdots, y_n\}$；

n：插值次数；

la：保存 Lagrange 插值多项式；

Ln：保存化简后的 Lagrange 插值多项式.

（4）计算实例.

Example 2.9.1 Given Table 2-12.Find Lagrange interpolation polynomial.

Table 2-12 The experimental data

x	1.2	2	3.5	4.8	5.5
y	3.2	1.8	5.3	3.5	4.9

① 将光标定位在要执行的 Cell 中，按小键盘的【Enter】键；

② 在弹出的两个对话框中分别输入插值点的横坐标序列与纵坐标序列

```
{1.2,2.,3.5,4.8,5.5},{3.2,1.8,5.3,3.5,4.9}
```

如图 2-8(a)所示；

③ 每次输入后单击 OK 命令按钮，系统给出 Lagrange 插值多项式及化简后的插值多项式；

④ 在新的 Cell 中输入 Ln[2.6]，则得 $f(x)$ 在 $x=2.6$ 处的近似值为 3.560 15，如图 2-8(b)所示.

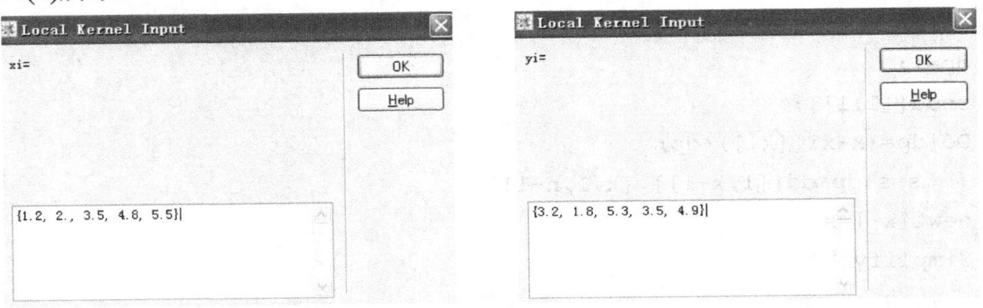

(a) 数据输入界面

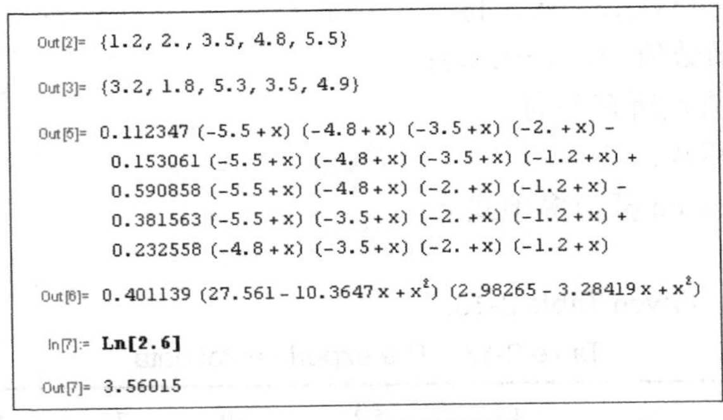

(b) Lagrange 插值计算结果

图 2-8 例 2.9.1 界面及执行结果

2. Newton 插值

（1）算法.

① 输入 n+1 个插值点(已知)：(x_i, y_i), i=0,1,2,⋯,n;

② 计算差商表；

③ 给出 n 次 Newton 插值多项式.

（2）程序清单.

```
(*Newton Interpolation*)
Clear[newt,s,x];
xi=Input["xi="]
yi=Input["yi="]
n=Length[xi];
(*计算差商表*)
dd=Table[0,{n},{n}];
Do[dd[[i,1]]=yi[[i]],{i,1,n}]
Do[dd[[i,j+1]]=(dd[[i,j]]-dd[[i+1,j]])/(xi[[i]]-xi[[i+j]]),
{j,1,n-1},{i,1,n-j}]
Print["差商表"]
Do[Print[xi[[i]],"   ",dd[[i]]],{i,1,n}]
(*计算Newton 插值多项式*)
dp=1;
s=dd[[1,1]];
Do[dp=(x-xi[[k]])*dp;
   s=s+dp*dd[[1,k+1]],{k,1,n-1}]
newt[x_]=s
Simplify[%]
```

（3）变量说明.

xi：存放插值基点$\{x_0, x_1,\cdots, x_n\}$；

yi：存放对应函数值$\{y_0, y_1,\cdots, y_n\}$；

dd：保存函数值及所有差商；

dp：保存差分累积；

newt：保存 Newton 插值多项式.

（4）计算实例.

Example 2.9.2 Given Table 2-13.

Table 2-13 The experimental data

x	3.1	4.8	5.2	6.0	7.3	8.5
y	2.4	1.9	5.7	4.3	2.8	6.2

（1）Find Newton interpolation polynomial of degree 5.

（2）Solve approximate value to $f(4.5)$.

操作步骤如下：

① 将光标定位在要执行的 Cell 中，按小键盘的【Enter】键；

② 在弹出的对话框中按提示分别输入插值点的横坐标序列与纵坐标序列

```
{3.1,4.8,5.2,6,7.3,8.5},{2.4,1.9,5.7,4.3,2.8,6.2}
```

③ 每次输入后单击 OK 命令按钮，系统输出差商表、Newton 插值多项式及化简后的插值多项式；

④ 在新的 Cell 中输入 newt[4.5]，则得 $f(x)$在 $x = 4.5$ 处的近似值为−2.995 95，如图 2-9 所示.

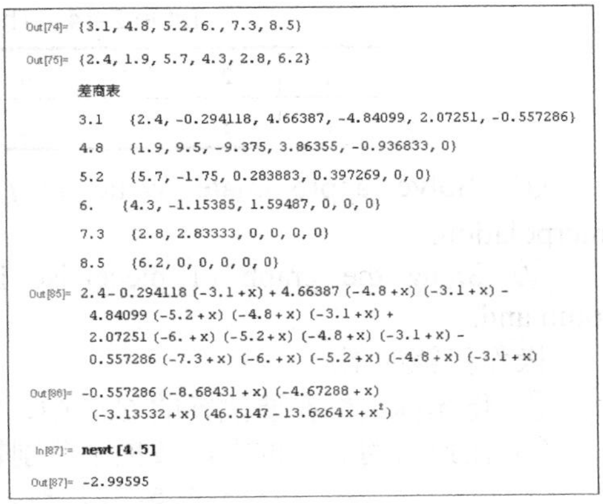

```
Out[74]= {3.1, 4.8, 5.2, 6., 7.3, 8.5}
Out[75]= {2.4, 1.9, 5.7, 4.3, 2.8, 6.2}
         差商表
         3.1   {2.4, -0.294118, 4.66387, -4.84099, 2.07251, -0.557286}
         4.8   {1.9, 9.5, -9.375, 3.86355, -0.936833, 0}
         5.2   {5.7, -1.75, 0.283883, 0.397269, 0, 0}
         6.    {4.3, -1.15385, 1.59487, 0, 0, 0}
         7.3   {2.8, 2.83333, 0, 0, 0, 0}
         8.5   {6.2, 0, 0, 0, 0, 0}
Out[85]= 2.4-0.294118 (-3.1+x) + 4.66387 (-4.8+x) (-3.1+x) -
          4.84099 (-5.2+x) (-4.8+x) (-3.1+x) +
          2.07251 (-6.+x) (-5.2+x) (-4.8+x) (-3.1+x) -
          0.557286 (-7.3+x) (-6.+x) (-5.2+x) (-4.8+x) (-3.1+x)
Out[86]= -0.557286 (-8.68431+x) (-4.67288+x)
          (-3.13532+x) (46.5147-13.6264x+x^2)
In[87]:= newt[4.5]
Out[87]= -2.99595
```

图 2-9　Newton 插值计算结果

3. 分段插值

（1）算法.

① 输入 n 个插值点：(x_i,y_i), $i=1,2,\cdots,n$；

② 输入要做近似计算的自变量 x_a；

③ 求出每个小区间上的线性函数；

④ 寻找包含 x_a 的小区间$[x_i, x_{i+1}]$；

⑤ 用$[x_i,x_{i+1}]$上的线性函数在 x_a 处的值作为 $f(x_a)$的近似值.

（2）程序清单.

```
(*Piecewise Interpolation*)
Clear[x,a,b];
li[a_,b_,x_]:=(b[[2]]-a[[2]])/(b[[1]]-a[[1]])*(x-a[[1]])+a[[2]]
xi=Input["xi="]
yi=Input["yi="]
xa=Input["xa="]
n=Length[xi];
pp=Table[li[{xi[[m]],yi[[m]]},{xi[[m+1]],yi[[m+1]]},x],{m,1,n-1}]
If[xa<xi[[1]]||xa>xi[[n]],
  Print["超限"] ;Break[ ],
  Do[If[xa>=xi[[k]],m=k],{k,1,n-1}];
  Print["m=",m, "    ", "x=",xi[[m]]];
  li[{xi[[m]],yi[[m]]},{xi[[m+1]],yi[[m+1]]},xa]
  ]
```

（3）变量说明.

xi：存放插值基点$\{x_0, x_1,\cdots, x_n\}$；

yi：存放对应函数值$\{y_0, y_1,\cdots, y_n\}$；

xa：存放要做近似计算的自变量；

n：插值基点个数；

li[a_,b_,x_]：定义经过$(a_1，a_2)$, $(b_1，b_2)$两点的直线方程；

pp：保存每个小区间上的线性函数.

（4）计算实例.

Example 2.9.3 Given Table 2-14.

Table 2-14 The experimental data

x	2	3	4	5	6	7	8
y	3	8	2	4	1	5	6

① Solve approximate value to $f(x)$ at $x=5.2$ by using piecewise linear interpolation.

② Show the graph of piecewise interpolation function by using Mathematica command.

操作步骤如下：

① 将光标定位在要执行的 Cell 中，按小键盘的【Enter】键；

② 在弹出的对话框中，按提示分别输入

```
{2,3,4,5,6,7,8},{3,8,2,4,1,5,6},5.2
```

③ 每次输入后单击 OK 命令按钮，输出结果如图 2-10(a)所示.

输出结果表明 $f(x)$ 在 $x=5.2$ 的近似值为 3.4，它对应的小区间为 $[x_4, x_5]=[5,6]$.

④ 在新的 Cell 中输入命令

```
g=Interpolation[{{2,3},{3,8},{4,2},{5,4},{6,1},{7,5},{8,6}},
                 InterpolationOrder->1]
Plot[g[x],{x,2,8}]
```

输出分段线性插值函数图像，如图 2-10(b)所示.

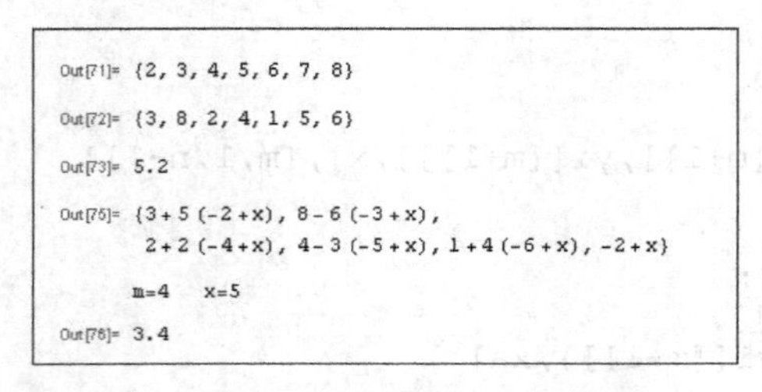

(a) 分段插值计算结果

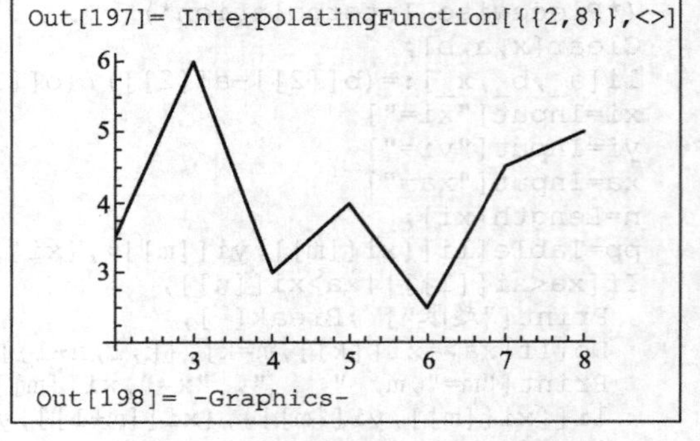

(b) 分段线性插值函数图像

图 2-10 分段插值函数

4. 三次样条插值

（1）算法.

① 判定插值类型，输入边界条件；

② 计算各值，得到正规方程组；

③ 解此正规方程组，求得 $M_0, M_1, M_2, \cdots, M_n$；

④ 将求得的M_i值代回样条函数$S(x)$.

（2）程序清单.

```
(*Cubic Spline Interpolation*)
type=Input["1," "2" "please" "select"];
f1=Input["f1="];
f2=Input["f2="];
xi=Input["xi="];
yi=Input["yi="];
n=Length[xi]-1;
h=Table[xi[[i+1]]-xi[[i]],{i,1,n}];
mu=Table[h[[i]]/(h[[i]]+h[[i+1]]),{i,1,n-1}];
lam=1-mu;
d=Table[6((yi[[i+1]]-yi[[i]])/h[[i]]-(yi[[i]]-yi[[i-1]])/h[[i-1]])/
(h[[i-1]]+h[[i]]),{i,2,n}];
If[type==2,
  d=Prepend[d,2f1];
  d=Append[d,2f2];
  mu=Append[Prepend[mu,0],0];
  lam=Append[Prepend[lam,0],0],
  d=Prepend[d,6*((yi[[2]]-yi[[1]])/h[[1]]-f1)/h[[1]]];
  d=Append[d,6*(f2-(yi[[n+1]]-yi[[n]])/h[[n]])];
  mu=Append[Prepend[mu,0],1];
  lam=Append[Prepend[lam,1],0]
  ] ;
c=lam;
b=Table[2,{n+1}];
a=mu;
n=n+1;
eps=0.000000001;
Do[a[[k]]=a[[k]]/b[[k-1]];
    b[[k]]=b[[k]]-a[[k]]*c[[k-1]];
    d[[k]]=d[[k]]-a[[k]]*d[[k-1]],
    {k,2,n}];
x=Table[0,{n}];
x[[n]]=d[[n]]/b[[n]];
Do[x[[k]]=(d[[k]]-c[[k]]*x[[k+1]])/b[[k]],{k,n-1,1,-1}];
Print["Mk的值为",x//N];
Clear[x1,x2,y1,y2,m1,m2,h1,t,sk]
sk[x1_,x2_,y1_,y2_,m1_,m2_,h1_,t_]:=m1*(x2-t)^3/(6*h1)+m2*(t-x1)^3/
(6*h1)+(y1-m1*h1^2/6)*(x2-t)/h1+(y2-m2*h1^2/6)*(t-x1)/h1;
sp=Table[sk[xi[[i]],xi[[i+1]],yi[[i]],yi[[i+1]],x[[i]],x[[i+1]],h[[
i]],t],{i,1,n-1}];
Print["所求三次样条函数S(t)为"]
Do[Print["S(t)=",Simplify[sp[[i]]]];
Print[" ",xi[[i]],"<=t<",xi[[i+1]]],{i,1,n-1}]
```

（3）变量说明.

type：边界条件类型；

xi：存放插值基点$\{x_0, x_1,\cdots, x_n\}$；

yi：存放对应函数值$\{y_0, y_1,\cdots, y_n\}$；

h：存放小区间长度；

mu：存放计算过程中的μ_k值；

lam：存放计算过程中的λ_k值；

x：存放计算过程中的M_k值, $k=0,1,\cdots,n$；

sp：存放所求出的三次样条函数；

sk：存放第k个小区间上的三次样条函数, $k=1,\cdots,n-1$.

（4）计算实例.

Example 2.9.4　Given Table 2-15.

Find the natural cubic spline, and give approximate value to $f(2.6)$.

操作步骤如下：

① 将光标定位在要执行的 Cell 中，按小键盘的【Enter】键；

② 在弹出的对话框中，按提示分别输入

```
{2,0,0},{0,1,2,3},{1,1,0,10}
```

③ 每次输入后单击OK 命令按钮，系统输出所求三次样条函数，再输入 sp[[3]]/.t->2.6 得 $f(2.6)$ 的近似值为 4.992，如图 2-11 所示.

Table 2-15　The experimental data

x	0	1	2	3
y	1	1	0	10

```
Mk的值为{0., -6., 18., 0.}
所求三次样条函数S(t)为
S(t)=1 + t - t^3
 0<=t<1
S(t)=-4 + 16 t - 15 t^2 + 4 t^3
 1<=t<2
S(t)=52 - 68 t + 27 t^2 - 3 t^3
 2<=t<3
In[27]:=
sp[[3]] /. t → 2.6
Out[27]= 4.992
```

图 2-11　三次样条插值计算结果

2.9.3　Experiments by Matlab

1．拉格朗日插值

（1）函数语句.

```
[P,yy]=lagrintp(x,y,xx)
```

（2）参数说明.

x：实向量，输入参数，节点向量；

y：实向量，输入参数，相应的函数值向量；

xx：实向量，输入参数，自变量的值；

P：符号函数，输出参数，拉格朗日插值多项式；

yy：实变量，输出参数，拉格朗日插值多项式在 xx 的函数值.

（3）lagrintp.m 程序.

```
function [P,yy]=lagrintp(x,y,xx)
n=length(x);m=length(y);
t=sym('x');
if m~=n
   error('向量x与y的长度不一致')
end
P=sym('0');
for k=1:n
   p=1;
   for j=1:n
      if j~=k
         p=p*(t-x(j))/(x(k)-x(j));
      end
   end
   P=P+p*y(k);
end
if nargin<3
   return
else
   yy=subs(P,'x',xx);
end
```

（4）计算实例.

Example 2.9.5 Given Table 2-16.

Table 2-16 The experimental data

x	1.2	2	3.5	4.8	5.5
y	3.2	1.8	5.3	3.5	4.9

Find Lagrange interpolation polynomial and solve approximate value to $f(2.6)$.

在命令窗口输入：

```
>> clear
>>
x=[1.2,2,3.5,4.8,5.5];y=[3.2,1.8,5.3,3.5,4.9];xx=2.6;lagrintp(x,y,xx)
P =
    -800/8901*(-5/4*x+5/2)*(x-7/2)*(x-24/5)*(x-11/2)-6/49*(5/4*x-3/2)
    *(x-7/2)*(x-24/5)*(x-11/2)+53/39*(10/23*x-12/23)*(x-2)*(x-24/5)
    *(x-11/2)-125/91*(5/18*x-1/3)*(x-2)*(x-7/2)*(x-11/2)+(10/43*x-12/
    43)*(x-2)*(x-7/2)*(x-24/5)
yy =
    3.5602
```

2. 牛顿插值

（1）函数语句.

```
[P,yy]=newtint(x,y,xx)
```

（2）参数说明.

x：实向量，输入参数，节点向量；

y：实向量，输入参数，相应的函数值向量；

xx：实向量，输入参数，自变量的值；

P：符号函数，输出参数，牛顿插值多项式；

yy：实变量，输出参数，插值多项式在 xx 的函数值.

（3）newtint.m 程序.

```
function [P,yy]=newtint(x,y,xx)
n=length(y);
if length(x)~=n
   error('x and y are not compatible');
end
%下面调用了求差商表程序 divdiff(x,y)
D=divdiff(x,y);
syms t
P=D(1,1);w=1;
for i=1:n-1
   w=w*(t-x(i));
   P=D(i+1,i+1)*w+P;
end
if nargin<3
   return
end
yy=subs(P,'t',xx);
```

divdiff.m 程序如下：

```
function D=divdiff(x,y)
%产生差商表,x 为插值节点,y 为相应的函数值,输出 D 为一矩阵.
n=length(y);
if length(x)~=n
   fprintf('出错信息: 节点向量与函数值向量的维数不相等.');
end
D=zeros(n,n);
D(:,1)=y(:);   %D 的第一列就是 y
for j=2:n;
   for i=j:n
      D(i,j)=(D(i,j-1)-D(i-1,j-1))/(x(i)-x(i-j+1));
   end
end
```

（4）计算实例.

Example 2.9.6 Given Table 2-17.

Table 2-17 The experimental data

x	3.1	4.8	5.2	6.0	7.3	8.5
y	2.4	1.9	5.7	4.3	2.8	6.2

① Find Newton interpolation of degree 5.

② Solve approximate value to $f(4.5)$.

```
>> clear
>> x=[3.1,4.8,5.2,6.0,7.3,8.5];y=[2.4,1.9,5.7,4.3,2.8,6.2];
xx=4.5;[P,yy]=newtint(x,y,xx)
 P =
    -5019584132683039/9007199254740992*(t-31/10)*(t-24/5)*(t-26/5)*
     (t-6)*(t-73/10)+4666877220606461/2251799813685248*(t-31/10)*
     (t-24/5)*(t-26/5)*(t-6)-66825/13804*(t-31/10)*(t-24/5)*
     (t-26/5)+555/119*(t-31/10)*(t-24/5)-5/17*t+563/170
 yy =
    -2.9960
```

3. 分段线性插值

（1）函数语句.

```
[P,yy]=PiecewiseLineInt(x,y,xx)
```

（2）参数说明.

x：实向量，输入参数，节点向量；

y：实向量，输入参数，相应的函数值向量；

xx：实向量，输入参数，自变量的值；

P：符号函数向量，输出参数，P(k)是在[x(k),x(k+1)]上的线性插值函数；

yy：实向量，输出参数，分段线性插值函数 P 在 xx 的函数值.

（3）PiecewiseLineInt.m 程序.

```
function [P,yy]=PiecewiseLineInt(x,y,xx)
n=length(x);
for i=1:n-1
    P(i)=lagrintp([x(i),x(i+1)],[y(i),y(i+1)]);%调用拉格朗日插值函数
end
hold on
for k=1:n-1
    ezplot(P(k),[x(k),x(k+1)])
    end
plot(x,y,'*')
axis([x(1)-1,x(n)+1,min(y),max(x)])%控制坐标轴的显示范围
hold off
 if nargin<3
    return
end
m=length(xx);
for k=1:m
  if xx(k)<x(1)
      yy(k)=subs(P(1),'x',xx(k));
```

```
        continue
    end
    if xx(k)>=x(n)
        yy(k)=subs(P(n-1),'x',xx(k));
        continue
    end
    for i=1:n-1
        if xx(k)<x(i+1)&xx(k)>=x(i)
            yy(k)=subs(P(i),'x',xx(k));
            break
        end
    end
end
```

（4）计算实例.

Example 2.9.7 Given Table 2-18.

Table 2-18 The experimental data

x	2	3	4	5	6	7	8
y	3	8	2	4	1	5	6

① Solve approximate value to $f(x)$ at x=5.2 by using piecexise linear interpolation.

② Show the graph of piecewise interpolation function.

在命令窗口输入：

```
clear
>> x=2:8;y=[3,8,2,4,1,5,6];xx=5.2;[P,yy]=PiecewiseLineInt(x,y,xx)
 P =
    [ 5*x-7,-6*x+26,2*x-6,-3*x+19,4*x-23,x-2]
 yy =
     3.4000
```

并输出分段线性插值函数图像及（x(i),y(i)）的图像，如图 2-12 所示.

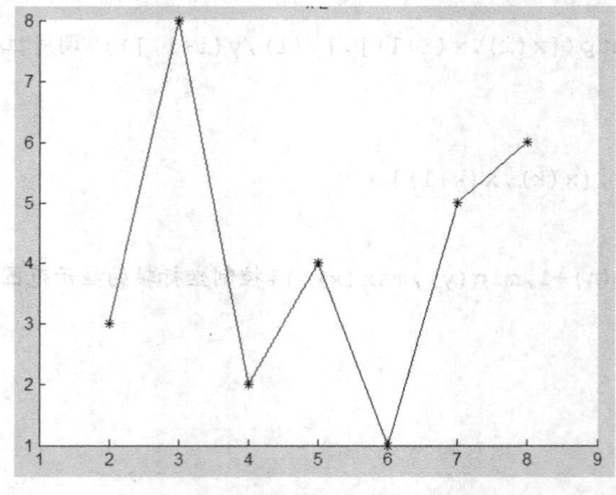

图 2-12 分段线性插值函数图像

4．三次样条插值函数

（1）函数语句.

```
[S,yy]=CubicSplineIntp(x,y,xx)
```

（2）参数说明.

x：实向量，输入参数，节点向量；

y：实向量，输入参数，相应的函数值向量；

xx：实向量，输入参数，自变量的值；

S：符号函数向量，输出参数，S(k)是在[x(k),x(k+1)]上的三次样条函数；

yy：实向量，输出参数，S 在 xx 的函数值.

（3）CubicSplineIntp.m 程序.

```
function [S,yy]=CubicSplineIntp(x,y,xx)
flag=input('请输入边界条件的类型(1 或 2 或 3): ')
if flag==1
   a=cubicspline1(x,y);
elseif flag==2
   a=cubicspline2(x,y);
elseif flag==3
   a=cubicspline3(x,y);
else
   fprintf('输入的参数有误')
end
hold on
plot(x,y,'*');
n=length(x);
t=sym('x');
hold on
for i=1:n-1
   S(i)=a(i,1)*t^3+a(i,2)*t^2+a(i,3)*t+a(i,4);
   ezplot(S(i),[x(i),x(i+1)])
end
axis([x(1)-1,x(n)+1,-10,15]);

if nargin<3
   return
end
m=length(xx);
for k=1:m
  if xx(k)<x(1)
     yy(k)=subs(S(1),'x',xx(k));
       continue
  end
  if xx(k)>=x(n)
     yy(k)=subs(S(n-1),'x',xx(k));
       continue
```

```
  end
    for i=1:n-1
        if xx(k)<x(i+1)&xx(k)>=x(i)
            yy(k)=subs(S(i),'x',xx(k));
            break
        end
    end
end
plot(xx,yy,'O');
hold off
```

子程序 1：cubicspline1.m 如下.

```
function a=cubicspline1(x,y,xx)
fp=input('请输入第一类边界条件，即在首尾两节点的一阶导数值：(fp1,fpn)=')
fp1=fp(1);
fpn=fp(2);
n=length(x);
h=diff(x);
divdif=diff(y)./h;
D=zeros(n,n);
D(1,1)=2;D(1,2)=1;D(n,n-1)=1;D(n,n)=2;
b=zeros(n,1);
b(1)=6*((y(2)-y(1))/h(1)-fp1)/h(1);
b(n)=6*((y(n-1)-y(n))/h(n-1)+fpn)/h(n-1);
for i=2:n-1
    D(i,i)=2;
    D(i,i-1)=h(i-1)/(h(i)+h(i-1));
    D(i,i+1)=1-D(i,i-1);
    b(i)=6*((y(i+1)-y(i))/h(i)-(y(i)-y(i-1))/h(i-1))/(h(i)+h(i-1));
end
M=D\b
for i=1:n-1
    a(i,1)=(M(i+1)-M(i))/(6*h(i));
    a(i,2)=(x(i+1)*M(i)-x(i)*M(i+1))/(2*h(i));
   a(i,3)=0.5*x(i)^2*M(i+1)/h(i)-0.5*x(i+1)^2*M(i)/h(i)+(y(i+1)-y(i))/h(i)
-(M(i+1)-M(i))*h(i)/6;
 a(i,4)=-x(i)^3*M(i+1)/(6*h(i))+x(i+1)^3*M(i)/(6*h(i))-x(i)*(y(i+1)
-y(i))/h(i)+x(i)*(M(i+1)-M(i))*h(i)/6+y(i)-M(i)*h(i)^2/6;
end
```

子程序 2：cubicspline2.m 如下.

```
function a=cubicspline2(x,y)
fp=input('请输入第二类边界条件，即在首尾两节点的二阶导数值：(fp1,fp2)=')
fp1=fp(1);
fpn=fp(2);
n=length(x);
h=diff(x);
D=zeros(n,n);
```

```
D(1,1)=2;D(1,2)=1;D(n,n-1)=1;D(n,n)=2;
 for i=2:n-1
    D(i,i)=2;
    D(i,i-1)=h(i-1)/(h(i)+h(i-1));
    D(i,i+1)=1-D(i,i-1);
    g(i-1)=6*((y(i+1)-y(i))/h(i)-(y(i)-y(i-1))/h(i-1))/(h(i)+h(i-1));
end
g(1)=g(1)-D(2,1)*fp1;
g(n-2)=g(n-2)-D(n-1,n)*fpn;
D1=D(2:n-1,2:n-1);
m=D1\g';
M(1)=fp1;M(n)=fpn;
M(2:n-1)=m;
for i=1:n-1
    a(i,1)=(M(i+1)-M(i))/(6*h(i));
    a(i,2)=(x(i+1)*M(i)-x(i)*M(i+1))/(2*h(i));
    a(i,3)=0.5*x(i)^2*M(i+1)/h(i)-0.5*x(i+1)^2*M(i)/h(i)+(y(i+1)-y(i))/h(i)
-(M(i+1)-M(i))*h(i)/6;
a(i,4)=-x(i)^3*M(i+1)/(6*h(i))+x(i+1)^3*M(i)/(6*h(i))-x(i)*(y(i+1)-
y(i))/h(i)+x(i)*(M(i+1)-M(i))*h(i)/6+y(i)-M(i)*h(i)^2/6;
End
```

子程序 3：cubicspline3.m 如下.

```
function a=cubicspline3(x,y)
n=length(x);
h=diff(x);
D=zeros(n,n);
D(1,1)=2;D(1,2)=1;D(n,n-1)=h(n-1)/(h(1)+h(n-1));D(n,n)=2;
 for i=2:n-1
    D(i,i)=2;
    D(i,i-1)=h(i-1)/(h(i)+h(i-1));
    D(i,i+1)=1-D(i,i-1);
    g(i-1)=6*((y(i+1)-y(i))/h(i)-(y(i)-y(i-1))/h(i-1))/(h(i)+h(i-1));
end
g(n-1)=6*((y(2)-y(1))/h(1)-(y(n)-y(n-1))/h(n-1))/(h(1)+h(n-1));
D1=D(2:n,2:n);
D1(1,n-1)=D(2,1);D1(n-1,1)=1-D(n,n-1);
m=D1\g';
M(1)=m(n-1);
M(2:n)=m;
for i=1:n-1
    a(i,1)=(M(i+1)-M(i))/(6*h(i));
    a(i,2)=(x(i+1)*M(i)-x(i)*M(i+1))/(2*h(i));
    a(i,3)=0.5*x(i)^2*M(i+1)/h(i)-0.5*x(i+1)^2*M(i)/h(i)+(y(i+1)-y(i))/h(i)
    -(M(i+1)-M(i))*h(i)/6;
    a(i,4)=-x(i)^3*M(i+1)/(6*h(i))+x(i+1)^3*M(i)/(6*h(i))-x(i)*
    (y(i+1)-y(i))/h(i)+x(i)*(M(i+1)-M(i))*h(i)/6+y(i)-M(i)*h(i)^
```

```
    2/6;
end
```

（4）计算实例.

Example 2.9.8 Given Table 2-19.

Table 2-19 The experimental data

x	0	1	2	3
y	1	1	0	10

Find the natural cubic spline,and give approximate value to $f(2.6)$.

在命令窗口输入：

```
clear
>> x=[0,1,2,3];y=[1,1,0,10];xx=2.6;[P,yy]=CubicSplineIntp(x,y,xx)
请输入边界条件的类型(1 或 2 或 3): 2
flag =
    2
请输入第二类边界条件，即在首尾两节点的二阶导数值: (fp1,fp2)=[0,0]
fp =
    0     0
P =
    [ -x^3+x+1,   4*x^3-15*x^2+16*x-4, -3*x^3+27*x^2-68*x+52]
 yy =
    4.9920
```

输出图像如图 2-13 所示.

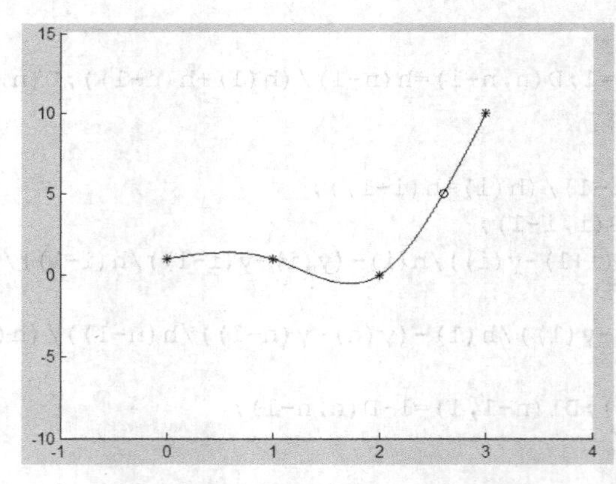

图 2-13 三次样条插值函数图像

Exercises 2

Questions

1. What are the base functions and Lagrange form of the interpolating polynomial for the data in the following table?

x	5	-7	-6	0
y	1	-23	-54	-954

2. When $x = 1,-1,2$, $f(x) = 0,-3,4$, solve the quadratic interpolation polynomial of the function $f(x)$.

3. If the function $f(x) = \sin x$ is approximated by a polynomial of degree 9 that interpolates f at ten points in the interval [0, 1], how large is the error on this interval?

4. Let $x_j (j = 0,1,\cdots,n)$ be distinct points, show that

(1) $\sum\limits_{j=0}^{n} x_j^k l_j(x) \equiv x^k \quad (k = 0,1,\cdots,n)$;

(2) $\sum\limits_{j=0}^{n} (x_j - x)^k l_j(x) \equiv 0 \quad (k = 0,1,\cdots,n)$.

5. Prove that the coefficient of x^n in the polynomial p of Lagrange interpolation is

$$\sum_{i=0}^{n} y_i \prod_{\substack{j=0 \\ j \neq i}}^{n} (x_i - x_j)^{-1}$$

6. Find a polynomial that assumes these values:

x	1	2	0	3
y	3	2	-4	5

7. Suppose a table is to be prepared for the function $f(x) = \mathrm{e}^x$, for x in [−4,4], and that the step size is h. What should h be for quadratic interpolation to give an absolute error of at most 10^{-6}?

8. If $y_n = 2^n$, compute $\Delta^4 y_n$ and $\delta^4 y_n$.

9. Prove that $\Delta(f_k g_k) = f_k \Delta g_k + g_{k+1} \Delta f_k$.

10. Prove that $\sum\limits_{j=0}^{n-1} \Delta^2 y_j = \Delta y_n - \Delta y_0$.

11. The polynomial p of degree $\leqslant n$ that interpolates a given function f at n+1 prescribed nodes is uniquely defined. Hence, there is a mapping $f \mapsto p$. Denote this mapping by L and show that

$$Lf = \sum_{i=0}^{n} f(x_i) l_i$$

Show that L is linear, that is, $L(af + bg) = aLf + bLg$.

12. The polynomial

$$p(x) = 2 - (x+1) + x(x+1) - 2x(x+1)(x-1)$$

interpolates the first four points in the table:

x	-1	0	1	2	3
y	2	1	2	-7	10

By adding one additional term to p, find a polynomial that interpolates the whole table.

13. Given $f(x)=x^7+x^4+3x+1$, compute $f[2^0,2^1,\cdots,2^7]$ and $f[2^0,2^1,\cdots,2^7,2^8]$.

14. Solve a piecewise linear interpolation function $I_h(x)$ of $f(x)=x^2$ on [0, 2] with $h=0.5$, and estimate its error.

15. Find a polynomial $P(x)$ whose degree is less than or equal to 4 and satisfies $P(0)=P'(0)=0,\quad P(1)=P'(1)=1,\quad P(2)=1$.

16. Determine the natural cubic spline S that interpolates the data $f(0)=0$, $f(1)=1$, and $f(2)=2$.

17. Determine all the values of a, b, c, d, e for which the following function is a cubic spline:

$$f(x)=\begin{cases} a(x-2)^2+b(x-1)^3 & x\in(-\infty,1] \\ c(x-2)^2 & x\in[1,3] \\ d(x-2)^2+e(x-3)^3 & x\in[3,\infty) \end{cases}$$

18. A natural cubic spline S on [0,2] is defined by

$$S(x)=\begin{cases} S_0(x)=1+2x-x^3 & 0\leqslant x<1 \\ S_1(x)=2+b(x-1)+c(x-1)^2+d(x-1)^3 & 1\leqslant x\leqslant 2 \end{cases}$$

Find b, c, and d.

19. A natural cubic spline S is defined by

$$S(x)=\begin{cases} S_0(x)=1+B(x-1)-D(x-1)^3 & 1\leqslant x<2 \\ S_1(x)=1+b(x-2)-\dfrac{3}{4}(x-2)^2+d(x-2)^3 & 2\leqslant x\leqslant 3 \end{cases}$$

If S interpolates the data (1,1), (2,1), and (3,0), find B, D, b, and d.

20. Determine whether this function is a natural cubic spline:

$$f(x)=\begin{cases} 2(x+1)+(x+1)^3 & x\in[-1,0] \\ 3+5x+3x^2 & x\in[0,1] \\ 11+11(x-1)+3(x-1)^2-(x-1)^3 & x\in[1,2] \end{cases}$$

Computer Questions

1. Find the Lagrange interpolating polynomial, Newton interpolating polynomial, and natural cubic spline function respectively for the following data set.

D={(0,5), (1,−2), (2,3), (3,4), (4,−1), (5,7), (6,5), (7,2)}

2. For $n=5,10$, and 15, find the Newton interpolating polynomial p_n for the function $f(x)=1/(1+x^2)$ on the interval [−5,5]. Use equally spaced nodes. In each case, compute $f(x)-p_n(x)$ for 30 equally spaced points in [−5,5] in order to see the divergence of p_n from f.

3. The following table lists the population of the United States, in thousands of people, from 1940 to 1990.In reviewing these data, whether you could be used to provide a reasonable estimate of the population, say, in 1965 or even in the year 2010.

Year	1940	1950	1960	1970	1980	1990
Population (in thousand)	132 165	151 326	179 323	203 302	226 542	249 633

4. Given the table:

x	0.25	0.32	0.40	0.46	0.55
y	0.520 0	0.547 8	0.625 0	0.680 8	0.728 8

Find the cubic spline interpolation $S(x)$, which satisfies the following condition:

(1) $S'(0.25)=1,\ S'(0.55)=0.68$;

(2) $S''(0.25)=S''(0.55)=0$.

3 Best Approximation

提　要

在实际问题中，常常遇到以各种方式给出的函数，其中有些函数的表达式用来计算函数值或分析函数的性态很不方便，因此，希望能用一个简单的函数 $P(x)$来代替给定的函数 $f(x)$，并使 $P(x)$与 $f(x)$之差在某种度量意义下为最小，这就是函数逼近问题．其中，$P(x)$称为逼近函数，$f(x)$称为被逼近函数．前一章所介绍的插值问题实际上也是函数逼近问题．

词　汇

iterative method	迭代法
normed space	赋范空间
norm	范数
homogeneity	齐次性
divergent	发散的
trivial solution	平凡解
weight function	权函数
orthogonal	正交
orthonormal	标准正交
lower bound	下界
deviation	偏差
best approximation	最佳逼近
invariant	不变的
monotonic	单调的
end point	端点
normal equation	正规方程
inequality	不等式
family of orthogonal functions	正交函数族
homogeneous system of equations	齐次方程组
least squares approximation	最小平方逼近
least square solution	最小二乘解
curve fitting	曲线拟合
nonsingular	非奇异
sequence	序列
axiom	公理
conjugate	共轭
empirical formula	经验公式
function of many variables	多元函数

3.1 Introduction

The study of approximation theory involves two general types of problems. One problem arises when a function is given explicitly, but we wish to find a “simpler” type of function, such as a polynomial, that can be used to determine approximate values of the given function. The other problem in approximation theory is concerned with fitting functions to given data and finding the "best" function in a certain class to represent the data.

Both problems have been touched upon in Chapter 2.The Lagrange interpolating polynomials, Newton interpolating polynomials, were discussed both as approximating polynomials and as polynomials to fit certain data. Cubic splines were also discussed in that chapter. In this chapter, limitations to these techniques are considered, and other **avenues** (方法) of approach are discussed.

We begin by introducing the notions of **norms of vectors and matrices** (向量范数和矩阵范数) and their elementary properties, where we assume that the reader is familiar with the concept of linear spaces or vector spaces and their basic properties.

When you finish this chapter, you will understand basic theory of normed space. You will know how to express an polynomial in terms of the data points using least squares method. You will know how to solve best linear approximation and best square approximation. These overall goals will be achieved by mastering the following chapter objectives.

Objectives

- Know how to construct best linear approximation and best squares approximation.
- Know how to fit a function using least squares approximation.
- Understand vector norm , matrix norm and their properties.

3.2 Norms

3.2.1 Vector Norms

Definition 3.2.1 **A vector norm** on $\mathbf{R}^n$ is a function, $\|\cdot\|$, from $\mathbf{R}^n$ into $\mathbf{R}$, with the following properties:

(1) $\|\boldsymbol{x}\|\geqslant 0$ for all $\boldsymbol{x}\in\mathbf{R}^n$ (positivity);

(2) $\|\boldsymbol{x}\|=0$ if and only if $\boldsymbol{x}=0$ (definiteness);

(3) $\|\alpha\boldsymbol{x}\|=|\alpha|\|\boldsymbol{x}\|$ for all $\alpha\in\mathbf{R}$ and $\boldsymbol{x}\in\mathbf{R}^n$ (homogeneity);

(4) $\|\boldsymbol{x}+\boldsymbol{y}\|\leqslant\|\boldsymbol{x}\|+\|\boldsymbol{y}\|$ for all $\boldsymbol{x},\boldsymbol{y}\in\mathbf{R}^n$ (triangle inequality).

A linear space equipped with a norm is called a **normed space** (赋范空间). Since vectors in $\mathbf{R}^n$ are column vectors, it is convenient to use the transpose notation when a vector is represented in terms of its components. For example, the vector

$$\boldsymbol{x}=\begin{pmatrix}x_1\\x_2\\\vdots\\x_n\end{pmatrix}$$

will be written $\boldsymbol{x}=(x_1,\cdots,x_n)^{\mathrm{T}}$.

Definition 3.2.2 The l_1, l_2 and l_∞ norms for the vector $\boldsymbol{x}=(x_1,\cdots,x_n)^{\mathrm{T}}$ are defined by

$$\|\boldsymbol{x}\|_1=\sum_{i=1}^{n}|x_i|,\quad \|\boldsymbol{x}\|_2=\left(\sum_{i=1}^{n}x_i^2\right)^{\frac{1}{2}},\quad \text{and}\quad \|\boldsymbol{x}\|_\infty=\max_{1\leqslant i\leqslant n}|x_i|$$

The l_2 norm is also called the **Euclidean norm**(欧几里得范数)of the vector x since it represents the usual notion of distance from the origin in case $\boldsymbol{x}$ is in $\mathbf{R}^1\equiv\mathbf{R}$, $\mathbf{R}^2$, or $\mathbf{R}^3$. For example, the l_2 norm of the vector $\boldsymbol{x}=(x_1,\cdots,x_n)^{\mathrm{T}}$ gives the length of the straight line joining the points (0,0,0) and $\boldsymbol{x}=(x_1,x_2,x_3)$. The l_∞ norm is called **the maximum norm.** The three norms are special cases of the l_p norm

$$\|\boldsymbol{x}\|_p=\left(\sum_{j=1}^{n}|x_j|^p\right)^{\frac{1}{p}} \tag{3.2.1}$$

defined for any real number $p\geqslant 1$. The l_∞ norm is the limiting case of (3.2.1) as $p\to\infty$.

It is an easy exercise for the reader to verify that the **norm axioms** (公理) (1)-(3)satisfied for l_1, l_2 and l_∞ norms . It is easy to show that the triangle inequality holds for the norms l_1 and l_∞ since they follow from similar results for absolute values. For example, if $\boldsymbol{x}=(x_1,\cdots,x_n)^{\mathrm{T}}$ and $\boldsymbol{y}=(y_1,\cdots,y_n)^{\mathrm{T}}$, then

$$\|\boldsymbol{x}+\boldsymbol{y}\|_1=\sum_{i=1}^{n}|x_i+y_i|\leqslant\sum_{i=1}^{n}(|x_i|+|y_i|)=\sum_{i=1}^{n}|x_i|+\sum_{i=1}^{n}|y_i|=\|\boldsymbol{x}\|_1+\|\boldsymbol{y}\|_1$$

$$\|\boldsymbol{x}+\boldsymbol{y}\|_\infty=\max_{1\leqslant i\leqslant n}|x_i+y_i|\leqslant\max_{1\leqslant i\leqslant n}(|x_i|+|y_i|)\leqslant\max_{1\leqslant i\leqslant n}|x_i|+\max_{1\leqslant i\leqslant n}|y_i|=\|\boldsymbol{x}\|_\infty+\|\boldsymbol{y}\|_\infty$$

To show that

$$\|\boldsymbol{x}+\boldsymbol{y}\|_2\leqslant\|\boldsymbol{x}\|_2+\|\boldsymbol{y}\|_2,\ \text{for each } \boldsymbol{x},\boldsymbol{y}\in\mathbf{R}^n$$

we need a famous inequality.

Theorem 3.2.1 (Cauchy-Buniakowsky-Schwarz Inequality for Sums)

For each $\boldsymbol{x}=(x_1,\cdots,x_n)^{\mathrm{T}}$ and $\boldsymbol{y}=(y_1,\cdots,y_n)^{\mathrm{T}}$ in $\mathbf{R}^n$,

$$\boldsymbol{x}^{\mathrm{T}}\boldsymbol{y}=\sum_{i=1}^{n}x_iy_i\leqslant\left(\sum_{i=1}^{n}x_i^2\right)^{1/2}\left(\sum_{i=1}^{n}y_i^2\right)^{1/2}=\|\boldsymbol{x}\|_2\cdot\|\boldsymbol{y}\|_2$$

Proof If $\boldsymbol{y}=\mathbf{0}$ or $\boldsymbol{x}=\mathbf{0}$, the result is immediate since both sides of the inequality are zero. Suppose $\boldsymbol{y}\neq\mathbf{0}$ and $\boldsymbol{x}\neq\mathbf{0}$. For each $\lambda\in\mathbf{R}$,

$$0\leqslant\|\boldsymbol{x}-\lambda\boldsymbol{y}\|_2^2=\sum_{i=1}^{n}(x_i-\lambda y_i)^2=\sum_{i=1}^{n}x_i^2-2\lambda\sum_{i=1}^{n}x_iy_i+\lambda^2\sum_{i=1}^{n}y_i^2$$

and

$$2\lambda\sum_{i=1}^{n}x_iy_i\leqslant\sum_{i=1}^{n}x_i^2+\lambda^2\sum_{i=1}^{n}y_i^2=\|\boldsymbol{x}\|_2^2+\lambda^2\|\boldsymbol{y}\|_2^2$$

Since $\|x\|_2 > 0$ and $\|y\|_2 > 0$, we can let $\lambda = \frac{\|x\|_2}{\|y\|_2}$ to give

$$\left(2\frac{\|x\|_2}{\|y\|_2}\right)\left(\sum_{i=1}^{n} x_i y_i\right) \leqslant \|x\|_2^2 + \frac{\|x\|_2^2}{\|y\|_2^2}\|y\|_2^2 = 2\|x\|_2^2$$

therefore

$$\sum_{i=1}^{n} x_i y_i \leqslant \|x\|_2 \|y\|_2$$

So

$$x^{\mathrm{T}} y = \sum_{i=1}^{n} x_i y_i \leqslant \|x\|_2 \cdot \|y\|_2 = \left(\sum_{i=1}^{n} x_i^2\right)^{1/2} \left(\sum_{i=1}^{n} y_i^2\right)^{1/2}$$

From this result we see that for each $x, y \in \mathbf{R}^n$.

$$\|x + y\|_2^2 = \sum_{i=1}^{n} (x_i + y_i)^2 = \sum_{i=1}^{n} x_i^2 + 2\sum_{i=1}^{n} x_i y_i + \sum_{i=1}^{n} y_i^2 \leqslant \|x\|_2^2 + 2\|x\|_2 \|y\|_2 + \|y\|_2^2$$

which gives the norm property

$$\|x + y\|_2 \leqslant (\|x\|_2^2 + 2\|x\|_2 \|y\|_2 + \|y\|_2^2)^{1/2} = \|x\|_2 + \|y\|_2$$ □

Since the norm of a vector gives a measure for the distance between an arbitrary vector and the zero vector, the distance between two vectors is defined as the norm of the difference of the vectors.

Definition 3.2.3 Let $x = (x_1, \cdots, x_n)^{\mathrm{T}}$ and $y = (y_1, \cdots, y_n)^{\mathrm{T}}$ be vectors in $\mathbf{R}^n$, the l_1, l_2 and l_∞ distances between x and y are defined by

$$\|x - y\|_1 = \sum_{i=1}^{n} |x_i - y_i|, \quad \|x - y\|_2 = \left(\sum_{i=1}^{n} (x_i - y_i)^2\right)^{\frac{1}{2}}$$

and

$$\|x - y\|_\infty = \max_{1 \leqslant i \leqslant n} |x_i - y_i|$$

Example 3.2.1 Using the norm $\|\cdot\|_1$, compare the lengths of the following three vectors in $\mathbf{R}^4$. Then repeat the calculation for the norms $\|\cdot\|_2$ and $\|\cdot\|_\infty$.

$$x = (4,4,-4,4)^{\mathrm{T}}, \quad v = (0,5,5,5)^{\mathrm{T}}, \quad w = (6,0,0,0)^{\mathrm{T}}$$

Solution The results are displayed here (see Table 3-1).

Table 3-1 The norms

vectors	$\|\cdot\|_1$	$\|\cdot\|_2$	$\|\cdot\|_\infty$
x	16	8	4
v	15	8.66	5
w	6	6	6

Remark 3.2.1 For each norm, the second triangle inequality

$$\big|\|x\| - \|y\|\big| \leqslant \|x - y\|$$

holds for all $x, y \in X$.

Proof From the triangle inequality we have

$$\|x\| = \|x - y + y\| \leqslant \|x - y\| + \|y\|$$

hence $\|x\| - \|y\| \leqslant \|x - y\|$ follows. Analogously, by interchanging the roles of x and y we have $\|y\| - \|x\| \leqslant \|y - x\|$. □

The concept of distance in $\mathbf{R}^n$ is also used to define a limit of a sequence of vectors in this space.

Definition 3.2.4 A sequence $\{x^{(k)}\}_{k=1}^{\infty}$ of vectors in $\mathbf{R}^n$ is called convergent if there exists an element $x \in \mathbf{R}^n$ such that

$$\lim_{k\to\infty} \|x^{(k)} - x\| = 0$$

i.e., if for every $\varepsilon > 0$ there exists an integer $N(\varepsilon)$, such that $\|x^{(k)} - x\| < \varepsilon$ for all $k \geqslant N(\varepsilon)$. The element x is called **the limit of the sequence** $\{x^{(k)}\}$, and we write

$$\lim_{k\to\infty} x^{(k)} = x$$

or

$$x^{(k)} \to x,\ k \to \infty$$

A sequence that does not converge is called **divergent** (发散).

Theorem 3.2.2 The sequence of vectors $\{x^{(k)}\}_{k=1}^{\infty}$ converges to x in $\mathbf{R}^n$ with respect to $\|\cdot\|_{\infty}$ if and only if $\lim\limits_{k\to\infty} x_i^{(k)} = x_i$, for each $i = 1,\cdots,n$.

Proof Suppose $\{x^{(k)}\}$ converges to x with respect to $\|\cdot\|_{\infty}$. Given any $\varepsilon>0$, there exists an integer $N(\varepsilon)$ such that for all $k \geqslant N(\varepsilon)$,

$$\max_{i=1,2,\cdots,n} \left| x_i^{(k)} - x_i \right| = \left\| x^{(k)} - x \right\|_{\infty} < \varepsilon$$

This result implies that $\left| x_i^{(k)} - x_i \right| < \varepsilon$, for each $i = 1,\cdots, n$, so $\lim\limits_{k\to\infty} x_i^{(k)} = x_i$ for each i.

Conversely, suppose that $\lim\limits_{k\to\infty} x_i^{(k)} = x_i$, for every $i = 1,\cdots,n$. For a given $\varepsilon>0$, there exist integers $N_i(\varepsilon)(i = 1,\cdots,n)$, such that

$$\left| x_i^{(k)} - x_i \right| < \varepsilon$$

whenever $k \geqslant N_i(\varepsilon)$.

Define $N(\varepsilon) = \max\limits_{i=1,2,\cdots,n} N_i(\varepsilon)$. If $k \geqslant N(\varepsilon)$, then

$$\max_{i=1,2,\cdots,n} \left| x_i^{(k)} - x_i \right| = \left\| x^{(k)} - x \right\|_{\infty} < \varepsilon$$

This implies that $\{x^{(k)}\}$ converges to x with respect to $\|\cdot\|_{\infty}$. □

Example 3.2.2 Let $x^{(k)} \in \mathbf{R}^4$ be defined by

$$x^{(k)} = (x_1^{(k)}, x_2^{(k)}, x_3^{(k)}, x_4^{(k)})^{\mathrm{T}} = \left(1 + \frac{1}{k}, \frac{1}{2^k}, 3\mathrm{e}^{-k}, \frac{1}{k^2}\right)^{\mathrm{T}}$$

Since

$$\lim_{k\to\infty}\left(1+\frac{1}{k}\right)=1,\ \lim_{k\to\infty}\frac{1}{2^k}=0,\ \lim_{k\to\infty}\left(3e^{-k}\right)=0, \text{ and } \lim_{k\to\infty}\left(\frac{1}{k^2}\right)=0$$

From Theorem 3.2.2, the given sequence $\{x^{(k)}\}$ converges to $(1,0,0,0)^{\mathrm{T}}$ with respect to $\|\cdot\|_\infty$.

Theorem 3.2.3 The limit of a convergent sequence is uniquely determined.

Proof Assume that $x_n \to x$ and $x_n \to y$ for $n\to\infty$. Then from the triangle inequality we obtain that

$$\|x-y\|=\|x-x_n+x_n-y\|\leqslant\|x-x_n\|+\|x_n-y\|\to 0,\quad n\to\infty$$

Therefore, $\|x-y\|=0$ and $x=y$. □

Definition 3.2.5 Two norms on $\mathbf{R}^n$ are equivalent if they have the same convergent sequence, i.e., if $\|\cdot\|_a$ and $\|\cdot\|_b$ are any two norms on $\mathbf{R}^n$, and $\{x^{(k)}\}_{k=1}^{\infty}$ has the limit x with respect to $\|\cdot\|_a$, then $\{x^{(k)}\}_{k=1}^{\infty}$ also has the limit x with respect to $\|\cdot\|_b$.

Theorem 3.2.4 Two norms $\|\cdot\|_a$ and $\|\cdot\|_b$ on $\mathbf{R}^n$ are equivalent if and only if there exist positive number c_1 and c_2, such that

$$c_1\|x\|_a\leqslant\|x\|_b\leqslant c_2\|x\|_a \tag{3.2.2}$$

for all $x\in\mathbf{R}^n$. The limits with respect to the two norms coincide.

Proof Provided that the condition (3.2.2) is satisfied, from $\|x_n-x\|_a\to 0$, $n\to\infty$, it follows that $\|x_n-x\|_b\to 0$, $n\to\infty$, and vice versa.

Conversely, let the two norms be equivalent and assume that there is no $c_2>0$, such that $\|x\|_b\leqslant c_2\|x\|_a$ for all $x\in\mathbf{R}^n$. Then there exists a sequence $\{x_n\}$ with $\|x_n\|_a=1$ and $\|x_n\|_b\geqslant n^2$. Now, the sequence $\{y_n\}$ with $y_n=x_n/n$ converges to zero with respect to $\|\cdot\|_a$, whereas with respect to $\|\cdot\|_b$ it is divergent because of $\|y_n\|_b\geqslant n$. □

Example 3.2.3 For each $x\in\mathbf{R}^n$, show that $\|x\|_\infty\leqslant\|x\|_1\leqslant n\|x\|_\infty$.

Proof For each $x\in(x_1,x_2,\cdots,x_n)^{\mathrm{T}}\in\mathbf{R}^n$,

$$\|x\|_\infty=\max_{i=1,2,\cdots,n}|x_i|\leqslant\sum_{i=1}^{n}|x_i|=\|x\|_1$$

so

$$\|x\|_\infty\leqslant\|x\|_1$$

and

$$\|x\|_1=\sum_{i=1}^{n}|x_i|\leqslant\sum_{i=1}^{n}(\max_{j=1,2,\cdots,n}|x_j|)=n\|x\|_\infty$$

so

$$\|x\|\leqslant n\|x\|_\infty$$

Theorem 3.2.4 implies that $\|\cdot\|_\infty$ and $\|\cdot\|_1$ are equivalent.

Theorem 3.2. 5 All norms on $\mathbf{R}^n$ are equivalent.

3.2.2 Matrix Norms

In the subsequent sections of later chapters, we will need methods for determining the distance between $n\times n$ matrices. This again requires the use of a norm.

Definition 3.2.6 **A matrix norm** on the set of all $n\times n$ matrices is a real-valued function, $\|\cdot\|$, defined on this set, satisfying for all $n\times n$ matrices $\boldsymbol{A}$ and $\boldsymbol{B}$ and all real numbers c:

(1) $\|\boldsymbol{A}\|\geqslant 0$ ($\|\boldsymbol{A}\|=0 \Leftrightarrow \boldsymbol{A}=0$) (positivity 正定性);

(2) $\|c\boldsymbol{A}\|=|c|\|\boldsymbol{A}\|,\ c\in\mathbf{R}$ (homogeneity 齐次性);

(3) $\|\boldsymbol{A}+\boldsymbol{B}\|\leqslant\|\boldsymbol{A}\|+\|\boldsymbol{B}\|$ (triangle inequality 三角不等式);

(4) $\|\boldsymbol{AB}\|\leqslant\|\boldsymbol{A}\|\|\boldsymbol{B}\|$.

The distances between $n\times n$ matrices $\boldsymbol{A}$ and $\boldsymbol{B}$ with respect to this matrix norm is $\|\boldsymbol{A}-\boldsymbol{B}\|$.

Definition 3.2.7 Let $\boldsymbol{x}\in\mathbf{R}^n$, $\boldsymbol{A}\in\mathbf{R}^{n\times n}$, given a vector norm $\|\cdot\|$, then the matrix norm

$$\|\boldsymbol{A}\|=\max_{x\neq 0}\frac{\|\boldsymbol{Ax}\|}{\|\boldsymbol{x}\|}$$

is called an **operator norm of matrix** $\boldsymbol{A}$ (矩阵的算子范数).

Theorem 3.2.6 For any vector $\boldsymbol{x}$, matrix $\boldsymbol{A}$ and any norm $\|\cdot\|$, we have

$$\|\boldsymbol{Ax}\|\leqslant\|\boldsymbol{A}\|\,\|\boldsymbol{x}\|$$

Theorem 3.2.7 Let $\boldsymbol{x}\in\mathbf{R}^n$, $\boldsymbol{A}\in\mathbf{R}^{n\times n}$, then

(1) $\|\boldsymbol{A}\|_\infty=\max\limits_{1\leqslant i\leqslant n}\sum\limits_{j=1}^{n}|a_{ij}|$, the l_∞ norm or row norm of $\boldsymbol{A}$;

(2) $\|\boldsymbol{A}\|_1=\max\limits_{1\leqslant j\leqslant n}\sum\limits_{i=1}^{n}|a_{ij}|$, the l_1 norm or column norm of $\boldsymbol{A}$;

(3) $\|\boldsymbol{A}\|_2=\sqrt{\lambda_{\max}(\boldsymbol{A}^{\mathrm{T}}\boldsymbol{A})}$, the l_2 norm.

where $\lambda_{\max}(\boldsymbol{A}^{\mathrm{T}}\boldsymbol{A})$ represent the biggest **eigenvalue** (特征值, See Section 3.3) of $\boldsymbol{A}^{\mathrm{T}}\boldsymbol{A}$.

Proof We only prove term (1).

Let $\boldsymbol{x}=(x_1,x_2,\cdots,x_n)^{\mathrm{T}}\neq 0$, and $\boldsymbol{A}\neq 0$. We denote

$$t=\|\boldsymbol{x}\|_\infty=\max_{1\leqslant i\leqslant n}|x_i|,\qquad \mu=\max_{1\leqslant i\leqslant n}\sum_{j=1}^{n}|a_{ij}|$$

then

$$\|Ax\|_\infty = \max_{1\leqslant i\leqslant n}\left|\sum_{j=1}^{n} a_{ij}x_j\right| \leqslant \max_{1\leqslant i\leqslant n}\sum_{j=1}^{n}|a_{ij}||x_j| \leqslant t\max_{1\leqslant i\leqslant n}\sum_{j=1}^{n}|a_{ij}|$$

This indicates that for any $x \in \mathbf{R}^n$, we have

$$\frac{\|Ax\|_\infty}{\|x\|_\infty} \leqslant \mu$$

In the following, we prove that there exists a vector $x_0 \neq \mathbf{0}$, such that

$$\frac{\|Ax\|_\infty}{\|x\|_\infty} = \mu$$

Let

$$\mu = \sum_{j=1}^{n}|a_{i_0 j}|, \quad \text{take} \quad x_0 = (x_1, x_2, \cdots, x_n)^{\mathrm{T}}$$

where

$$x_j = \begin{cases} 1 & a_{i_0 j} \geqslant 0 \\ -1 & a_{i_0 j} < 0 \end{cases}$$

Obviously, $\|x_0\|_\infty = 1$ and $\sum_{j=1}^{n} a_{i_0 j}x_j = \sum_{j=1}^{n}|a_{i_0 j}| = \mu$, this indicates

$$\|Ax_0\|_\infty = \max_{1\leqslant i\leqslant n}\left|\sum_{j=1}^{n} a_{ij}x_j\right| = \sum_{j=1}^{n}|a_{i_0 j}| = \mu$$

□

Example 3.2.4 Given $A = \begin{pmatrix} 1 & 3 & -5 \\ -4 & 7 & 0 \\ 6 & 2 & -2 \end{pmatrix}$ compute $\|A\|_\infty$ and $\|A\|_1$.

Solution Since $\sum_{j=1}^{3}|a_{1j}| = 9, \ \sum_{j=1}^{3}|a_{2j}| = 11, \ \sum_{j=1}^{3}|a_{3j}| = 10$

$$\sum_{i=1}^{3}|a_{i1}| = 11, \ \sum_{i=1}^{3}|a_{i2}| = 12, \ \sum_{i=1}^{3}|a_{i3}| = 7$$

so

$$\|A\|_\infty = \max\{9, 11, 10\} = 11$$

$$\|A\|_1 = \max\{11, 12, 7\} = 12$$

Analogously, for continuous functional space $C[a,b]$, if $f \in \mathbf{C}[a,b]$, three common norms are defined by

$$\|f\|_\infty = \max_{a\le x\le b}|f(x)| \quad (\infty\text{-norm or maximum norm})$$

$$\|f\|_1 = \int_a^b |f(x)|\mathrm{d}x \quad (1\text{-norm})$$

$$\|f\|_2 = \left(\int_a^b f^2(x)\mathrm{d}x\right)^{\frac{1}{2}} \quad (2\text{-norm or Euclidean norm})$$

3.3 Spectral Radius（谱半径）

Definition 3.3.1 If A is a square matrix, the **characteristic polynomial** (特征多项式) and **characteristic equation** (特征方程) of A are defined respectively by

$$p(\lambda) = \det(A - \lambda I)$$

and

$$\det(A - \lambda I) = 0$$

It is not difficult to show that p is an nth-degree polynomial and consequently, has at most n distinct zeros, some of which may be complex.

Definition 3.3.2 If p is the characteristic polynomial of the matrix A, the zeros of p are eigenvalues, or characteristic values, of the matrix A. If λ is an eigenvalue of A and $x\neq 0$ satisfies $(A-\lambda I)x=0$, then x is an eigenvector, or characteristic vector, of A corresponding to the eigenvalue λ.

If x is an eigenvector associated with the eigenvalue λ, then , $Ax=\lambda x$**,** so the matrix A takes the vector x into a scalar multiple of itself, as illustrated in Figure 3-1.

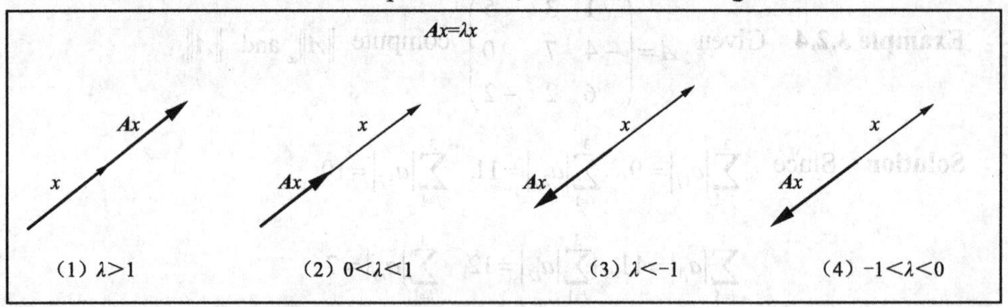

Figure 3-1 Eigenvalues and eigenvectors

Example 3.3.1 Let

$$A = \begin{pmatrix} 1 & 0 & 1 \\ 2 & 2 & 1 \\ -1 & 0 & 0 \end{pmatrix}$$

compute the eigenvalues of A.

Solution The characteristic equation:

$$\det(\boldsymbol{A}-\lambda\boldsymbol{I})=\begin{vmatrix}1-\lambda & 0 & 1\\ 2 & 2-\lambda & 1\\ -1 & 0 & -\lambda\end{vmatrix}=(2-\lambda)(\lambda^2-\lambda+1)=0$$

the eigenvalues are

$$\lambda_1=2,\quad \lambda_2=\frac{1}{2}+\frac{\sqrt{3}}{2}\mathrm{i},\ \text{and}\ \lambda_3=\frac{1}{2}-\frac{\sqrt{3}}{2}\mathrm{i}$$

An eigenvector x of $\boldsymbol{A}$ associated with λ_1 is a solution of the system $(\boldsymbol{A}-\lambda_1\boldsymbol{I})x=0$:

$$\begin{pmatrix}-1 & 0 & 1\\ 2 & 0 & 1\\ -1 & 0 & -2\end{pmatrix}\begin{pmatrix}x_1\\ x_2\\ x_3\end{pmatrix}=\begin{pmatrix}0\\ 0\\ 0\end{pmatrix}$$

Thus

$$x_1-x_3=0,\ \ 2x_1+x_3=0,\ \text{and}\ x_1+2x_3=0$$

which implies that

$$x_1=x_3=0,\ \text{and}\ x_2\ \text{is arbitrary}$$

The choice $x_2=1$ produces the eigenvector $(0,1,0)^{\mathrm{T}}$ corresponding to the eigenvalue $\lambda_1=2$, for this choice, we have $\|(0,1,0)^{\mathrm{T}}\|_\infty=1$.

Analogously, we can obtain eigenvectors of $\boldsymbol{A}$ associated with λ_2 and λ_3 respectively.

Definition 3.3.3 The spectral radius of a matrix $\boldsymbol{A}$ is defined by

$$\rho(\boldsymbol{A})=\max_{1\leqslant i\leqslant n}|\lambda_i|,$$

where λ_i is an eigenvalue of $\boldsymbol{A}$. Recall that for complex $\lambda=\alpha+\mathrm{i}\beta$, we have $|\lambda|=(\alpha^2+\beta^2)^{1/2}$.

For the matrix considered in Example 3.3.1,

$$\rho(\boldsymbol{A})=\max\{2,|(1+\sqrt{3}\,\mathrm{i})/2|,\ |(1-\sqrt{3}\,\mathrm{i})/2|\}=2$$

Theorem 3.3.1 Let $\boldsymbol{A}\in\mathbf{R}^{n\times n}$, then

$$\rho(\boldsymbol{A})\leqslant\|\boldsymbol{A}\|$$

Proof Let λ be an arbitrary eigenvalue of $\boldsymbol{A}$, $\boldsymbol{x}$ is an eigenvector associated with λ, then $\boldsymbol{Ax}=\lambda\boldsymbol{x}$, and

$$|\lambda|\|\boldsymbol{x}\|=\|\lambda\boldsymbol{x}\|=\|\boldsymbol{Ax}\|\leqslant\|\boldsymbol{A}\|\|\boldsymbol{x}\|,$$

notice that $\|\boldsymbol{x}\|\neq0$, therefore $|\lambda|\leqslant\|\boldsymbol{A}\|$. □

Theorem 3.3.2 If matrix $\boldsymbol{A}\in\mathbf{R}^{n\times n}$ is symmetric , then $\|\boldsymbol{A}\|_2=\rho(\boldsymbol{A})$.

Example 3.3.2 If

$$A = \begin{pmatrix} 2 & 1 & 0 \\ 1 & 2 & 0 \\ 0 & 0 & 3 \end{pmatrix}$$

compute $\rho(A)$ and $\|A\|_2$.

Solution To compute $\rho(A)$, consider

$$0 = \det(A - \lambda I) = \det\begin{pmatrix} 2-\lambda & 1 & 0 \\ 1 & 2-\lambda & 0 \\ 0 & 0 & 3-\lambda \end{pmatrix} = -(\lambda-3)^2(\lambda-1),$$

We have

$$\lambda_1 = 1, \quad \lambda_2 = \lambda_3 = 3$$

so

$$\rho(A) = \max_{1\leqslant i\leqslant 3}|\lambda_i| = \max\{1,3,3\} = 3$$

$$A^{\mathrm{T}}A = \begin{pmatrix} 2 & 1 & 0 \\ 1 & 2 & 0 \\ 0 & 0 & 3 \end{pmatrix}\begin{pmatrix} 2 & 1 & 0 \\ 1 & 2 & 0 \\ 0 & 0 & 3 \end{pmatrix} = \begin{pmatrix} 5 & 4 & 0 \\ 4 & 5 & 0 \\ 0 & 0 & 9 \end{pmatrix}$$

To compute $\|A\|_2$ we need the eigenvalues of $A^{\mathrm{T}}A$. If

$$0 = \det(A^{\mathrm{T}}A - \lambda I) = \det\begin{pmatrix} 5-\lambda & 4 & 0 \\ 4 & 5-\lambda & 0 \\ 0 & 0 & 9-\lambda \end{pmatrix} = -(\lambda-9)^2(\lambda-1),$$

then

$$\lambda_1 = 1, \quad \lambda_2 = \lambda_3 = 9$$

so

$$\|A\|_2 = \sqrt{\lambda_{\max}(A^{\mathrm{T}}A)} = \sqrt{\lambda_{\max}\{1,9\}} = 3$$

From Example 3.3.2, Theorem 3.3.2 is verified.

Theorem 3.3.3 If $\|B\| < 1$, then $I \pm B$ is nonsingular, and

$$\|(I \pm B)^{-1}\| \leqslant \frac{1}{1 - \|B\|}$$

where $\|\cdot\|$ refers to an operator norm of A.

Proof If det $(I-B)=0$ then the homogeneous system of equations $(I-B)x=0$ has nonzero solution, i.e., there exists $x_0 \neq 0$ such that $Bx_0 = x_0$, $\frac{\|Bx_0\|}{\|x_0\|} = 1$, therefore $\|B\| \geqslant 1$, this is a contradiction with the assumption.

Since $(I-B)(I-B)^{-1}=I$, therefore

$$(I-B)^{-1} = I + B(I-B)^{-1}$$

hence

$$\|(I-B)^{-1}\| \leqslant \|I\| + \|B\|\|(I-B)^{-1}\|$$

that is

$$\|(I-B)^{-1}\| \leqslant \frac{1}{1-\|B\|}$$

□

Definition 3.3.4 We call an $n \times n$ matrix A **convergent** if

$$\lim_{k\to\infty}(A^k)_{ij} = 0$$

for each $i = 1,2,\cdots,n$ and $j = 1,2,\cdots,n$.

Example 3.3.3 Let

$$A = \begin{pmatrix} \frac{1}{2} & 0 \\ \frac{1}{4} & \frac{1}{2} \end{pmatrix}$$

Computing powers of A. We have

$$A^2 = \begin{pmatrix} \frac{1}{4} & 0 \\ \frac{1}{4} & \frac{1}{4} \end{pmatrix}, \quad A^3 = \begin{pmatrix} \frac{1}{8} & 0 \\ \frac{3}{16} & \frac{1}{8} \end{pmatrix}, \quad A^4 = \begin{pmatrix} \frac{1}{16} & 0 \\ \frac{1}{8} & \frac{1}{16} \end{pmatrix}$$

and in general

$$A^k = \begin{pmatrix} \left(\frac{1}{2}\right)^k & 0 \\ \frac{k}{2^{k+1}} & \left(\frac{1}{2}\right)^k \end{pmatrix}$$

Since

$$\lim_{k\to\infty}\left(\frac{1}{2}\right)^k = 0 \text{ and } \lim_{k\to\infty}\frac{k}{2^{k+1}} = 0$$

A is a convergent matrix.

3.4 Best Linear Approximation

3.4.1 Basic Concepts and Theories

Let $f \in \mathbf{C}[a,b]$, we seek a polynomial $P_n^*(x)$ in

$$H_n = \text{span}\{1, x, \cdots, x^n\}$$

such that

$$\|f-P_n^*\|_\infty=\max_{a\leqslant x\leqslant b}|f(x)-P_n^*(x)|=\min_{P_n\in H_n}\|f(x)-P_n(x)\|_\infty$$

We first give the following definition.

Definition 3.4.1 Let $P_n(x)\in H_n, f(x)\in \mathbf{C}[a,b]$, we call

$$\Delta(f,P_n)=\|f-P_n\|_\infty=\max_{a\leqslant x\leqslant b}|f(x)-P_n(x)| \tag{3.4.1}$$

deviation (偏差) of $f(x)$ and $P_n(x)$ on $[a,b]$.

Obviously $\Delta(f,P_n)\geqslant 0$, denote $\{\Delta(f,P_n)\}$ is a set of all deviations, it has **lower bound** (下界) 0, we call

$$E_n=\inf_{P_n\in H_n}\{\Delta(f,P_n)\}=\inf_{P_n\in H_n}\max_{a\leqslant x\leqslant b}|f(x)-P_n(x)| \tag{3.4.2}$$

minimum deviation of $f(x)$ on $[a,b]$.

Definition 3.4.2 Let $f(x)\in\mathbf{C}[a,b]$, if there exists $P_n^*(x)\in H_n$, such that

$$\Delta(f,P_n^*)=E_n \tag{3.4.3}$$

then $P_n^*(x)$ is called **polynomial of the best uniform approximation** (最佳一致逼近多项式) or **polynomial of the minimum deviation approximation** (最小偏差逼近多项式), it is simply called polynomial of the best approximation.

Theorem 3.4.1 Let $f(x)\in\mathbf{C}[a,b]$, and there exists $P_n^*(x)\in H_n$, such that

$$\|f(x)-P_n^*(x)\|_\infty=E_n$$

Definition 3.4.3 Let $f(x)\in\mathbf{C}[a,b], P(x)\in H_n$, if on $x=x_0$, we have

$$|P(x_0)-f(x_0)|=\max_{a\leqslant x\leqslant b}|P(x)-f(x)|=\mu$$

then x_0 is called **deviation point** (偏差点) of $P(x)$.

If $P(x_0)-f(x_0)=\mu$, then x_0 is called a **positive deviation point.**

If $P(x_0)-f(x_0)=-\mu$, then x_0 is called a **negative deviation point.**

Since $P(x)-f(x)$ is continuous on $[a,b]$, therefore, there exists a point $x_0\in[a,b]$ at least, such that

$$|P(x_0)-f(x_0)|=\mu$$

Theorem 3.4.2 $P(x)\in H_n$ is the polynomial of the best approximation of $f(x)\in\mathbf{C}[a,b]$ if and only if $P(x)$ has at least $n+2$ deviation points which are positive and negative by turns, that is there exists $n+2$ points $a\leqslant x_1<x_2<\cdots<x_{n+2}\leqslant b$, such that

$$P(x_k)-f(x_k)=(-1)^k\sigma\|P(x)-f(x)\|_\infty,\quad \sigma=\pm 1 \tag{3.4.4}$$

This set of points is called **the set of Chebyshev alternate points** (切比雪夫交错点).

Corollary 3.4.1 If $f(x)\in\mathbf{C}[a,b]$, then there exists a unique polynomial of the best approximation in H_n.

3.4.2 Best Linear Approximation

Theorem 3.4.2 gives the property of the polynomial $P(x)$ of the best approximation, and it is very difficult to solve $P(x)$. Now, we discuss the case when $n=1$. Assume $f(x)\in \mathbf{C}^2[a,b]$, and the sign of $f''(x)$ is invariant in (a,b), we will seek the polynomial $P_1(x)=a_0+a_1x$ of the best approximation. According to Theorem 3.4.2, there exist at least three points $a\leqslant x_1<x_2<x_3\leqslant b$, such hat

$$P_1(x_k)-f(x_k)=(-1)^k\sigma\max_{a\leqslant x\leqslant b}|P_1(x)-f(x)| \quad (\sigma=\pm1,k=1,2,3)$$

Since the sign of $f''(x)$ is invariant on $[a,b]$, therefore $f'(x)$ is **monotonic** (单调的), $f'(x)-a_1$ has only one zero in (a,b), denoted by x_2, hence

$$P_1'(x_2)-f'(x_2)=a_1-f'(x_2)=0, \quad \text{i.e.,} \quad f'(x_2)=a_1$$

and the other two deviation points must be the **end points** (端点) of the interval, i.e., $x_1=a$, $x_3=b$, and satisfy

$$P_1(a)-f(a)=P_1(b)-f(b)=-(P_1(x_2)-f(x_2))$$

we obtain

$$\begin{cases} a_0+a_1a-f(a)=a_0+a_1b-f(b) \\ a_0+a_1a-f(a)=f(x_2)-(a_0+a_1x_2) \end{cases} \tag{3.4.5}$$

This implies that

$$a_1=\frac{f(b)-f(a)}{b-a}=f'(x_2) \tag{3.4.6}$$

From (3.4.5) and (3.4.6), we have

$$a_0=\frac{f(a)+f(x_2)}{2}-\frac{f(b)-f(a)}{b-a}\frac{a+x_2}{2} \tag{3.4.7}$$

Thus, we can obtain the polynomial $P_1(x)$ of the best approximation, see Figure 3-2 of its geometric significance.

The straight line $y_1=P_1(x)$ **parallels** (平行) the **chord** (弦) MN and passes through the middle point D of MQ, whose equation is of the form

$$y=\frac{1}{2}(f(a)+f(x_2))+a_1\left(x-\frac{a+x_2}{2}\right)$$

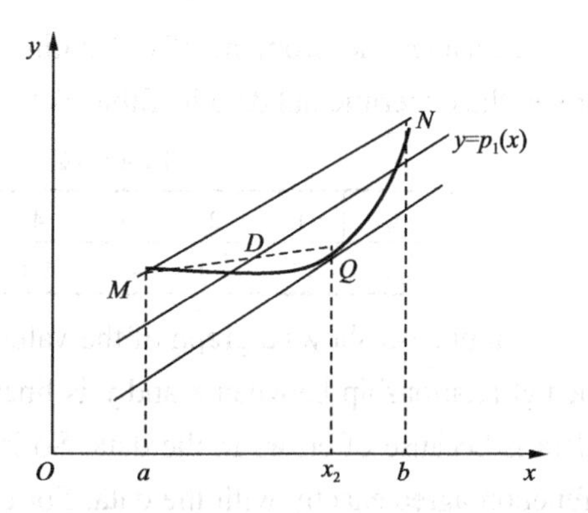

Figure 3-2 Best linear approximation

Example 3.4.1 Find the polynomial of the best linear approximation to $f(x)=\sqrt{1+x^2}$ on [0,1].

Solution From (3.4.6), we have

$$a_1 = \sqrt{2} - 1 \approx 0.414$$

and $f'(x) = \dfrac{x}{\sqrt{1+x^2}}$, therefore $\dfrac{x_2}{\sqrt{1+x_2^2}} = \sqrt{2} - 1$, this implies

$$x_2 = \sqrt{\frac{\sqrt{2}-1}{2}} \approx 0.4511, \quad f(x_2) = \sqrt{1+x_2^2} \approx 1.0986$$

From (3.4.7), we obtain

$$a_0 = \frac{1+\sqrt{1+x_2^2}}{2} - a_1 \frac{x_2}{2} \approx 0.955$$

Hence we obtain the polynomial of the best linear approximation

$$P_1(x) = 0.955 + 0.414x$$

$$\sqrt{1+x^2} \approx 0.955 + 0.414x, \quad 0 \leqslant x \leqslant 1 \tag{3.4.8}$$

Its limit of the error

$$\max_{0 \leqslant x \leqslant 1} \left| \sqrt{1+x^2} - P_1(x) \right| \leqslant 0.045$$

If let $x = \dfrac{b}{a} \leqslant 1$ in (3.4.8), then we obtain a formula

$$\sqrt{a^2 + b^2} \approx 0.955a + 0.414b$$

3.5 Discrete Least Squares Approximation（离散型最小平方逼近）

Consider the problem of estimating the values of a function at nontabulated points, given the experimental data in Table 3-2.

Table 3-2 The experimental data

x_i	1	2	3	4	5	6	7	8	9	10
y_i	1.3	3.4	4.3	5.0	7.0	8.8	10.2	12.5	13.0	15.6

Figure 3-3 shows a graph of the values in Table 3-2. From this graph, it appears that the actual relationship between x and y is linear. The likely reason that no line precisely fits the data is because of errors in the data. So it is unreasonable to require that the approximating function agree exactly with the data. For example, the ninth degree interpolating polynomial on the data shown in Figure 3-4 is obtained in mathematica using the commands.

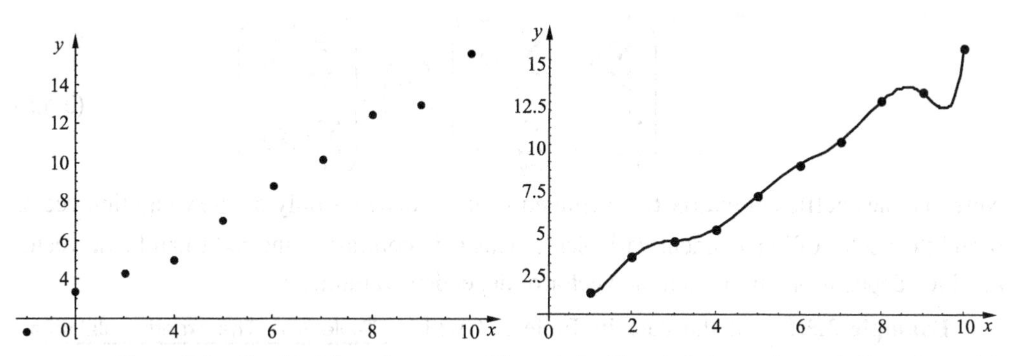

Figure 3-3 Graph of the values Figure 3-4 The ninth degree interpolating polynomial

This polynomial is clearly a poor predictor of information between a number of the data points. A better approach would be to find the "best" (in some sense) approximating line, even if it does not agree precisely with the data at any point.

For the given m+1 points $(x_i,y_i)(i=0,1,\cdots,m)$, the simplest trend curve is a linear polynomial

$$P_1(x)=a_0+a_1x \tag{3.5.1}$$

The least squares approach to this problem involves determining the best approximating line when the error involved is the sum of the squares of the differences between the y-values on the approximating line and the given y-values. Hence, constants a_0 and a_1 must be found that minimize the least error:

$$\varphi(a_0,a_1)=\sum_{i=0}^{m}w_i(y_i-(a_0+a_1x_i))^2 \tag{3.5.2}$$

Here $w_k>0$ denotes a **weight** (权) associated with the kth data sample, with the most common case being uniform weighting:

$$w_i=1,\quad 0\leqslant i\leqslant m \tag{3.5.3}$$

The polynomial P_1 whose coefficients minimize $\varphi(a_0,a_1)$ is referred to as the weighted least squares fit or simply the least squares fit. Although alternative error criteria can be used, the least squares objective in Equation (3.5.2) is preferred because a solution can be easily found. Indeed, if we differentiate $\varphi(a_0,a_1)$ with respect to the unknown components of the coefficient vector a, this yields

$$\begin{cases}\dfrac{\partial\varphi}{\partial a_0}=2\sum\limits_{i=0}^{m}w_i(y_i-p_1(x_i))\\[2ex]\dfrac{\partial\varphi}{\partial a_1}=2\sum\limits_{i=0}^{m}w_i(y_i-p_1(x_i))x_i\end{cases} \tag{3.5.4}$$

Setting $\partial\varphi/\partial a=0$ then results in two equations in the unknowns a_0 and a_1. After rearrangement, these equations can be expressed as a linear algebraic system of the form $Ca=b$.

$$\begin{pmatrix} \sum_{i=0}^{m} w_i & \sum_{i=0}^{m} w_i x_i \\ \sum_{i=0}^{m} w_i x_i & \sum_{i=0}^{m} w_i x_i^2 \end{pmatrix} \begin{pmatrix} a_0 \\ a_1 \end{pmatrix} = \begin{pmatrix} \sum_{i=0}^{m} w_i y_i \\ \sum_{i=0}^{m} w_i x_i y_i \end{pmatrix} \tag{3.5.5}$$

Note that the coefficient matrix $\boldsymbol{C}$ is symmetric, and it depends only on the weighting vector w and the vector of independent variables x. This is in contrast to the right-hand-side vector b, which depends on w, x, and the vector of dependent variables y.

Example 3.5.1 Fit the data in Table 3-3 with the discrete least squares polynomial of degree 1.

Table 3-3 The experimental data

x_i	1	2	3	4	5
f_i	4	4.5	6	8	8.5
ω_i	2	1	3	1	1

Solution $p_1(x)=a_0+a_1x$, where $m=4$.

The normal equations

$$\begin{cases} 8a_0 + 22a_1 = 47 \\ 22a_0 + 74a_1 = 145.5 \end{cases}$$

hence $a_0=2.77$, $a_1=1.13$

The fitting curve

$$S_1^*(x) = 2.77 + 1.13x$$

See Figure 3-5.

Figure 3-5 Least squares curve fitting

The general problem of approximating a set of data, $(x_i,y_i)(i=0,1,\cdots,m)$, with an algebraic polynomial

$$P_n(x) = a_0 + a_1 x + \cdots + a_n x^n$$

of degree $n<m$, using the least squares procedure is handed in a similar manner. We choose the constants $a_0, a_1, \cdots, a_n$ to minimize the least squares error

$$\begin{aligned} \varphi &\equiv \varphi(a_0, a_1, \cdots, a_n) \\ &= \sum_{i=0}^{m} w_i (y_i - p_n(x_i))^2 \\ &= \sum_{i=0}^{m} w_i y_i^2 - 2\sum_{i=0}^{m} w_i p_n(x_i) y_i + \sum_{i=0}^{m} w_i (p_n(x_i))^2 \\ &= \sum_{i=0}^{m} w_i y_i^2 - 2\sum_{i=0}^{m} w_i \left(\sum_{j=0}^{n} a_j x_i^j \right) y_i + \sum_{i=0}^{m} w_i \left(\sum_{j=0}^{n} a_j x_i^j \right)^2 \end{aligned} \tag{3.5.6}$$

As in the linear case, for $\varphi(a_0, a_1, \cdots, a_n)$ to be minimized, it is necessary that $\partial\varphi / \partial a_j = 0,\ j = 0,1,\cdots,n$.

$$0 = \frac{\partial \varphi}{\partial a_j} = -2\sum_{i=0}^{m} w_i y_i x_i^j + 2\sum_{k=0}^{n} a_k \sum_{i=0}^{m} w_i x_i^{j+k}$$

This gives n+1 **normal equations** (正规方程) in the n+1 unknowns a_j.

$$\sum_{k=0}^{n} a_k \sum_{i=0}^{m} w_i x_i^{j+k} = \sum_{i=0}^{m} w_i y_i x_i^j, \quad \text{for each} \quad j = 0,1,\cdots,n$$

It is helpful to write the equations as follows:

$$a_0 \sum_{i=0}^{m} w_i x_i^0 + a_1 \sum_{i=0}^{m} w_i x_i^1 + a_2 \sum_{i=0}^{m} w_i x_i^2 + \cdots + a_n \sum_{i=0}^{m} w_i x_i^n = \sum_{i=0}^{m} w_i y_i x_i^0$$

$$a_0 \sum_{i=0}^{m} w_i x_i^1 + a_1 \sum_{i=0}^{m} w_i x_i^2 + a_2 \sum_{i=0}^{m} w_i x_i^3 + \cdots + a_n \sum_{i=0}^{m} w_i x_i^{n+1} = \sum_{i=0}^{m} w_i y_i x_i^1$$

…

$$a_0 \sum_{i=0}^{m} w_i x_i^n + a_1 \sum_{i=0}^{m} w_i x_i^{n+1} + a_2 \sum_{i=0}^{m} w_i x_i^{n+2} + \cdots + a_n \sum_{i=0}^{m} w_i x_i^{2n} = \sum_{i=0}^{m} w_i y_i x_i^n$$

These equations can be cast as a linear algebraic system $\boldsymbol{Ca}=\boldsymbol{b}$.

$$\begin{pmatrix} \sum_{i=0}^{m} w_i & \sum_{i=0}^{m} w_i x_i & \cdots & \sum_{i=0}^{m} w_i x_i^n \\ \sum_{i=0}^{m} w_i x_i & \sum_{i=0}^{m} w_i x_i^2 & \cdots & \sum_{i=0}^{m} w_i x_i^{n+1} \\ \vdots & \vdots & & \vdots \\ \sum_{i=0}^{m} w_i x_i^n & \sum_{i=0}^{m} w_i x_i^{n+1} & \cdots & \sum_{i=0}^{m} w_i x_i^{2n} \end{pmatrix} \begin{pmatrix} a_0 \\ a_1 \\ \vdots \\ a_n \end{pmatrix} = \begin{pmatrix} \sum_{i=0}^{m} w_i y_i \\ \sum_{i=0}^{m} w_i x_i y_i \\ \vdots \\ \sum_{i=0}^{m} w_i x_i^n y_i \end{pmatrix} \tag{3.5.7}$$

These normal equations have a unique solution provided that the x_i are distinct.

Example 3.5.2 Given the data of $(x_i,y_i)(i=0,1,2,3,4)$ as Table 3-4, the mathematical model is $y=ae^{bx}$, determine a and b with least squares method.

Table 3-4 The experimental data

i	0	1	2	3	4
x_i	1.00	1.25	1.50	1.75	2.00
y_i	5.10	5.79	6.53	7.45	8.46
$\overline{y_i}$	1.629	1.756	1.876	2.008	2.135

Solution For $y=ae^{bx}$, take logarithm of the two sides, we have

$$\ln y = \ln a + bx, \tag{3.5.8}$$

Let

$$\overline{y} = \ln y, \quad A = \ln a,$$

then the Equation (3.5.8) can be rewritten

$$\overline{y} = A + bx$$

Extending the table and computing as follows:

$$\sum_{i=0}^{4} x_i = 7.50, \quad \sum_{i=0}^{4} x_i^2 = 11.875, \quad \sum_{i=0}^{4} \overline{y_i} = 9.404, \quad \sum_{i=0}^{4} x_i \overline{y_i} = 14.422.$$

The normal equations

$$\begin{cases} 5A + 7.50b = 9.404 \\ 7.50A + 11.875b = 14.422 \end{cases}$$

We have

A=1.122, b=0.505, $a=e^A$=3.071

Hence $y=3.071e^{0.505x}$

The graph shown in Figure 3-6 is obtained in mathematica using the commands.

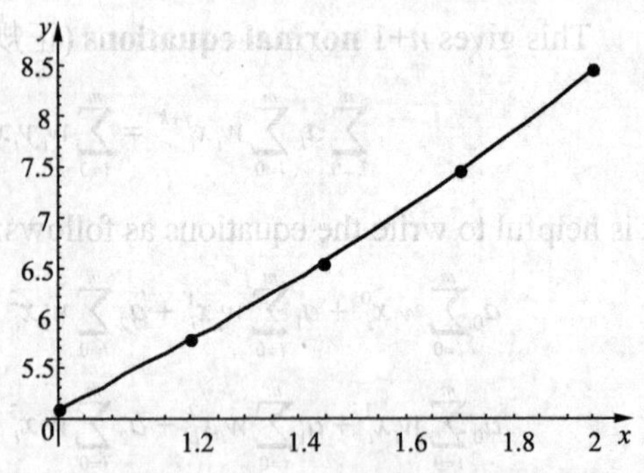

Figure 3-6 Least squares curve fittings

Example 3.5.3 Given the table as Table 3-5, if $y=a+bx$, determine the coefficients a and b with the least squares method.

Solution $\sum_{i=1}^{4}\omega_i = 54$, $\sum_{i=1}^{4}\omega_i x_i = 28 + 108 + 72 + 8 = 216$

$$\sum_{i=1}^{4}\omega_i x_i^2 = 984,\quad \sum_{i=1}^{4}\omega_i y_i = 701,\quad \sum_{i=1}^{4}\omega_i x_i y_i = 3\,580$$

Normal equations

$$\begin{cases} 54a + 216b = 701 \\ 216a + 984b = 3580 \end{cases}$$

We have

$$a=-12.885,\quad b=6.467$$

Example 3.5.4 Given Table 3-6.

Table 3-5 The experimental data

i	1	2	3	4
x_i	2	4	6	8
y_i	2	11	28	40
ω_i	14	27	12	1

Table 3-6 The experimental data

x	1	2	3	4
y	0.8	1.5	1.8	2.0

and **empirical formula** (经验公式) $g(x)=ax+bx^2$, determine the coefficients a and b with the least square method .

Solution The coefficient of x^0 in the empirical formula is zero, we can not use the Formula (3.5.7) directly, consider

$$\varphi(a,b) = \sum_{i=1}^{4}(ax_i + bx_i^2 - y_i)^2$$

Let

$$\begin{cases} 0 = \dfrac{\partial\varphi}{\partial a} = 2\sum_{i=1}^{4}(ax_i + bx_i^2 - y_i)x_i \\ 0 = \dfrac{\partial\varphi}{\partial b} = 2\sum_{i=1}^{4}(ax_i + bx_i^2 - y_i)x_i^2 \end{cases}$$

That is

$$\begin{cases} a\sum_{i=1}^{4}x_i^2+b\sum_{i=1}^{4}x_i^3=\sum_{i=1}^{4}y_ix_i \\ a\sum_{i=1}^{4}x_i^3+b\sum_{i=1}^{4}x_i^4=\sum_{i=1}^{4}y_ix_i^2 \end{cases}$$

where

$$\sum_{i=1}^{4}x_i^2=30,\ \sum_{i=1}^{4}x_i^3=100,\ \sum_{i=1}^{4}x_i^4=354,\ \sum_{i=1}^{4}y_ix_i=17.2,\ \sum_{i=1}^{4}y_ix_i^2=55.$$

Normal equations

$$\begin{cases} 30a+100b=17.2 \\ 100a+354b=55 \end{cases}$$

We have

$$a=\frac{58.88}{62},\quad b=-\frac{7}{62}$$

Therefore

$$g(x)=0.949\,7x-0.112\,9x^2$$

3.6 Least Squares Approximation and Orthogonal Polynomials（最小平方逼近和正交多项式）

The previous section considered the problem of least squares approximation to fit a collection of data. The other approximation problem mentioned in the introduction concerns the approximation of functions.

Suppose $f\in \mathbf{C}[a,b]$ and that a polynomial $P_n(x)$ of degree at most n is required that will minimize the error

$$\int_a^b (f(x)-P_n(x))^2\,\mathrm{d}x$$

To determine a least squares approximating polynomial, that is, a polynomial to minimize this expression, let

$$P_n(x)=a_0+a_1x+\cdots+a_nx^n=\sum_{k=0}^{n}a_kx^k$$

and define

$$\varphi\equiv\varphi(a_0,a_1,\cdots,a_n)=\int_a^b\left(f(x)-\sum_{k=0}^{n}a_kx^k\right)^2\mathrm{d}x$$

The problem is to find real coefficients $a_0,a_1,\cdots,a_n$ that will minimize φ. A necessary condition for the numbers $a_0,a_1,\cdots,a_n$ to minimize φ is that

$$\frac{\partial \varphi}{\partial a_j}=0, \quad \text{for each } j=0,1,\cdots,n$$

Since

$$\varphi=\int_a^b (f(x))^2 \mathrm{d}x-2\sum_{k=0}^{n} a_k \int_a^b x^k f(x)\mathrm{d}x+\int_a^b\left(\sum_{k=0}^{n} a_k x^k\right)^2 \mathrm{d}x$$

we have

$$\frac{\partial \varphi}{\partial a_j}=-2\int_a^b x^j f(x)\mathrm{d}x+2\sum_{k=0}^{n} a_k \int_a^b x^{j+k}\mathrm{d}x$$

Hence, to find $P_n(x)$, the n+1 linear normal equations

$$\sum_{k=0}^{n} a_k \int_a^b x^{j+k}\mathrm{d}x=\int_a^b x^j f(x)\mathrm{d}x, \quad \text{for each } j=\text{d}0,1,\cdots,n \tag{3.6.1}$$

must be solved for the n+1 unknowns a_j. The normal equations always have a unique solution provided $f\in \mathbf{C}[a,b]$.

Example 3.6.1 Find the least squares approximating polynomial of degree 1 for the function $f(x)=\sqrt{1+x^2}$ on [0,1].

Solution The normal equations for $P_1(x)=a_0+a_1x$ are

$$a_0\int_0^1 1\mathrm{d}x+a_1\int_0^1 x\mathrm{d}x=\int_0^1\sqrt{1+x^2}\,\mathrm{d}x$$

$$a_0\int_0^1 x\mathrm{d}x+a_1\int_0^1 x^2\mathrm{d}x=\int_0^1 x\sqrt{1+x^2}\,\mathrm{d}x$$

where

$$\int_0^1 x\mathrm{d}x=\frac{1}{2}, \quad \int_0^1 x^2\mathrm{d}x=\frac{1}{3}$$

$$\int_0^1\sqrt{1+x^2}\,\mathrm{d}x=\frac{1}{2}\ln(1+\sqrt{2})+\frac{\sqrt{2}}{2}\approx 1.147$$

$$\int_0^1 x\sqrt{1+x^2}\,\mathrm{d}x=\frac{1}{3}(1+x^2)^{3/2}\Big|_0^1=\frac{2\sqrt{2}-1}{3}\approx 0.609$$

The normal equations are

$$a_0+\frac{1}{2}a_1=1.147, \quad \frac{1}{2}a_0+\frac{1}{3}a_1=0.609$$

We have

$$a_0=0.934, \quad a_1=0.426$$

Therefore

$$P_1(x)=0.934+0.426x$$

Example 3.6.2 Find the least squares approximating polynomial of degree 2 for the function $f(x)=\sin \pi x$ on the interval[0,1].

Solution The normal equations for $P_2(x)=a_0+a_1x+a_2x^2$ are

$$a_0\int_0^1 1\mathrm{d}x+a_1\int_0^1 x\mathrm{d}x+a_2\int_0^1 x^2\mathrm{d}x=\int_0^1 \sin\pi x\mathrm{d}x,$$

$$a_0\int_0^1 x\mathrm{d}x+a_1\int_0^1 x^2\mathrm{d}x+a_2\int_0^1 x^3\mathrm{d}x=\int_0^1 x\sin\pi x\mathrm{d}x$$

$$a_0\int_0^1 x^2\mathrm{d}x+a_1\int_0^1 x^3\mathrm{d}x+a_2\int_0^1 x^4\mathrm{d}x=\int_0^1 x^2\sin\pi x\mathrm{d}x$$

Performing the integration yields

$$a_0+\frac{1}{2}a_1+\frac{1}{3}a_2=\frac{2}{\pi}$$

$$\frac{1}{2}a_0+\frac{1}{3}a_1+\frac{1}{4}a_2=\frac{1}{\pi}$$

$$\frac{1}{3}a_0+\frac{1}{4}a_1+\frac{1}{5}a_2=\frac{\pi^2-4}{\pi^3}$$

These three equations in three unknowns can be solved to obtain

$$a_0=\frac{12\pi^2-120}{\pi^3}\approx -0.050\,465$$

$$a_1=-a_2=\frac{720-62\pi^2}{\pi^3}\approx 4.122\,51$$

Consequently, the least squares approximating polynomial of degree 2 for the function $f(x)=\sin\pi x$ on the interval$[0,1]$ is

$$P_2(x)=-0.050\,465+4.122\,51x-4.122\,51x^2$$

Example 3.6.2 illustrates the difficulty in obtaining a least squares polynomial approximation. A $(n+1)\times(n+1)$ linear system for the unknowns $a_0,a_1,\cdots,a_n$ must be solved, and the coefficients in the linear system are of the form

$$\int_a^b x^{j+k}\mathrm{d}x=\frac{b^{j+k+1}-a^{j+k+1}}{j+k+1}$$

The Matrix in the linear system

$$\boldsymbol{H}=\begin{pmatrix}1 & 1/2 & \cdots & 1/(n+1)\\ 1/2 & 1/3 & \cdots & 1/(n+2)\\ \vdots & \vdots & & \vdots\\ 1/(n+1) & 1/(n+2) & \cdots & 1/(2n+1)\end{pmatrix}$$

is known as a Hilbert matrix.

A different technique to obtain least squares approximation will now be considered. This turns out to be computationally efficient, and once $P_n(x)$ is known, it is easy to determine $P_{n+1}(x)$. To facilitate the discussion, we need some new concepts.

Definition 3.6.1 The set of functions $\{\varphi_0,\cdots,\varphi_n\}$ is said to be linearly independent on $[a,b]$, whenever

$$c_0\varphi_0(x)+c_1\varphi_1(x)+\cdots c_n\varphi_n(x)=0, \quad \text{for all } x\in[a,b]$$

we have $c_0=c_1=\cdots=c_n=0$. Otherwise the set of functions is said to be **linearly dependent.**

Theorem 3.6.1 If $\varphi_j(x)$ is a polynomial of degree j, for each j=0,1,$\cdots$,n, then $\{\varphi_0,\cdots,\varphi_n\}$is linearly independent on any interval$[a,b]$.

Proof Suppose $c_0,\cdots,c_n$ are real numbers for which

$$P(x)=c_0\varphi_0(x)+c_1\varphi_1(x)+\cdots c_n\varphi_n(x)=0, \quad \text{for all } x\in[a,b]$$

The polynomial $P(x)$ vanishes on $[a,b]$, so it must be zero polynomial, and the coefficients of all the powers of x are zero. In particular, the coefficient of x^n is zero. Since $c_n\varphi_n(x)$ is the only term in $P(x)$ that contains x^n, we must have c_n=0 and

$$P(x)=\sum_{j=0}^{n-1}c_j\varphi_j(x)$$

In this representation of $P(x)$, the only term that contains a power of x^{n-1} is $c_{n-1}\varphi_{n-1}(x)$, so this term must also be zero and

$$P(x)=\sum_{j=0}^{n-2}c_j\varphi_j(x)$$

In like manner, the remaining constants $c_{n-2},c_{n-3},\cdots,c_1,c_0$ are all zero, which implies that $\{\varphi_0,\varphi_1,\cdots,\varphi_n\}$ is linearly independent. □

Example 3.6.3 Let $\varphi_0(x)$=3, $\varphi_1(x)=x-2$, and $\varphi_2(x)=x^2+4x+7$. By Theorem 3.6.1, $\{\varphi_0, \varphi_1, \varphi_2\}$is linearly independent on any interval $[a,b]$. Suppose $q(x)=a_0+a_1x+a_2x^2$, Show that there exist constants c_0, c_1, and c_2, such that $q(x)=c_0\varphi_0(x)+c_1\varphi_1(x)+c_2\varphi_2(x)$.

Solution First note that

$$1=\frac{1}{3}\varphi_0(x),\quad x=\varphi_1(x)+2=\varphi_1(x)+\frac{2}{3}\varphi_0(x),$$

and

$$\begin{aligned}x^2&=\varphi_2(x)-4x-7=\varphi_2(x)-4\left(\varphi_1(x)+\frac{2}{3}\varphi_0(x)\right)-7\left(\frac{1}{3}\varphi_0(x)\right)\\&=\varphi_2(x)-4\varphi_1(x)-5\varphi_0(x)\end{aligned}$$

Hence

$$\begin{aligned}q(x)&=a_0\left(\frac{1}{3}\varphi_0(x)\right)+a_1\left(\varphi_1(x)+\frac{2}{3}\varphi_0(x)\right)+a_2(\varphi_2(x)-4\varphi_1(x)-5\varphi_0(x))\\&=\left(\frac{1}{3}a_0+\frac{2}{3}a_1-5a_2\right)\varphi_0(x)+(a_1-4a_2)\varphi_1(x)+a_2\varphi_2(x)\end{aligned}$$

Let Π_n be the set of all polynomials of degree at most n. The following result is used extensively in many applications of linear algebra.

Theorem 3.6.2 If $\{\varphi_0, \varphi_1, \cdots, \varphi_n\}$ is a collection of linearly independent polynomials in Π_n, then any polynomial in Π_n can be written uniquely as a linear combination of $\{\varphi_0, \varphi_1, \cdots, \varphi_n\}$.

Suppose $\{\varphi_0, \varphi_1, \cdots, \varphi_n\}$ is a set of linearly independent functions on $[a,b]$, $w(x)$ is a weight function for $[a,b]$, and for $f \in \mathbf{C}[a,b]$, a linear combination

$$P(x) = \sum_{k=0}^{n} a_k \varphi_k(x)$$

is sought to minimize the error

$$\varphi(a_0, a_1, \cdots, a_n) = \int_a^b w(x)(f(x) - \sum_{k=0}^{n} a_k \varphi_k(x))^2 \mathrm{d}x$$

This problem reduces to the situation considered at the beginning of this section in the special case when $w(x) \equiv 1$ and $\varphi_k(x) = x^k$, for each $k = 0, 1, \cdots, n$.

The normal equations associated with this problem are derived from the fact that for each $j=0,1,\cdots,n$,

$$0 = \frac{\partial \varphi}{\partial a_j} = 2\int_a^b w(x)\left(f(x) - \sum_{k=0}^{n} a_k \varphi_k(x)\right)\varphi_j(x)\mathrm{d}x$$

The system of normal equations can be written

$$\int_a^b w(x) f(x) \varphi_j(x)\mathrm{d}x = \sum_{k=0}^{n} a_k \int_a^b w(x)\varphi_k(x)\varphi_j(x)\mathrm{d}x, \quad \text{for } j = 0, 1, \cdots, n$$

If the function $\varphi_0(x), \varphi_1(x), \cdots, \varphi_n(x)$ can be chosen, so that

$$\int_a^b w(x)\varphi_k(x)\varphi_j(x)\mathrm{d}x = \begin{cases} 0 & j \neq k \\ m_j > 0 & j = k \end{cases} \tag{3.6.2}$$

then the normal equations reduce to

$$\int_a^b w(x) f(x)\varphi_j(x)\mathrm{d}x = a_j \int_a^b w(x)(\varphi_j(x))^2 \mathrm{d}x = a_j m_j$$

for each $j = 0, 1, \cdots, n$, and are easily solved to give

$$a_j = \frac{1}{m_j}\int_a^b w(x) f(x)\varphi_j(x)\mathrm{d}x$$

Hence the least squares approximation problem is greatly simplified when the functions φ_0, φ_1, $\cdots$, φ_n are chosen to satisfy the orthogonality condition in (3.6.2). The remainder of this section is devoted to studying collections of this type.

Definition 3.6.2 $\{\varphi_0, \varphi_1, \cdots, \varphi_n\}$ is said to be an **orthogonal set of functions** (正交函数集) for the interval $[a,b]$ with respect to the weight function w if

$$\int_a^b w(x)\varphi_k(x)\varphi_j(x)\mathrm{d}x=\begin{cases}0 & j\neq k\\ m_j>0 & j=k\end{cases}$$

If, in addition , m_j=1 for each j=0,1⋯,n, the set is said to be **orthonormal** (标准正交).

This definition, together with the remarks preceding it, produces the following theorem.

Theorem 3.6.3 If $\{\varphi_0, \varphi_1,\cdots, \varphi_n\}$ is an orthogonal set of functions on an interval $[a,b]$ with respect to the weight function w, the least squares approximation to f on $[a,b]$ respect to w is

$$P(x)=\sum_{k=0}^{n}a_k\varphi_k(x)$$

where, for each k=0,1,⋯,n.

$$a_k=\frac{\int_a^b w(x)f(x)\varphi_k(x)\mathrm{d}x}{\int_a^b w(x)(\varphi_k(x))^2\mathrm{d}x}=\frac{1}{m_k}\int_a^b w(x)f(x)\varphi_k(x)\mathrm{d}x \tag{3.6.3}$$

The next theorem, which is based on the Gram-Schmidt process, describes how to construct orthogonal polynomials on $[a,b]$ with respect to a weight function w.

Theorem 3.6.4 The set of polynomial functions $\{\varphi_0,\varphi_1, \cdots ,\varphi_n\}$ defined in the following way is orthogonal on $[a,b]$ with respect to the weight function w.

$$\varphi_0(x)\equiv 1,\quad \varphi_1(x)=x-B_1,\quad \text{for each } x \text{ in } [a,b]$$

where

$$B_1=\frac{\int_a^b xw(x)(\varphi_0(x))^2\mathrm{d}x}{\int_a^b w(x)(\varphi_0(x))^2\mathrm{d}x}$$

and when $k\geqslant 2$,

$$\varphi_k(x)=(x-B_k)\varphi_{k-1}(x)-C_k\varphi_{k-2}(x),\quad \text{for each } x \text{ in } [a,b]$$

where

$$B_k=\frac{\int_a^b xw(x)(\varphi_{k-1}(x))^2\mathrm{d}x}{\int_a^b w(x)(\varphi_{k-1}(x))^2\mathrm{d}x}$$

and

$$C_k=\frac{\int_a^b xw(x)\varphi_{k-1}(x)\varphi_{k-2}(x)\mathrm{d}x}{\int_a^b w(x)(\varphi_{k-2}(x))^2\mathrm{d}x}$$

Theorem 3.6.4 provides a recursive procedure for constructing a set of orthogonal polynomials. The proof of this theorem follows by applying mathematical introduction to the degree of the polynomial $\varphi_n(x)$.

Corollary 3.6.1 For any $n>0$, the set of polynomial functions $\{\varphi_0, \varphi_1,\cdots, \varphi_n\}$ given in Theorem 3.6.3 is linearly independent on $[a,b]$ and

$$\int_a^b w(x)\varphi_n(x)Q_k(x)\mathrm{d}x = 0$$

for any polynomial $Q_k(x)$ of degree $k<n$.

Proof Since $\varphi_n(x)$ is a polynomial of degree n, Theorem 3.6.1 implies that $\{\varphi_0, \varphi_1,\cdots, \varphi_n\}$ is a linearly independent set.

Let $Q_k(x)$ be a polynomial of degree k. By Theorem 3.6.2, there exist numbers $c_0,\cdots,c_k$, such that

$$Q_k(x) = \sum_{j=0}^{k} c_j\varphi_j(x)$$

Thus

$$\int_a^b w(x)\varphi_n(x)Q_k(x)\mathrm{d}x = \sum_{j=0}^{k} c_j \int_a^b w(x)\varphi_j(x)\varphi_n(x)\mathrm{d}x = \sum_{j=0}^{k} c_j \cdot 0 = 0$$

since φ_n is orthogonal to φ_j for each $j = 0,1,\cdots,k$. □

Example 3.6.4 The set of Legendre polynomials, $\{P_n(x)\}$, is orthogonal on $[-1,1]$ with respect to the weight function $w(x)\equiv 1$. The classical definition of the Legendre polynomials requires that $P_n(1)=1$ for each n, and a recursive relation is used to generate the polynomials when $n\geqslant 2$.

Using the recursive procedure of Theorem 3.6.4 with $P_0(x)\equiv 1$ gives

$$B_1 = \frac{\int_{-1}^{1} x\mathrm{d}x}{\int_{-1}^{1}\mathrm{d}x} = 0 \quad \text{and} \quad P_1(x) = x - B_1 = x$$

Also

$$B_2 = \frac{\int_{-1}^{1} x^3\mathrm{d}x}{\int_{-1}^{1} x^2\mathrm{d}x} = 0 \quad \text{and} \quad C_2 = \frac{\int_{-1}^{1} x^2\mathrm{d}x}{\int_{-1}^{1} 1\mathrm{d}x} = \frac{1}{3}$$

So

$$P_2(x) = (x - B_2)P_1(x) - C_2P_0(x) = (x-0)x - \frac{1}{3}\cdot 1 = x^2 - \frac{1}{3}$$

Analogously, we have

$$B_3 = 0 \quad \text{and} \quad C_3 = \frac{4}{15}$$

thus

$$P_3(x) = xP_2(x) - \frac{4}{15}P_1(x) = x^3 - \frac{1}{3}x - \frac{4}{15}x = x^3 - \frac{3}{5}x$$

$$P_4(x) = x^4 - \frac{6}{7}x^2 + \frac{3}{35}$$

$$P_5(x) = x^5 - \frac{10}{9}x^3 + \frac{5}{21}x$$

3.7 Rational Function Approximation

The class of algebraic polynomial has some distinct advantages for use in approximation:

(1) there are a sufficient number of polynomials to approximate any continuous function on a closed interval to within an arbitrary tolerance;

(2) polynomials are easily evaluated at arbitrary values;

(3) the derivatives and integrals of polynomials exist and are easily determined.

3.7.1 Continued Fractions

Many of the special functions that occur in the applications of mathematics are defined by infinite processes, such as series, integrals, and iterations. The continued fraction is one of these infinite processes. An example of a continued fraction is this one, due to Lambert in 1770:

$$\arctan x = \cfrac{x}{1+\cfrac{x^2}{3+\cfrac{4x^2}{5+\cfrac{9x^2}{7+\cfrac{16x^2}{9+\cdots}}}}} \quad (|x|<1) \tag{3.7.1}$$

This can also be written as

$$\arctan x = \frac{x}{1+}\,\frac{x^2}{3+}\,\frac{4x^2}{5+}\,\frac{9x^2}{7+}\,\frac{16x^2}{9+}\cdots \quad (|x|<1) \tag{3.7.2}$$

The right side of this equation represents a limit in the following way: The expression in equation (3.7.2) is terminated after n terms to give a well-defined function f_n:

$$f_n(x) = \frac{x}{1+}\,\frac{x^2}{3+}\,\frac{4x^2}{5+}\,\frac{9x^2}{7+}\cdots\frac{(n-1)^2x^2}{2n-1} \quad (n \geqslant 2) \tag{3.7.3}$$

This is called the nth convergent of the continued fraction. Equation (3.7.2) is defined to mean

$$\arctan x = \lim_{n\to\infty} f_n(x)\cdots \quad (|x|<1) \tag{3.7.4}$$

If we assume the validity of this equation, we have an alternative method of computing values of the arctangent function. Whether the method is practical depends on the rapidity of convergence in Equation (3.7.4). To judge this numerically, let us compute $\arctan(1/\sqrt{3}) = \pi/6 \approx 0.523\,598\,775\,6$ by means of the sequence $f_n(1/\sqrt{3})$ for $n \geqslant 2$. Here are the results:

n	$f_n(1/\sqrt{3})$
2	0.519 615
3	0.523 892
4	0.523 577
5	0.523 600
6	0.523 599
7	0.523 599

This list shows that six decimal places of precision have been obtained at the sixth convergent.

Another example, for function $\ln(1+x)$, the Taylor expansion is

$$\ln(1+x)=\sum_{k=1}^{\infty}(-1)^{k-1}\frac{x^k}{k}, \qquad x\in[-1,1], \tag{3.7.5}$$

take the partial sum

$$S_n(x)=\sum_{k=1}^{n}(-1)^{k-1}\frac{x^k}{k}\approx\ln(1+x)\cdot$$

On the other hand, using division to obtain a continued fraction

$$\ln(1+x)=\cfrac{x}{1+\cfrac{1\cdot x}{2+\cfrac{1\cdot x}{3+\cfrac{2^2\cdot x}{4+\cfrac{2^2\cdot x}{5+\cdots}}}}}$$

This can also be written as the following compact format (紧凑格式)

$$\ln(1+x)=\frac{x}{1+}\,\frac{1\cdot x}{2+}\,\frac{1\cdot x}{3+}\,\frac{2^2\cdot x}{4+}\,\frac{2^2\cdot x}{5+}\cdots$$

Take the second, fourth, sixth and eighth as the approximations of $\ln(1+x)$ respectively, then

$$R_{11}(x)=\frac{2x}{2+x}$$

$$R_{22}(x)=\frac{6x+3x^2}{6+6x+x^2}$$

$$R_{33}(x)=\frac{60x+60x^2+11x^3}{60+90x+36x^2+3x^3}$$

$$R_{44}(x)=\frac{420x+630x^2+260x^3+25x^4}{420+480x+540x^2+120x^3+6x^4}$$

We compare the errors of $S_{2n}(x)$ and errors of $R_{2n}(x)$ at x=1, the results are displayed in Table 3-7.

Table 3-7 Errors of $S_{2n}(x)$ and $R_{2n}(x)$

n	S_{2n}	$\varepsilon_S = \lvert\ln 2 - S_{2n}(1)\rvert$	$S_{2n}(1)$	$\varepsilon_R = \lvert\ln 2 - R_{2n}(1)\rvert$
1	0.50	0.19	0.667	0.026
2	0.58	0.11	0.692 31	0.000 84
3	0.617	0.076	0.693 122	0.000 025
4	0.634	0.058	0.693 146 42	0.000 000 76

Clearly, Rational function approximation is much better than that of polynomial.

Recursive Formulas

The task of evaluating a continued fraction is not as easy as evaluating a series, In the case of an infinite series, say $\sum_{k=1}^{\infty} a_k$, we compute the partial sum $S=\sum_{k=1}^{n} a_k$ by means of the formula $S_{n+1}=S_n+a_{n+1}$. Let us consider the analogous problem for a continued fraction:

$$C=\frac{a_1}{b_1+}\ \frac{a_2}{b_2+}\ \frac{a_3}{b_3+}\ \cdots \tag{3.7.6}$$

We want to discover a recursive algorithm for computing the successive convergence:

$$C_n=\frac{a_1}{b_1+}\ \frac{a_2}{b_2+}\ \frac{a_3}{b_3+}\ \cdots\frac{a_{n-1}}{b_{n-1}+}\ \frac{a_n}{b_n} \tag{3.7.7}$$

We introduce for temporary use the functions

$$f_n(x)=\frac{a_1}{b_1+}\ \frac{a_2}{b_2+}\ \frac{a_3}{b_3+}\ \cdots\frac{a_{n-1}}{b_{n-1}+}\ \frac{a_n}{b_n+x} \tag{3.7.8}$$

Obviously then, we obtain

$$C_n=f_n(0)$$

To obtain $f_n(x)$ from $f_{n-1}(x)$, we observe first

$$f_n(x)=f_{n-1}\left(\frac{a_n}{b_n+x}\right)$$

By direct calculation, we have

$$f_1(x)=\frac{a_1}{b_1+x}$$

$$f_2(x)=\frac{a_1b_2+a_1x}{b_1b_2+a_2+b_1x}$$

$$f_3(x)=\frac{a_1b_2b_3+a_1a_3+(a_1b_2)x}{b_1b_2b_3+a_2b_3+b_1a_3+(b_1b_2+a_2)x}$$

The pattern that emerges here suggests that

$$f_n(x)=\frac{A_n+A_{n-1}x}{B_n+B_{n-1}x}\qquad (n\geqslant 2) \tag{3.7.9}$$

where

$$\begin{cases} A_0 = 0, \ A_1 = a_1 \\ A_n = b_n A_{n-1} + a_n A_{n-2} \qquad (n \geqslant 2) \end{cases} \tag{3.7.10}$$

and

$$\begin{cases} B_0 = 1, \ B_1 = b_1 \\ B_n = b_n B_{n-1} + a_n B_{n-2} \qquad (n \geqslant 2) \end{cases} \tag{3.7.11}$$

Theorem 3.7.1 (Theorem on Continued Fraction) If the sequences $\{a_n\}_{n=1}^{\infty}$ and $\{b_n\}_{n=1}^{\infty}$ are given, and if the sequences $\{A_n\}_{n=0}^{\infty}$ and $\{B_n\}_{n=0}^{\infty}$ are defined by Formulas (3.7.10), then

$$C_n = \frac{A_n}{B_n} \qquad (n \geqslant 1)$$

Proof The validity of Equation (3.7.9) for indices 1, 2 and 3 has already been verified. Now suppose that Equation (3.7.9) is true for indices $1, 2, \dots, n-1$. Then

$$\begin{aligned} f_n(x) = f_{n-1}\left(\frac{a_n}{b_n + x}\right) &= \frac{A_{n-1} + A_{n-2}a_n/(b_n + x)}{B_{n-1} + B_{n-2}a_n/(b_n + x)} \\ &= \frac{A_{n-1}(b_n + x) + A_{n-2}a_n}{B_{n-1}(b_n + x) + B_{n-2}a_n} = \frac{A_{n-1}b_n + A_{n-2}a_n + A_{n-1}x}{B_{n-1}b_n + B_{n-2}a_n + B_{n-1}x} \\ &= \frac{A_n + A_{n-1}x}{B_n + B_{n-1}x} \end{aligned}$$

This establishes the Equation (3.7.9) by induction. Then, of course

$$C_n = f_n(0) = \frac{A_n}{B_n} \qquad \square$$

The recursive formulas developed here form the basis for an efficient algorithm. For example, the numerical results for $\tan^{-1}(1/\sqrt{3})$, at the beginning of this section, were computed using this recursive relation with $a_n = (n-1)^2 x^2$ and $b_n = 2n-1$, starting with $a_1 = x$ and $b_1 = 1$.

3.7.2 Padé Approximation

Suppose R_{nm} is a rational function of degree $N = n + m$ of the form

$$R_{nm} = \frac{p(x)}{q(x)} = \frac{p_0 + p_1 x + \cdots p_n x^n}{q_0 + q_1 x + \cdots q_m x^m}$$

That is used to approximate a function f on a closed interval I containing zero. For R_{nm} to be defined at zero requires that $q_0 \neq 0$. In fact, we can assume that $q = 1$, for if this is not the case we simply replace $p(x)$ by $p(x)/q_0$ and $q(x)$ by $q(x)/q_0$. Consequently, there are $N+1$ parameters $q_1, q_2, \dots, q_m$, $p_0, p_1, p_2, \dots, p_n$ available for the approximation of f by R_{nm}.

The **Padé approximation technique,** which is the extension of Taylor polynomial approximation to rational functions, chooses the $N+1$ parameters so that $f^{(k)}(0)=R_{nm}{}^{(k)}(0)$, for each $k=0,1,\ldots,N$. When $n=N$ and $m=0$, the Padé approximation is just the Nth **Maclaurin polynomial**.

Consider the difference

$$f(x)-R_{nm}=f(x)-\frac{p(x)}{q(x)}=\frac{f(x)q(x)-p(x)}{q(x)}=\frac{f(x)\sum_{i=0}^{m}q_ix^i-\sum_{i=0}^{n}p_ix^i}{q(x)}$$

and suppose f has the Maclaurin polynomial series expansion $f(x)=\sum_{i=0}^{\infty}a_ix^i$. Then

$$f(x)-R_{nm}=\frac{\sum_{i=0}^{\infty}a_ix^i\sum_{i=0}^{m}q_ix^i-\sum_{i=0}^{n}p_ix^i}{q(x)} \tag{3.7.12}$$

The object is to choose the constants $q_1,q_2,\ldots,q_m$, and $p_0,p_1,p_2,\ldots,p_n$, so that

$$f^{(k)}(0)=R_{nm}{}^{(k)}(0),\quad \text{for each } k=0,1,\ldots,N.$$

We find that this is equivalent to $f-R_{nm}$ having a zero of multiplicity $N+1$ at $x=0$. As a consequence, we choose $q_1,q_2,\ldots,q_m$, and $p_0,p_1,p_2,\ldots,p_n$, so that the numerator on the right side of Eq. (3.7.12),

$$(a_0+a_1x+\cdots)(1+q_1x+\cdots q_mx^m)-(p_0+p_1x+\cdots p_nx^n) \tag{3.7.13}$$

has no terms of degree less than or equal to N.

To simplify notation, we define

$$p_{n+1}=p_{n+2}=\ldots=p_N=0 \text{ and } q_{m+1}=q_{m+2}=\ldots=q_N=0$$

We can then express the coefficient of x^k in expression (3.7.13) as

$$\left(\sum_{i=0}^{k}a_iq_{k-i}\right)-p_k$$

So, the rational function for Padé approximation results from the solution of the $N+1$ linear equations

$$\sum_{i=0}^{k}a_iq_{k-i}=p_k,\quad k=0,1,\ldots,N$$

in the $N+1$ unknowns $q_1,q_2,\ldots,q_m$, $p_0,p_1,p_2,\ldots,p_n$.

Example 3.7.1 The Maclaurin series expansion for e^{-x} is

$$\sum_{i=0}^{\infty}\frac{(-1)^i}{i!}x^i$$

To find the Padé approximation to e^{-x} of degree 5 with $n=3$ and $m=2$. requires choosing, p_0,p_1,p_2,p_3,q_1, and q_2, so that the coefficients of x^k for $k=0,1,\ldots,5$ are zero in the expression

$$\left(1-x+\frac{x^2}{2}-\frac{x^3}{6}+\cdots\right)(1+q_1x+q_2x^2)-(p_0+p_1x+p_2x^2+p_3x^3).$$

Expanding and collecting terms produces

$$x^5: -\frac{1}{120}+\frac{1}{24}q_1-\frac{1}{6}q_2=0; \qquad x^2: \frac{1}{2}-q_1+q_2=p_2$$

$$x^4: \frac{1}{24}-\frac{1}{6}q_1+\frac{1}{2}q_2=0; \qquad x^1: -1+q_1=p_1$$

$$x^3: -\frac{1}{6}+\frac{1}{2}q_1-q_2=p_3; \qquad x^0: 1=p_2$$

Giving

$$p_0=1,\quad p_1=-\frac{3}{5},\quad p_2=\frac{3}{20},\quad p_3=-\frac{1}{60},\quad q_1=\frac{2}{5},\quad \text{and}\quad q_2=\frac{1}{20}$$

So the Padé approximation is

$$R_{32}=\frac{1-\frac{3}{5}x+\frac{3}{20}x^2-\frac{1}{60}x^3}{1+\frac{2}{5}x+\frac{1}{20}x^2}$$

The fifth Maclaurin polynomial is

$$P_5=1-x+\frac{x^2}{2}-\frac{x^3}{6}+\frac{x^4}{24}-\frac{x^5}{120}$$

Table 3-8 lists values of $R_{32}(x)$ and P_5. The Padé approximation is clearly superior in example.

Table 3-8 Values of $R_{32}(x)$ and P_5

x	e^{-x}	$P_5(x)$	$\lvert e^{-x}-P_5(x)\rvert$	$R_{32}(x)$	$\lvert e^{-x}-R_{32}(x)\rvert$
0.2	0.818 730 75	0.818 730 67	8.64×10^{-8}	0.818 730 75	7.55×10^{-9}
0.4	0.670 320 05	0.670 314 67	5.38×10^{-6}	0.670 319 63	4.11×10^{-7}
0.6	0.548 811 64	0.548 752 00	5.96×10^{-5}	0.548 807 63	4.00×10^{-6}
0.8	0.449 328 96	0.449 002 67	3.26×10^{-4}	0.449309 66	1.93×10^{-5}
1.0	0.367 879 44	0.366 666 67	1.21×10^{-3}	0.367816 09	6.33×10^{-5}

3.8 Computer Experiments

3.8.1 Functions Needed in The Experiments by Mathematica

1. Plot——绘制函数图像函数

格式：`Plot[f,{x,xmin,xmax}]`

功能：绘制函数$f(x)$在区间[$x$min,$x$max]上的图像.

2. ListPlot——绘制散点图函数

格式 1：`ListPlot [{y1,y2,…}]`

功能：绘制横坐标 x 为 1,2,⋯，对应的纵坐标 y 为 y1,y2,⋯的散点图.

格式 2：`ListPlot [{{x1,y1},{x2,y2},…}]`

功能：绘制离散点(x1,y1),(x2,y2),⋯.

3．Show——图形重绘与组合函数

格式 1：`Show[fig]`

功能：重绘变量 fig 中保存的图形.

格式 2：`Show[fig1,fig2,…,fign]`

功能：组合绘制多个图形.

4．PlotStyle——设置图形的样式（绘图函数的可选项）

格式：`PlotStyle ->{{Style1},{Style2},…}`

功能：设置图形的线条粗细、点的大小、颜色变化等. 例如：

```
PlotStyle -> PointSize[0.04]]
```

设置点的尺寸：

```
PlotStyle ->RGBCorlor[0.9,0,0]
```

设置图形颜色为红色.

5．DisplayFunction——设置图形的样式（绘图函数的可选项）

格式 1：`DisplayFunction->Identity`

功能：禁止图形显示.

格式 2：`DisplayFunction->$DisplayFunction`

功能：恢复图形显示.

6．Solve——解方程（组）函数

格式：`Solve[eqns,vars]`

功能：从方程（组）eqns 中解出变量 vars.

7．MatrixForm——转换为矩阵形式函数

格式：`MatrixForm[list]]`

功能：按矩阵形式输出表 list.

3.8.2 Experiments by Mathematica

1．用最小二乘法求 m 次多项式拟合

（1）算法.

① 输入 n+1 个拟合点(x_i, y_i)，i=0,1,2,⋯,n;

② 输出散点图;

③ 根据散点图确定拟合多项式的次数 m;

④ 计算正规方程组的系数和右端常数项;

⑤ 解正规方程组，求得 $a_0^*, a_1^*, \cdots, a_m^*$;

⑥ 得出拟合多项式 $\varphi^*(x) = a_0^* + a_1^* x + a_2^* x^2 + \cdots + a_m^* x^m$;

⑦ 输出正规方程组系数矩阵、常数列、拟合多项式及散点图与拟合曲线画在一起的图形.

（2）程序清单.

```
(*mth-Interpolating Polynomial by Least Square Method*)
Clear[xi,xx,yi];
xi=Input["xi="]
yi=Input["yi="]
n=Length[xi];
d=ListPlot[Table[{xi[[i]],yi[[i]]},{i,1,n}],
   PlotStyle->PointSize[0.04]]
m=Input["m="多项式次数]
s=Table[Sum[xi[[k]]^i,{k,1,n}],{i,0,2m}];
a=Table[s[[i+j-1]],{i,1,m+1},{j,1,m+1}];
a[[1,1]]=n;
Print["a=",MatrixForm[a]];
b=Table[Sum[xi[[k]]^i*yi[[k]],{k,1,n}],{i,0,m}];
b[[1]]=Sum[yi[[k]],{k,1,n}];
Print["b=",MatrixForm[b]];
xx=Table[x[i],{i,1,m+1}];
g=Solve[a.xx==b,xx];
f=Sum[x[i]*t^(i-1),{i,1,m+1}]/.g[[1]];
p=f//N
fig=Plot[p,{t,xi[[1]],xi[[n]]},DisplayFunction->Identity];
Show[{fig,d}, DisplayFunction->$DisplayFunction];
```

（3）变量说明.

xi：保存拟合基点$\{x_0,x_1,\cdots,x_n\}$；

yi：保存对应函数值$\{y_0,y_1,\cdots,y_n\}$；

m：保存多项式次数；

a：保存正规方程组系数矩阵；

b：保存正规方程组常数项；

xx：定义正规方程组变量，保存 m 次拟合多项式的系数；

p：保存 m 次拟合多项式；

d：保存散点图；

fig：保存拟合函数图形.

（4）计算实例.

Example 3.8.1 Given Table 3-9.

Find a polynomial to fit them by using least square method.

Table 3-9 The experimental data

x	−3	−1	2	4	5
y	3	1	1.5	3	4.5

操作步骤如下：

① 将光标定位在要执行的 Cell 中，按小键盘的【Enter】键；

② 在弹出的对话框中按提示分别输入

```
{-3,-1,2,4,5},{3,1,1.5,3,4.5}
```

③ 每次输入后单击 OK 命令按钮，输出散点图，如图 3-7(a)所示；

④ 根据散点图确定拟合多项式的次数 $m=2$，在输入窗口输入：2，单击 OK 命令按钮后得图 3-7(b)所示的输出结果.

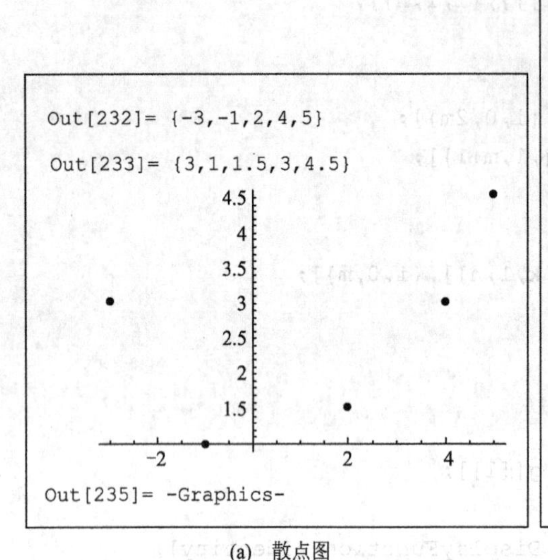

(a) 散点图

(b) 最小二乘拟合

图 3-7 例 3.8.1 运行结果

2. 用最小二乘法求线性模型拟合

（1）算法.

① 输入 $n+1$ 个拟合点(x_i, y_i)，$i=0,1,2,\cdots,n$；

② 输出散点图；

③ 根据散点图确定拟合基函数：$\{\varphi_0(x),\varphi_1(x),\cdots,\varphi_m(x)\}$；

④ 计算正规方程组的系数和右端常数项；

⑤ 解正规线性方程组，求得 $a_0^*,a_1^*,\cdots,a_m^*$；

⑥ 得出拟合多项式

$$\varphi^*(x)=a_0^*\varphi_0(x)+a_1^*\varphi_1(x)+a_2^*\varphi_2(x)+\cdots+a_m^*\varphi_m(x)$$

（2）程序清单.

```
(* General Linear Fitting by Least Square Method *)
Clear[x,xi,xx,yi];
xi=Input["xi="]
yi=Input["yi="]
n=Length[xi];
d=ListPlot[Table[{xi[[i]],yi[[i]]},{i,1,n}],PlotStyle PointSize[0.04]]
m1=Input["m1="拟合基函数组];
```

```
m=Length[m1];
p=Table[m1/.x xi[[k]],{k,1,n}];
a=Table[Sum[p[[k,i]]*p[[k,j]],{k,1,n}],{i,1,m},{j,1,m}]//N;
Print["a=",MatrixForm[a]];
b=Table[Sum[p[[k,i]]*yi[[k]],{k,1,n}],{i,1,m}]//N;
Print["b=",MatrixForm[b]];
xx=Table[xt[i],{i,1,m}];
g=Solve[a.xx b,xx];
f=Sum[xt[i]*m1[[i]],{i,1,m}]/.g[[1]]
pp=f//N;
fig=Plot[pp,{x,xi[[1]],xi[[n]]},DisplayFunction Identity];
Show[{fig,d},DisplayFunction $DisplayFunction]
```

（3）变量说明.

xi：保存拟合基点$\{x_0,x_1,\cdots,x_n\}$；

yi：保存对应函数值$\{y_0,y_1,\cdots,y_n\}$；

m：保存拟合基函数的个数；

m1：保存拟合基函数组；

a：保存正规方程组系数矩阵；

b：保存正规方程组常数项；

xx：定义正规方程组变量，保存线性模型拟合系数；

p：保存拟合基函数组在拟合基点$\{x_0,x_1,\cdots,x_n\}$的函数值；

pp：保存求出的线性模型拟合函数；

d：保存散点图；

fig：保存拟合函数图形.

（4）计算实例.

Example 3.8.2 Given Table 3-10.

Table 3-10 The experimental data

x	1	2	3	4	5	6	7	8	9	10	11	12
y	2.892	2.863	0.473	−2.252	−2.870	−0.836	1.972	2.968	1.236	−1.632	−1.784	−0.565

Find a function to fit them by using least square method.

操作步骤如下：

① 将光标定位在要执行的 Cell 中，按小键盘的【Enter】键；

② 在弹出的对话框中按提示分别输入

```
{1,2,3,4,5,6,7,8,9,10,11,12},
{2.892,2.863,0.473, -2.252, -2.870, -0.836,1.972,2.968,1.236, -1.632,
-1.784, -0.565}
```

③ 每次输入后单击 OK 命令按钮，输出散点图，如图 3-8(a)所示；

④ 散点图的分布与正弦函数曲线相似，所以确定拟合基函数为 1,sinx，在输入窗

口输入：{1,Sin[x]}，单击 OK 命令按钮后得图 3-8(b)所示的输出结果.

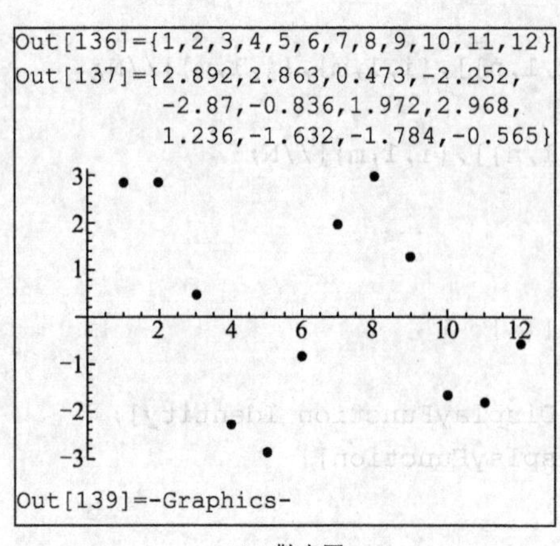

(a) 散点图

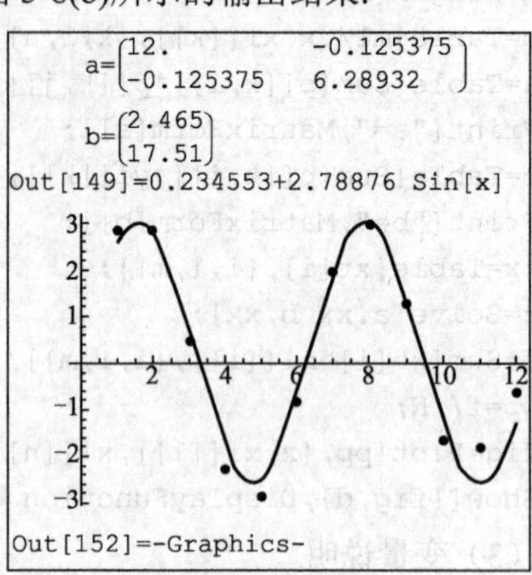

(b) 线性模型拟合

图 3-8 例 3.8.2 运行结果

3．求形如 $y=e^{a+bx}$ 的拟合曲线

求形如 $y=e^{a+bx}$ 的拟合曲线，可通过两边取对数：$\ln y=a+bx$，令 $z=\ln y$，则转化为 $z=a+bx$，即为线性模型拟合问题.此时拟合基函数组为$\{1,x\}$.

（1）程序清单.

```
(*Curve Fitting With the Form y=e^(a+bx)*)
xi=Input["xi="]
yi=Input["yi="]
yi=Log[yi];
n=Length[xi];
d=ListPlot[Table[{xi[[i]],yi[[i]]},{i,1,n}],PlotStyle->PointSize[0.04]]
s=Table[Sum[xi[[k]]^i,{k,1,n}],{i,0,2}];
a=Table[s[[i+j-1]],{i,1,2},{j,1,2}];
a[[1,1]]=n;
Print["a=",MatrixForm[a]];
b=Table[Sum[xi[[k]]^i*yi[[k]],{k,1,n}],{i,0,1}];
b[[1]]=Sum[yi[[k]],{k,1,n}];
Print["b=",MatrixForm[b]];
xx=Table[x[i],{i,1,2}];
g=Solve[a.xx==b,xx];
f=Sum[x[i]*x^(i-1),{i,1,2}]/.g[[1]];
p=f//N
fig=Plot[p,{x,xi[[1]],xi[[n]]},DisplayFunction->Identity];
Show[{fig,d},DisplayFunction->$DisplayFunction];
```

（2）计算实例.

Example 3.8.3 Given Table 3-11.

Table 3-11 The experimental data

x	1	1.25	1.5	1.75	2	2.5	3
y	5.12	5.79	6.63	7.37	8.31	9.62	10.9

Find a fitted curve with the form $y=e^{a+bx}$.

操作步骤如下：

① 将光标定位在要执行的 Cell 中，按小键盘的【Enter】键；

② 在弹出的对话框中按提示分别输入

```
{1,1.25,1.5,1.75,2,2.5,3}, {5.12,5.79,6.63,7.37,8.31,9.62,10.9}
```

③ 每次输入后单击 OK 命令按钮，输出(x_i,ln y_i)的散点图如图 3-9(a)所示及散点图的线性拟合曲线如图 3-9(b)所示.

所求拟合函数为：$y=e^{1.301\,84+0.379\,655x}$.拟合函数与离散点画在一起得图 3-9(c)所示结果.

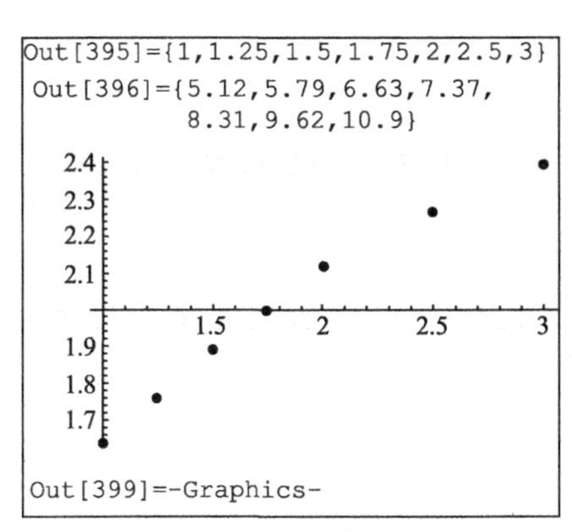

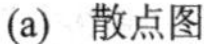

(a) 散点图

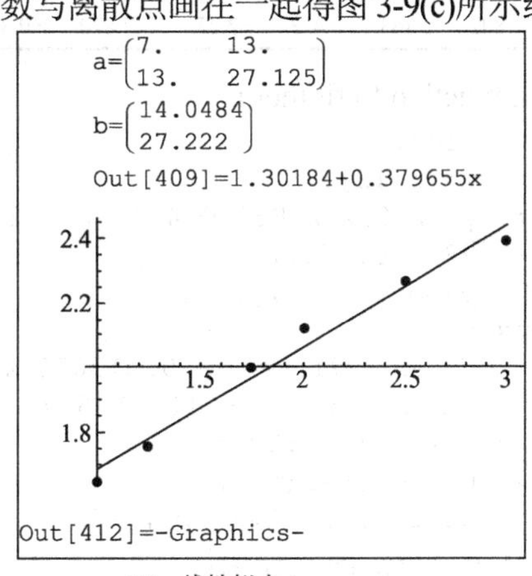

(b) 线性拟合

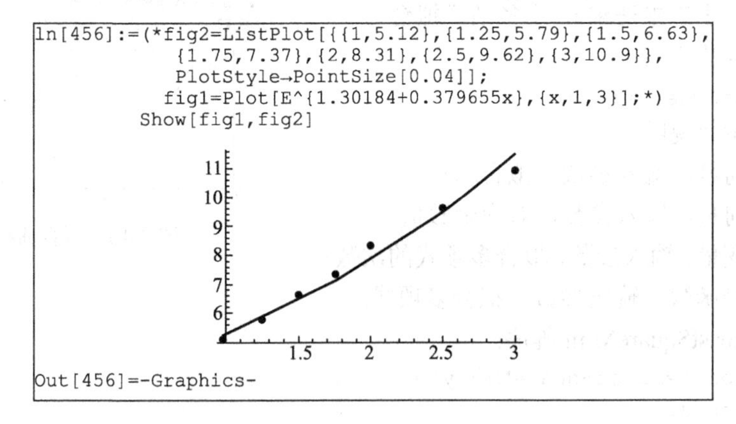

(c) 拟合函数与离散点

图 3-9 例 3.8.3 运行结果

3.8.3 Functions Needed in The Experiments by Matlab

实验中需用到的 Matlab 函数见表 3-12.

表 3-12 常用函数

函　数	意　义
polyfit(x,y,n)	给定数据的 n 次多项式拟合
polyval(pol)	计算多项式的值
poly2str(pol,'x')	以习惯方式显示多项式
plot(x,y)	绘制平面线图

3.8.4 Experiments by Matlab

Example 3.8.4 Given Table 3-13.

Table 3-13 The experimental data

x	1	2	3	4	5	6	7	8	9	10	11	12
y	2.892	2.863	0.473	−2.252	−2.870	−0.836	1.972	2.968	1.236	−1.632	−1.784	−0.565

Find a function to fit them.

```
>> clear
>> x=[1:12];
>> y=[2.892 2.863 0.473 -2.252 -2.870 -0.836 1.972 2.968 1.236 -1.632
-1.784 -0.565];
>> p=polyfit(x,y,7);
px =
   -0.00029715 x^7 + 0.013055 x^6 - 0.2242 x^5 + 1.895 x^4 - 8.1655 x^3
   + 17.0473 x^2 - 16.3146 x + 8.6477
>> xx=0.5:0.1:12.5;
>> pv=polyval(p,xx);
>> plot(x,y,'o',xx,pv,'k')
```

输出结果如图 3-10 所示.

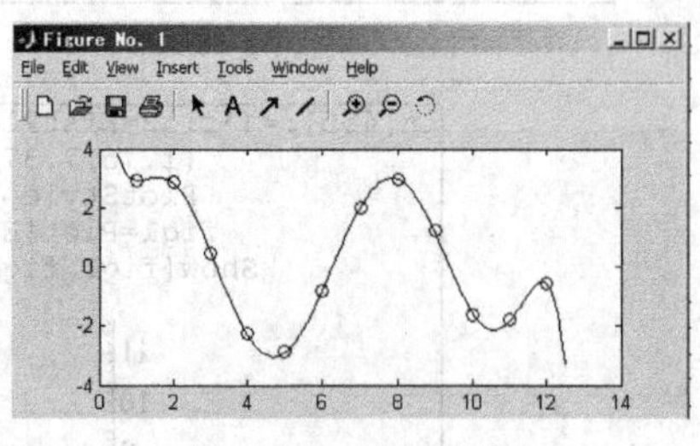

图 3-10 拟合曲线

1. 用最小二乘法求 m 次多项式拟合

（1）函数语句.

```
P=LeastSquareM(x,y)
```

（2）参数说明.

x：实向量，输入参数，拟合基点；

y：实向量，输入参数，对应函数值；

m：实变量，输入参数，拟合多项式的次数；

P：符号函数，输出参数，拟合多项式.

（3）LeastSquareM.m 程序.

```
function P=LeastSquareM(x,y)
n=length(x);
hold on
plot(x,y,'ro');
```

```
m=input('请输入拟合多项式的次数: ')
b=zeros(1:m+1);
f=zeros(n,m+1);
for k=1:m+1
    f(:,k)=x'.^(k-1);
end
a=f'*f;
b=f'*y';
c=a\b;
P=0;
syms t
for i=1:m+1
    d(i)=t^(i-1);
end
P=d*c;
ezplot(P,[x(1)-1,x(n)+1]);
hold off
```

（4）计算实例.

Example 3.8.5 Given Table 3-14

Find a polynomial to fit them by using least square method.

Table 3-14 The experimental data

x	−3	−1	2	4	5
y	3	1	1.5	3	4.5

在命令窗口输入：

```
>> clear
>> x=[-3,-1,2,4,5];y=[3,1,1.5,3,4.5];P=LeastSquareM(x,y)
请输入拟合多项式的次数:
```

输入 2，结果如下：

```
m =
    2
P =
    6263/7034-6299/42204*t+7361/42204*t^2
```

输出图形如图 3-11(a)和图 3-11(b)所示.

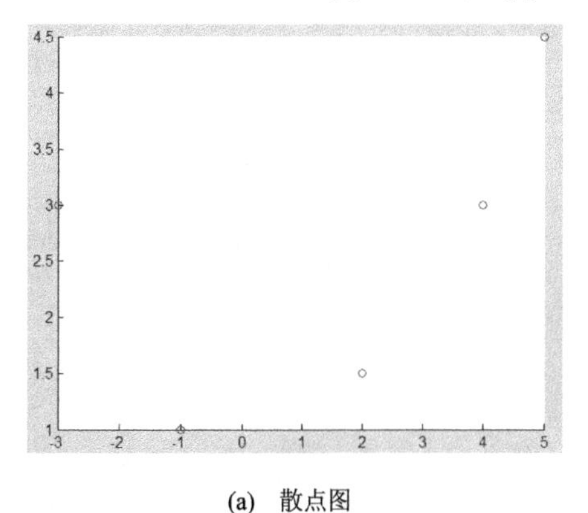

(a) 散点图

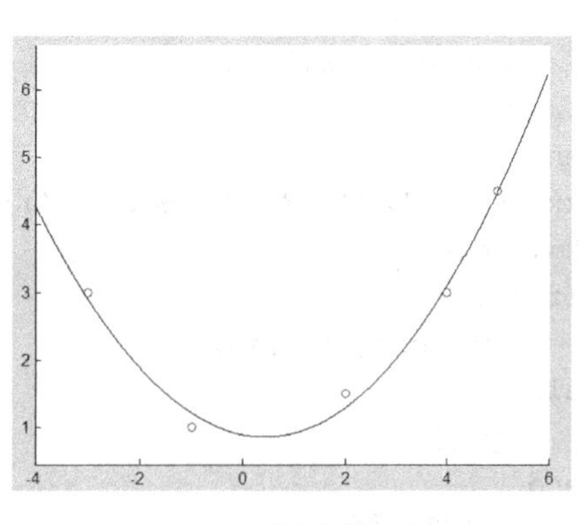

(b) 拟合曲线

图 3-11 例 3.8.5 运行结果

2. 用最小二乘法求线性模型拟合

（1）函数语句.

```
[g,yy]=GLeastSquareM(x,y)
```

（2）参数说明.

x：实向量，输入参数，拟合基点；

y：实向量，输入参数，对应函数值；

f：符号向量 ，输入参数，拟合基函数；

xx：实向量，输入参数，自变量取值；

g：符号函数，输出参数，拟合函数；

yy：实向量，输出参数，拟合函数在 xx 的值.

（3）GleastSquareM.m 程序.

```
function [g,yy]=GLeastSquareM(x,y)
plot(x,y,'*r');
syms t
f=input('请输入基函数向量(自变量用 t 表示): ')
m=length(f);
n=length(x);
for i=1:m
    for j=1:n
        a(i,j)=subs(f(i),'t',x(j));
    end
end
A=a*a';
b=a*y';
c=A\b;
g=f*c;
flag=input('是否画出拟合函数的图像? ('是'输入 1, '否'输入 0): ');
if flag==1
    hold on
    ezplot(g,[x(1),x(n)])
    hold off
end
xx=input('请输入插值点的横坐标: ');
k=length(xx);
if k>0
    yy=subs(g,'t',xx);
else
    yy=[];
end
```

（4）计算实例.

Example 3.8.6　Given Table 3-15.

Table 3-15 The experimental data

x	1	2	3	4	5	6	7	8	9	10	11	12
y	2.892	2.863	0.473	−2.252	−2.870	−0.836	1.972	2.968	1.236	−1.632	−1.784	−0.565

Find a function to fit them by using least method.

在命令窗口输入：

```
>> clear
>>
x=1:1:12;y=[2.892,2.863,0.473,-2.252,-2.870,-0.836,1.972,2.968,1.23
6,-1.632,-1.784,
-0.565];[g,yy]=GLeastSquareM(x,y)
```

输出：

```
请输入基函数向量(自变量用 t 表示):
```

输出图形如图 3-12(a)所示。

散点图的分布与正弦函数曲线相似，所以确定拟合基函数为 1，sin x，故在命令窗口提示光标位置输入：[1,sin(t)]，再按【Enter】键，出现结果：

```
f =
[ 1, sin(t)]
是否画出拟合函数的图像？('是'输入 1，'否'输入 0):
```

输入 1，再按【Enter】键，输出图 3-12(b)所示结果.

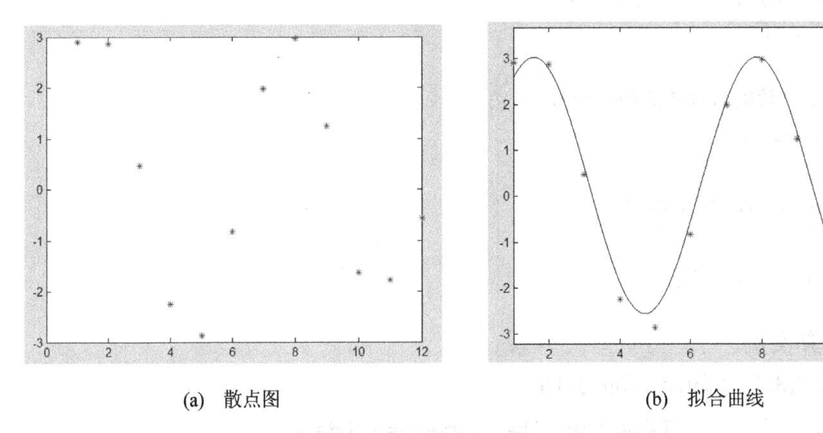

(a) 散点图　　(b) 拟合曲线

图 3-12 例 3.8.6 运行结果

并出现提示：

```
请输入插值点的横坐标:
```

按【Enter】键，出现结果：

```
g =
4225336895034461/18014398509481984+784965454453979/281474976710656*
sin(t)
yy =
     []
```

3．求形如 $y=\mathrm{e}^{a+bx}$ 的拟合曲线

（1）函数语句.

```
function [g,yy]=LSMexp(x,y)
```

（2）参数说明.

x：实向量，输入参数，拟合基点；

y：实向量，输入参数，对应函数值；

xx：实向量，输入参数，自变量取值；

g：符号函数，输出参数，拟合函数；

yy：实向量，输出参数，拟合函数在 xx 的值.

（3）LSMexp.m 程序.

```
function [g,yy]=LSMexp(x,y)
y1=log(y);
P=LeastSquareM(x,y1);
g=exp(P);
flag=input('是否画出拟合函数的图像? ('是'输入 1, '否'输入 0): ');
if flag==1
   hold on
   plot(x,y,'or');
   n=length(x);
   ezplot(g,[x(1),x(n)]);
   hold off
end
xx=input('请输入插值点的横坐标: ');
k=length(xx);
if k>0
   yy=subs(g,'t',xx);
else
   yy=[];
 end
```

（4）计算实例.

Example 3.8.7 Given Table 3-16.

Table 3-16 The experimental data

x	1	1.25	1.5	1.75	2	2.5	3
y	5.12	5.79	6.63	7.37	8.31	9.62	10.9

Find a fitted curve with the form $y=\mathrm{e}^{a+bx}$.

在命令窗口输入：

```
>> clear
>>
x=[1,1.25,1.5,1.75,2,2.5,3];y=[5.12,5.79,6.63,7.37,8.31,9.62,10.9];
```

```
[g,yy]=LSMexp(x,y)
请输入拟合多项式的次数: 1
m =
     1
是否画出拟合函数的图像? ('是'输入1, '否'输入0): 0
是否画出拟合函数的图像? ('是'输入1, '否'输入0): 1
请输入插值点的横坐标:
 g =
 exp(5862954323692635/4503599627370496+1709813863056559/4503599627370496*t)
 yy =
     []
```

输出图形如图 3-13(a)和图 3-13(b)所示.

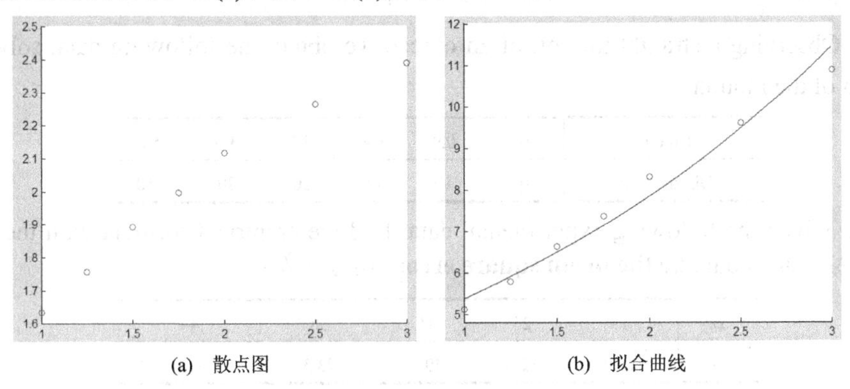

(a) 散点图　　(b) 拟合曲线

图 3-13　例 3.8.7 运行结果

Exercises 3

Questions

1. For each $x \in \mathbf{R}^n$, show that $\|x\|_\infty \leqslant \|x\|_2 \leqslant \sqrt{n}\|x\|_\infty$.

2. Compute the norms $\|f\|_\infty$, $\|f\|_1$ and $\|f\|_2$ of the following function $f(x)$ on $C[0,1]$:

(1) $f(x)=(x-1)^3$;　　(2) $f(x)=\left|x-\frac{1}{2}\right|$.

3. Find $\|\cdot\|_\infty$ and $\|\cdot\|_1$ for the following matrices.

(1) $\begin{pmatrix} 2 & -1 & 0 \\ -1 & 2 & -1 \\ 0 & -1 & 2 \end{pmatrix}$;　　(2) $\begin{pmatrix} 4 & -1 & 7 \\ -1 & 4 & 0 \\ -9 & 0 & 4 \end{pmatrix}$.

4. Compute the eigenvalues and associated eigenvectors of the following matrices and solve their spectral radius.

(1) $\begin{pmatrix} 1 & 1 \\ -2 & -2 \end{pmatrix}$; (2) $\begin{pmatrix} 3 & 2 & -1 \\ 1 & -2 & 3 \\ 2 & 0 & 4 \end{pmatrix}$.

5. Solve the best linear approximation of $f(x)=\sin x$ on $[0,\pi/2]$, and estimate the error.

6. Solve the best linear approximation of $f(x)=e^x$ on $[0,1]$.

7. Solve the polynomial of the best squares approximation of $f(x)=|x|$ on $[-1, 1]$ with $\Phi=\text{span}\{1,x^2,x^4\}$.

8. Solve the polynomial of the best squares approximation of the following $f(x)$ on the given interval with $\Phi=\text{span}\{1,x\}$.

(1) $f(x)=\frac{1}{x}$, $[1,3]$; (2) $f(x)=e^x$, $[0,1]$.

9. Observing a straight motion of an object, we obtain the following data, solve the equation of the motion.

Time t/s	0	0.9	1.9	3.0	3.9	5.0
Distance s/m	0	10	30	50	80	110

10. Given the following experimental data, find the empirical formula with the form $y=a+bx^2$, and compute **the mean square error** (均方误差).

x_i	19	25	31	38	44
y_i	19.0	32.3	49.0	73.3	97.8

11. Given the table

x_i	−2	−1	0	1	2
y_i	−0.1	0.1	0.4	0.9	1.6

Solve a quadratic and cubic fitting functions respectively with the least squares method.

12. Measure some length for n times, and obtain n approximate values $x_1, x_2, \cdots, x_n$, we usually take $x=\frac{1}{n}(x_1+x_2+\cdots+x_n)$, please illustrate this reason.

13. See the figure, the length of AB *is* x_1 and the length of BC *is* x_2. The results of someone obtained are $x_1=15.5$ m, $x_2=6.1$ m, $AC=x_1+x_2=20.9$ m, please determine x_1 and x_2 reasonably.

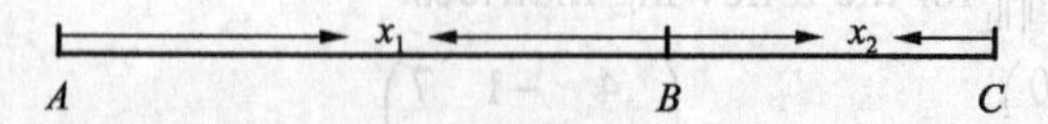

14. Express the following rational functions in continued fractions form:

(1) $\frac{3x^2+6x}{x^2+6x+6}$; (2) $\frac{2x^3-3x^2+4x-5}{x^2+2x+4}$.

15. Show that

$$\frac{1}{1+}\frac{1}{1+}\frac{1}{1+}\cdots=\frac{1}{2}(\sqrt{5}-1)$$

Hint: Set x equal to the continued fraction , and look at $1/x$.

16. Assuming that the following continued fraction converges, find its value

$$\frac{1}{6+}\frac{1}{6+}\frac{1}{6+}\frac{1}{6+}\cdots$$

17. Determine the Padé approximation of degree n=m=3 for function

$$f(x)=\sin x\ .$$

18. Determine the Padé approximation of degree n=m=1 for function

$$f(x)=\frac{1}{x}\ln(1+x)\cdot$$

19. Determine the Padé approximation of degree n=m=2 for function $f(x)=e^{2x}$. Compare the result at $x_i=0.2i$, for $i=1,2,3,4,5$, with the actual value $f(x_i)$.

Computer Questions

1. Find the least squares polynomials of degree1,2 and 3 for the data in the following table. Compute the sum of the squares error φ in each case. Graph the data and the polynomials.

x_i	1.0	1.1	1.3	1.5	1.9	2.1
y_i	1.84	1.96	2.21	2.45	2.94	3.18

2. Given the data:

x_i	4.0	4.2	4.5	4.7	5.1	5.5	5.9	6.3	6.8	7.1
y_i	102.56	113.18	130.11	142.05	167.53	195.14	224.87	256.73	299.50	326.72

(1) Construct the least squares polynomials of degree1,2 and 3, and compute the errors.

(2) Construct the least squares approximation of the form be^{ax}, and compute the error.

4 Numerical Integration and Differentiation

提 要

计算定积分一般可用 Newton-Leibniz 公式

$$\int_a^b f(x)\mathrm{d}x = F(b) - F(a)$$

而这一公式仅仅能解决一部分函数的积分问题．一方面，如果 $f(x)$ 的原函数不能用初等函数表出，或即使 $f(x)$ 的原函数可用初等函数表出，但其形式过于复杂不便于计算其数值；另一方面，如果函数关系是用表格形式给出的，这些都不便使用 Newton-Leibniz 公式．因此，有必要研究积分的数值计算问题．

本章主要介绍 Newton-Cotes 求积公式、Romberg 算法、Gauss 求积公式及数值微分方法等．

词 汇

quadrature formulae	求积公式	antiderivative	原函数
weight	权	remainder	余项
symmetry	对称	sequence	序列
derive	求导数	derivative	导数
degree of accuracy	精度	trapezoidal rule	梯形公式
Simpson's rule	辛普森公式	Cotes's rule	柯特斯公式
composite rule	复合公式	extrapolation method	外插法
Romberg's algorithm	龙贝格算法	numerical differentiation	数值微分
mean value theorem for integral	积分中值定理	continuously differentiable	连续可微
polynomial interpolation operator	多项式插值算子	integrand	被积函数
		elementary function	初等函数

4.1 Introduction

The need to compute derivatives and integrals is commonplace in engineering applications. The following example is representative.

Assume an environmental engineer is asked to estimate the cost of constructing a dam on a river. One of the preliminary steps is to determine the cross-sectional area of the river at the point were the dam is to be constructed as shown in Figure 4-1.

In order to estimate the cross-sectional area S, depth measurements are performed with the results summarized in Table 4-1.The cross-sectional area is obtained by evaluating the following integral. Again, since the measurements are available only in tabulated form, this calculation must be performed numerically.

$$S = \int_0^w y(x)\mathrm{d}x$$

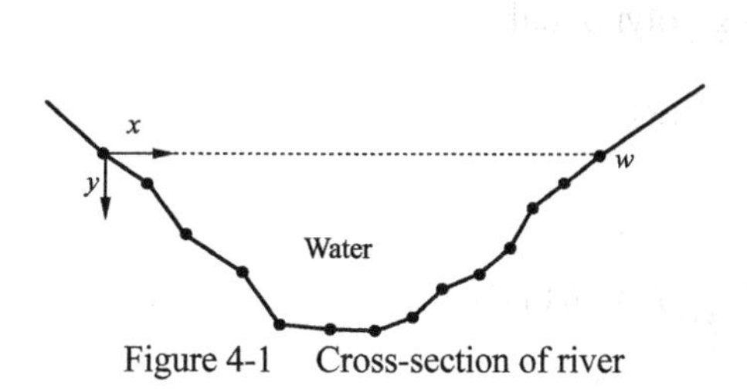

Figure 4-1 Cross-section of river

Table 4-1 Cross-section of river

x/m	y/m	x/m	y/m
0	0.0	70	13.4
10	2.2	80	10.9
20	5.9	90	9.8
30	8.5	100	7.6
40	13.1	110	5.1
50	14.3	120	3.2
60	14.6	130	0.0

When you finish this chapter, you will be able compute numerical derivatives and integrals using discrete data points. You will know how to estimate the size of the truncation error for numerical differentiation and integration formulas. You will know to perform numerical integration using integration formulas, including the trapezoid, Simpson, and Cotes's rules. You will use Richardson extrapolation to perform accurate numerical integration using Romberg's method. These overall goals will be achieved by mastering the following chapter objectives.

Objectives

- Be able to perform trapezoid, Simpson, and Cotes's rules.
- Be able to perform composite quadrature formulae.
- Be able to apply Richardson extrapolation to improve the accuracy of numerical derivatives.
- Know how to solve degree of accuracy of quadrature formulas.

- Know how to solve numerical differentiation.
- Understand Gauss quadrature formula.

4.2 Interpolatory Quadratures（插值型求积公式）

4.2.1 Interpolatory Quadratures

The most common quadrature formulae approximate the definite integral

$$\int_a^b f(x)\mathrm{d}x \tag{4.2.1}$$

of a continuous function f over the interval $[a,b]$ with $a<b$ by a weighted sum

$$\sum_{k=0}^{n} a_k f(x_k) \tag{4.2.2}$$

with n+1 distinct **quadrature points** (求积节点) $x_0,\cdots,x_n \in [a,b]$ and **quadrature weights** (求积权) $a_0,\cdots,a_n \in \mathbf{R}$.

The methods of quadrature in this section are based on the interpolation polynomials given in Chapter 2. We first select a set of distinct nodes $\{x_0,\cdots,x_n\}$ from the interval $[a,b]$. Then we integrate the Lagrange interpolating polynomial

$$L_n(x)=\sum_{k=0}^{n} f(x_k)l_k(x)$$

and its truncation error term over $[a,b]$ to obtain

$$\begin{aligned}\int_a^b f(x)\mathrm{d}x &\approx \int_a^b L_n(x)\mathrm{d}x = \int_a^b \sum_{k=0}^{n} f(x_k)l_k(x)\mathrm{d}x \\ &= \sum_{k=0}^{n} f(x_k)\int_a^b l_k(x)\mathrm{d}x = \sum_{k=0}^{n} a_k f(x_k)\end{aligned}$$

The quadrature formula is, therefore

$$\int_a^b f(x)\mathrm{d}x \approx \sum_{k=0}^{n} a_k f(x_k) \equiv I_n \tag{4.2.3}$$

and

$$a_k = \int_a^b l_k(x)\mathrm{d}x = \int_a^b \prod_{\substack{j=0\\ j\neq k}}^{n} \frac{x-x_j}{x_k-x_j}\mathrm{d}x, \quad k=0,1,\cdots,n \tag{4.2.4}$$

I_n is called **interpolatory quadrature,** with the error.

$$\begin{aligned}R(f) &= \int_a^b f(x)\mathrm{d}x - \int_a^b L_n(x)\mathrm{d}x \\ &= \int_a^b (f(x)-L_n(x))\mathrm{d}x = \int_a^b \frac{f^{(n+1)}(\xi)}{(n+1)!}\omega_{n+1}(x)\mathrm{d}x\end{aligned} \tag{4.2.5}$$

where $\omega_{n+1}(x)=(x-x_0)(x-x_1)\cdots(x-x_n)$.

Theorem 4.2.1 Given $n+1$ distinct quadrature points $x_0,\cdots,x_n\in[a,b]$, if $f(x)$ is a polynomial of degree n, then

$$\int_a^b f(x)\mathrm{d}x=\sum_{k=0}^{n}a_k f(x_k) \tag{4.2.6}$$

Proof If $f(x)$ is a polynomial of degree n and $P_n(x)$ interpolates $f(x)$ with $n+1$ distinct points $x_0,\cdots,x_n\in[a,b]$, then $f(x)=P_n(x)$, it follows that

$$\int_a^b f(x)\mathrm{d}x=\int_a^b P_n(x)\mathrm{d}x=\sum_{k=0}^{n}a_k f(x_k)$$

□

4.2.2 Degree of Accuracy（精度）

Definition 4.2.1 **The degree of accuracy**, or **precision**, of a quadrature formula is the largest positive integer m, such that the formula is exact for each $x^k, k=0,1,\cdots,m$.

Integration and summation are linear operations, that is,

$$\int_a^b(\alpha f(x)+\beta g(x))\mathrm{d}x=\alpha\int_a^b f(x)\mathrm{d}x+\beta\int_a^b g(x)\mathrm{d}x$$

and

$$\sum_{i=0}^{m}(\alpha f(x_i)+\beta g(x_i))=\alpha\sum_{i=0}^{m}f(x_i)+\beta\sum_{i=0}^{m}g(x_i)$$

for each pair of **integrable** (可积) functions f and g and each pair of real constants α and β. This implies that the degree of precision of a quadrature formula is m if and only if the error $R(P(x))=0$ for all polynomials $P(x)$ of degree $k=0,1,\cdots,m$, but $R(P(x))\neq 0$ for some polynomial $P(x)$ of degree $m+1$.

In generally, in order to make (4.2.3) have mth degree of accuracy at least, we assume (4.2.3) holds exactly for $f(x)=1,x,\cdots,x^m$ respectively, i.e.,

$$\begin{cases}\sum_{i=0}^{n}a_k=b-a\\ \sum_{i=0}^{n}a_k x_k=\frac{1}{2}(b^2-a^2)\\ \cdots\cdots\\ \sum_{i=0}^{n}a_k x_k^m=\frac{1}{m+1}(b^{m+1}-a^{m+1})\end{cases}$$

then solve the system of equations and obtain a_k $(k=0,1,\cdots,m)$.

Example 4.2.1 Find the constants A_{-1},A_0 and A_1 so that the quadrature formula

$$\int_{-2h}^{2h}f(x)\mathrm{d}x\approx A_{-1}f(-h)+A_0 f(0)+A_1 f(h)$$

has the highest possible degree of precision.

Solution Let the quadrature formula hold exactly for $f(x)=1,x,x^2$ respectively, then

$$\begin{cases} 4h = A_{-1} + A_0 + A_1 \\ 0 = -hA_{-1} + 0 + hA_1 \\ \dfrac{16}{3}h^3 = h^2 A_{-1} + 0 + h^2 A_1 \end{cases}$$

we obtain

$$A_{-1} = A_1 = \frac{8}{3}h, \quad A_0 = -\frac{4}{3}h$$

The quadrature formula is, therefore

$$\int_{-2h}^{2h} f(x)\mathrm{d}x \approx \frac{8}{3}hf(-h) - \frac{4}{3}hf(0) + \frac{8}{3}hf(h) \tag{4.2.7}$$

When $f(x)=x^3$, the Equation (4.2.7) holds exactly. when $f(x)=x^4$, the equation does not hold, so, the quadrature Formula (4.2.7) has degree of precision three.

4.3 Newton-Cotes Quadrature Formula（牛顿-柯特斯公式）

Theorem 4.3.1 The polynomial interpolatory quadrature of order n with equidistant quadrature points $x_k = a+kh,\ k=0,1,\cdots,n$, and step width $h=(b-a)/n$ is called the **Newton-Cotes quadrature formula** of order n. Its weights are given by

$$a_k = h\frac{(-1)^{n-k}}{k!(n-k)!}\int_0^n \prod_{\substack{j=0\\ j\neq k}}^{n}(z-j)\mathrm{d}z, \quad k=0,1,\cdots,n \tag{4.3.1}$$

and **symmetric** (对称), i.e., $a_k = a_{n-k},\ k=0,1,\cdots,n$.

Proof The weights are obtained from (4.2.4) by substituting $x=x_0+hz$, $x_j=x_0+jh$, and observing that

$$\begin{aligned} a_k &= h\frac{(-1)^{n-k}}{k!(n-k)!}\int_0^n z(z-1)\cdots(z-k+1)(z-k-1)\cdots(z-n)\mathrm{d}z \\ &= h\frac{(-1)^{n-k}}{k!(n-k)!}\int_0^n \prod_{\substack{j=0\\ j\neq k}}^{n}(z-j)\mathrm{d}z \quad (k=0,1,\cdots,n) \end{aligned}$$

The symmetry $a_k = a_{n-k}$ follows by substituting $z=n-y$. □

Since $h=(b-a)/n$, Newton-Cotes quadrature formula can be written as

$$\begin{aligned} I_n &= \sum_{k=0}^{n} a_k f(x_k) = (b-a)\sum_{k=0}^{n} f(x_k)\cdot\frac{(-1)^{n-k}}{nk!(n-k)!}\int_0^n \prod_{\substack{j=0\\ j\neq k}}^{n}(z-j)\mathrm{d}z \\ &= (b-a)\sum_{k=0}^{n} \mathrm{C}_k^{(n)} f(x_k) \end{aligned}$$

where

$$C_k^{(n)} = \frac{(-1)^{n-k}}{nk!(n-k)!}\int_0^n \prod_{\substack{j=0 \\ j\neq k}}^{n}(z-j)\mathrm{d}z \quad (k=0,1,\cdots,n) \tag{4.3.2}$$

are called **Cotes coefficients**.

The Newton-Cotes quadrature formula of order n = 1 is known as the **trapezoidal rule** (梯形公式), its Cotes coefficients can be obtained from evaluating (4.3.2) , we obtain

$$C_0^{(1)} = C_1^{(1)} = \frac{1}{2}$$

the trapezoidal rule has the form

$$T = \frac{b-a}{2}(f(a)+f(b)) \tag{4.3.3}$$

Geometrically speaking, the trapezoidal rule approximates the integral of f by the integral of the straight line connecting the two points $(a,f(a))$ and $(b,f(b))$. Hence, the approximate value coincides with the area of the trapezoid with the four corners $(a,0)$, $(b,0)$, $(a,f(a))$, and $(b,f(b))$ (see Figure 4-2).

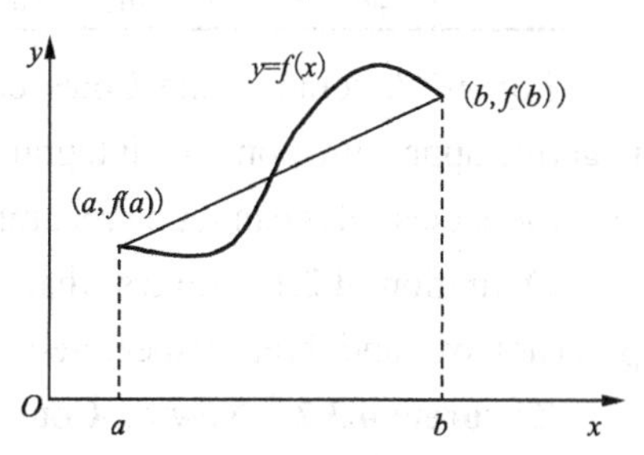

Figure 4-2 Trapezoidal rule

The Newton-Cotes quadrature formula of order n = 2 is called **Simpson's rule** (辛普森公式)**,** Its Cotes coefficients can be obtained from Evaluating (4.3.2).

$$C_0^{(2)} = \frac{1}{4}\int_0^2 (z-1)(z-2)\mathrm{d}z = \frac{1}{6}$$

$$C_1^{(2)} = -\frac{1}{2}\int_0^2 z(z-2)\mathrm{d}z = \frac{4}{6}$$

$$C_2^{(2)} = \frac{1}{4}\int_0^2 z(z-1)\mathrm{d}z = \frac{1}{6}$$

Simpson's rule is given by

$$S = \frac{b-a}{6}\left(f(a)+4f\left(\frac{a+b}{2}\right)+f(b)\right) \tag{4.3.4}$$

Geometrically speaking, the Simpson's rule approximates the integral of f by the integral of the parabola through the three points $(a,f(a))$, $\left(\frac{a+b}{2}, f\left(\frac{a+b}{2}\right)\right)$ and $(b,f(b))$.

When n = 4, we obtain **Cotes's rule** (柯特斯公式)

$$C = \frac{b-a}{90}\left(7f(x_0)+32f(x_1)+12f(x_2)+32f(x_3)+7f(x_4)\right) \tag{4.3.5}$$

where $x_k = a+kh, h=\frac{b-a}{4}$.

Table 4-2 gives some Cotes coefficients.

Table 4-2 Cotes coefficients

n	$C_k^{(n)}$	names
1	$\frac{1}{2}$ $\frac{1}{2}$	Trapezoidal rule
2	$\frac{1}{6}$ $\frac{2}{3}$ $\frac{1}{6}$	Simpson's rule
3	$\frac{1}{8}$ $\frac{3}{8}$ $\frac{3}{8}$ $\frac{1}{8}$	Newton's three-eights rule
4	$\frac{7}{90}$ $\frac{16}{45}$ $\frac{2}{15}$ $\frac{16}{45}$ $\frac{7}{90}$	Cotes's rule
5	$\frac{19}{288}$ $\frac{25}{96}$ $\frac{25}{144}$ $\frac{25}{144}$ $\frac{25}{96}$ $\frac{19}{288}$	
6	$\frac{41}{840}$ $\frac{9}{35}$ $\frac{9}{280}$ $\frac{34}{105}$ $\frac{9}{280}$ $\frac{9}{35}$ $\frac{41}{840}$	

For $n \geqslant 8$, some of the Cotes coefficients become negative. Since this might lead to negative approximations for integrals with positive **integrands** (被积函数), the high order Newton-Cotes rules cannot be recommended for numerical purposes.

Definition 4.2.1 implies that Trapezoidal and Simpson's rules have degrees of precision one and three, respectively.

Theorem 4.3.2 Newton-Cotes formula has $n+1$ degree of accuracy at least if n is an even number (偶数).

Proof We need only prove that if n is an even number, then the remainder of Newton-Cotes formula is zero for $f(x)=x^{n+1}$.

Since

$$f^{(n+1)}(x)=(n+1)!$$

therefore

$$R(f)=\int_a^b \frac{f^{(n+1)}(\xi)}{(n+1)!}\omega_{n+1}(x)\mathrm{d}x=\int_a^b \prod_{j=0}^{n}(x-x_j)\mathrm{d}x$$

By substituting $x=a+th$, and notice that $x_j=a+jh$, we have

$$R(f)=h^{n+2}\int_0^n \prod_{j=0}^{n}(t-j)\mathrm{d}t$$

If n is an even, then $\frac{n}{2}$ is an integer, let $t=u+\frac{n}{2}$, consequently we have

$$R(f)=h^{n+2}\int_{-\frac{n}{2}}^{\frac{n}{2}} \prod_{j=0}^{n}\left(u+\frac{n}{2}-j\right)\mathrm{d}u$$

Since the integrand

$$H(u)=\prod_{j=0}^{n}\left(u+\frac{n}{2}-j\right)=\prod_{j=-n/2}^{n/2}(u-j)$$

is an **odd function**(奇函数), therefore $R(f)=0$. □

We will carry out the error analysis for the Newton-Cotes formulae only for the two most important cases, $n=1$ and $n=2$, i.e., the trapezoidal rule and Simpson's rule.

Theorem 4.3.3 Let $f \in C^2[a,b]$(二次连续可微). Then the error for the trapezoidal rule can be represented in the form

$$\int_a^b f(x)\mathrm{d}x - \frac{b-a}{2}(f(a)+f(b)) = -\frac{h^3}{12}f''(\xi) \tag{4.3.6}$$

with some $\xi \in [a,b]$ and $h=b-a$.

Proof Let $P_1(x)$ denote the linear interpolation of f at the interpolation points $x_0=a$ and $x_1=b$. By construction of the trapezoidal rule, we have that the error

$$R_T = \int_a^b f(x)\mathrm{d}x - \frac{b-a}{2}(f(a)+f(b))$$

is given by

$$R_T = \int_a^b (f(x)-P_1(x))\mathrm{d}x = \int_a^b \frac{f''(\eta)}{2}(x-a)(x-b)\mathrm{d}x$$

Since $(x-a)(x-b)$ is nonpositive on $[a,b]$, from **the mean value theorem for integrals** (积分中值定理) we obtain that

$$R_T = \frac{f''(\xi)}{2}\int_a^b (x-a)(x-b)\mathrm{d}x = -\frac{f''(\xi)}{12}h^3, \quad \xi \in (a,b)$$

where $\int_a^b (x-a)(x-b)\mathrm{d}x = -\frac{(b-a)^3}{6} = -\frac{h^3}{6}$. □

Theorem 4.3.4 Let $f \in C^4[a,b]$. Then the error for the Simpson's rule can be represented in the form

$$\int_a^b f(x)\mathrm{d}x - \frac{b-a}{6}\left(f(a)+4f\left(\frac{a+b}{2}\right)+f(b)\right) = -\frac{h^5}{90}f^{(4)}(\xi) \tag{4.3.7}$$

for some $\xi \in [a,b]$ and $h=(b-a)/2$.

Proof Let $P_2(x)$ denote the quadratic interpolation polynomial for f at the interpolation points $x_0=a$, $x_1=(a+b)/2$, and $x_2=b$. By construction of Simpson's rule , we have that the error

$$R_S = \int_a^b f(x)\mathrm{d}x - \frac{b-a}{6}\left(f(a)+4f\left(\frac{a+b}{2}\right)+f(b)\right) \tag{4.3.8}$$

is given by

$$R_S = \int_a^b (f(x)-P_2(x))\,\mathrm{d}x$$

Consider the cubic polynomial $H(x)$, such that

$$\begin{aligned} &H(a)=f(a), \quad H(b)=f(b) \\ &H(c)=f(c), \quad H'(c)=f'(c) \end{aligned} \tag{4.3.9}$$

where $c=(a+b)/2$. Since Simpson's rule has three degree of accuracy, therefore

$$\int_a^b H(x)\mathrm{d}x = \frac{b-a}{6}(H(a)+4H(c)+H(b))$$

Notice that

$$\frac{b-a}{6}(H(a)+4H(c)+H(b))=\frac{b-a}{6}\left(f(a)+4f\left(\frac{a+b}{2}\right)+f(b)\right)=S_n$$

From the condition (4.3.9) and (2.6.11), we have the remainder

$$f(x)-H(x)=\frac{f^{(4)}(\eta)}{4!}(x-a)(x-c)^2(x-b)$$

Since $(x-a)(x-c)^2(x-b)$ is nonpositive, from the mean value theorem for integrals we obtain that

$$\begin{aligned} R_s &= \int_a^b [f(x)-H(x)]\mathrm{d}x \\ &= \int_a^b \frac{f^{(4)}(\eta)}{4!}(x-a)(x-c)^2(x-b)\mathrm{d}x \\ &= \frac{f^{(4)}(\xi)}{4!}\int_a^b (x-a)(x-c)^2(x-b)\mathrm{d}x \\ &= -\frac{h^5}{90}f^{(4)}(\xi) \end{aligned}$$

We conclude the statement of the theorem. □

Analogously, we have the remainder of Cotes's rule

$$R_c=-\frac{2(b-a)}{945}\left(\frac{b-a}{4}\right)^6 f^{(6)}(\eta) \tag{4.3.10}$$

Example 4.3.1 compute the integral $\int_{0.5}^1 \sqrt{x}\mathrm{d}x$ with trapezoidal rule, Simpson's rule and Newton-Cotes rule ($n=4$) respectively.

Solution Trapezoidal rule

$$\int_{0.5}^1 \sqrt{x}\mathrm{d}x \approx \frac{0.5}{2}\left(\sqrt{0.5}+1\right)=0.426\,776\,7$$

Simpson's rule

$$\int_{0.5}^1 \sqrt{x}\mathrm{d}x \approx \frac{0.5}{6}\left(\sqrt{0.5}+4\sqrt{0.75}+1\right)=0.430\,934\,03$$

Newton-Cotes rule ($n=4$)

$$\int_{0.5}^1 \sqrt{x}\mathrm{d}x \approx \frac{1-0.5}{90}\left(7\sqrt{0.5}+32\sqrt{0.625}+12\sqrt{0.75}+32\sqrt{0.875}+7\right)$$

$$=0.430\,964\,07$$

$$\int_{0.5}^1 \sqrt{x}\mathrm{d}x=\frac{2}{3}x^{\frac{3}{2}}\Big|_{0.5}^1=0.430\,964\,41$$

Example 4.3.2 The approximation of

$$\ln 2=\int_0^1 \frac{\mathrm{d}x}{1+x}$$

by the trapezoidal rule yields

$$\ln 2 \approx \frac{1}{2}\left(1+\frac{1}{2}\right)=0.75$$

For $f(x)=1/(x+1)$, we have

$$\frac{h^3}{12}\left\|f''\right\|_\infty=\frac{1}{6}$$

and hence, from Theorem 4.3.3, we obtain the estimate $|\ln 2-0.75|\leqslant 0.167$ as compared to the true error ln 2−0.75=−0.056⋯.

Simpson's rule yields

$$\ln 2 \approx \frac{1}{6}\left(1+\frac{4}{1+\frac{1}{2}}+\frac{1}{2}\right)=\frac{25}{36}=0.694\,4\cdots$$

and from Theorem 4.3.4 and

$$\frac{h^5}{90}\left\|f^{(4)}\right\|_\infty=\frac{1}{120}$$

we find the estimate $|\ln 2-0.694\,4|\leqslant 0.008\,4$ as compared to the true error $\ln 2-25/35=$ $-0.001\,2\cdots$.

4.4 Composite Quadrature Formula（复化求积公式）

In order to increase the accuracy, instead of using higher order Newton-Cotes rule, it is more practical to use so-called **composite rules**. These are obtained by subdividing the interval of integration and then applying a fixed rule with low interpolation order to each of the subintervals. The most frequently used quadrature rules of this type are the composite trapezoidal rule and the composite Simpson's rule.

4.4.1 Composite Trapezoidal Rule（复化梯形公式）

Subdivided the interval $[a,b]$ into n subintervals, and apply trapezoidal rule on each subinterval (see Figure 4-3), with $h=(b-a)/n$ and $x_k=a+kh, k=0,\cdots,n,$ we obtain

$$I=\int_a^b f(x)\mathrm{d}x=\sum_{k=1}^{n}\int_{x_{k-1}}^{x_k} f(x)\mathrm{d}x\approx\frac{h}{2}\sum_{k=1}^{n}\left(f(x_{k-1})+f(x_k)\right)$$

Denote

$$T_n=\frac{h}{2}\left(f(a)+2\sum_{k=1}^{n-1}f(x_k)+f(b)\right) \tag{4.4.1}$$

this is called **composite trapezoidal rule**.

Theorem 4.4.1 Let $f\in \mathbf{C}^2(a,b)$, $h=(b-a)/n$. Then the error for the composite trapezoidal rule is given by

$$\int_a^b f(x)\mathrm{d}x - T_n = -\frac{b-a}{12}h^2 f''(\xi) \tag{4.4.2}$$

for some $\xi \in [a,b]$.

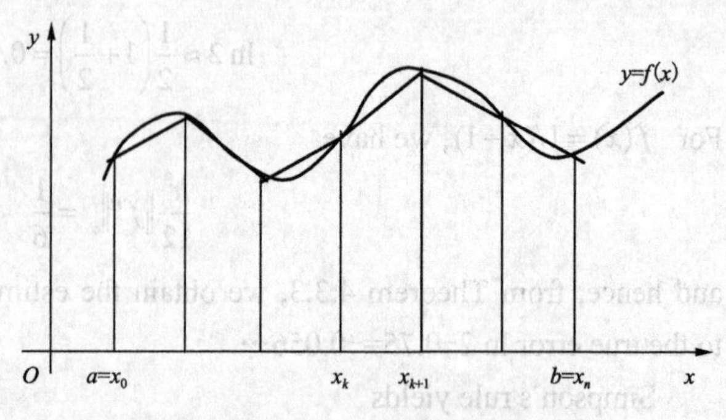

Figure 4-3 Composite trapezoidal rule

Proof From Theorem 4.3.3, the error on each subinterval $[x_{k-1}, x_k]$ is given by

$$-\frac{1}{12}h^3 f''(\xi_k) \quad (\xi_k \in [x_{k-1}, x_k])$$

We have that

$$\int_a^b f(x)\mathrm{d}x - T_n = -\frac{1}{12}h^3\left(f''(\xi_1) + f''(\xi_2) + \cdots + f''(\xi_n)\right)$$

That is

$$\int_a^b f(x)\mathrm{d}x - T_n = -\frac{b-a}{12}\cdot h^2 \cdot \frac{1}{n}\left(f''(\xi_1) + f''(\xi_2) + \cdots + f''(\xi_n)\right)$$

Since

$$\min_{x\in[a,b]} f''(x) \leqslant \frac{1}{n}[f''(\xi_1) + f''(\xi_2) + \cdots + f''(\xi_n)] \leqslant \max_{x\in[a,b]} f''(x)$$

From **the intermediate value theorem** (介值定理), there exists $\xi \in [a,b]$, such that

$$f''(\xi) = \frac{1}{n}\left(f''(\xi_1) + f''(\xi_2) + \cdots + f''(\xi_n)\right)$$

and the proof is finished. □

4.4.2 Composite Simpson's Rule

Subdivided the interval $[a,b]$ into n subintervals, and apply Simpson's rule on each subinterval (see Figure 4-4), with $h = (b-a)/n$ and $x_k = a + kh$, k=0,1, ⋯, n, denote $x_{k+\frac{1}{2}} = x_k + \frac{h}{2}$, we obtain

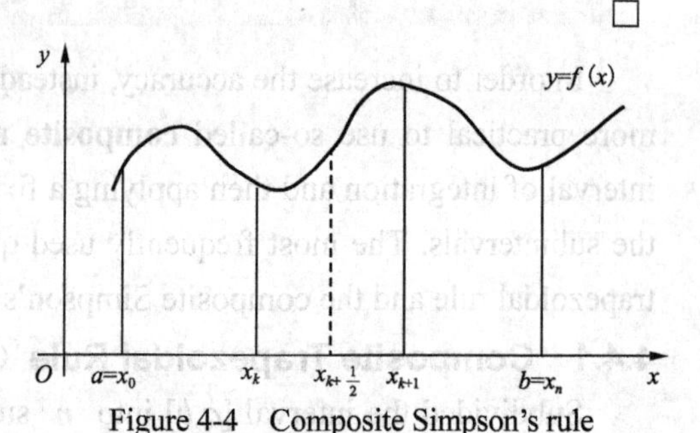

Figure 4-4 Composite Simpson's rule

$$I = \int_a^b f(x)\mathrm{d}x = \sum_{k=0}^{n-1}\int_{x_k}^{x_{k+1}} f(x)\mathrm{d}x$$

$$= \frac{h}{6}\sum_{k=0}^{n-1}\left(f(x_k) + 4f\left(x_{k+\frac{1}{2}}\right) + f(x_{k+1})\right) + R_n(f)$$

Denote

$$S_n = \frac{h}{6}\sum_{k=0}^{n-1}\left(f(x_k) + 4f(x_{k+\frac{1}{2}}) + f(x_{k+1})\right) \tag{4.4.3}$$

$$= \frac{h}{6}\left(f(a) + 4\sum_{k=0}^{n-1} f\left(x_{k+\frac{1}{2}}\right) + 2\sum_{k=1}^{n-1} f(x_k) + f(b)\right)$$

This is called **composite Simpson's rule.**

Theorem 4.4.2 Let $f \in \mathbf{C}^4(a,b)$, $h=(b-a)/n$. Then the error for the composite Simpson's rule is given by

$$R_n(f)=\int_a^b f(x)\mathrm{d}x - S_n = -\frac{b-a}{180}\left(\frac{h}{2}\right)^4 f^{(4)}(\xi)$$

for some $\xi \in [a,b]$.

Clearly, if the number n of quadrature points is doubled, i.e., if the step size h is halved, then the error for the trapezoidal rule is reduced by the factor 1/4 and for Simpson's rule by the factor 1/16, as predicted in Theorems 4.4.1 and 4.4.2.

Example 4.4.1 Compute $\pi = \int_0^1 \frac{4}{1+x^2}\mathrm{d}x$ with Newton-Cotes formulae.

Solution (1) Let $n=8$, then $h=\frac{1-0}{8}=\frac{1}{8}$, by using trapezoidal rule , we obtain

$$\pi \approx T_8 = \frac{1}{16}\left(f(0)+2\left(\left(f\left(\frac{1}{8}\right)+f\left(\frac{2}{8}\right)+f\left(\frac{3}{8}\right)+f\left(\frac{4}{8}\right)+f\left(\frac{5}{8}\right)+f\left(\frac{6}{8}\right)+f\left(\frac{7}{8}\right)\right)+f(1)\right)\right)$$

$$=3.138\,988$$

(2) Let $n=4$, then $h=\frac{1}{4}$, by using Simpson's rule , we obtain

$$\pi \approx S_4 = \frac{1}{4\times 6}\left(f(0)+4\left(f\left(\frac{1}{8}\right)+f\left(\frac{3}{8}\right)+f\left(\frac{5}{8}\right)+f\left(\frac{7}{8}\right)\right)+2\left(f\left(\frac{1}{4}\right)+f\left(\frac{1}{2}\right)+f\left(\frac{3}{4}\right)\right)+f(1)\right)$$

$$=3.141\,593$$

4.5 Romberg Integration（龙贝格积分）

4.5.1 Recursive Trapezoidal Rule（梯形公式的递推）

Recall the composite trapezoidal rule

$$T_n = \frac{h}{2}\left(f(a)+2\sum_{k=1}^{n-1} f(x_k)+f(b)\right),\quad h=\frac{b-a}{n}$$

If the number n is doubled, i.e., $2n$, then

$$\begin{aligned}
T_{2n} &= \frac{b-a}{4n}\left(f(a)+2\sum_{k=1}^{n-1} f\left(a+2k\frac{b-a}{2n}\right)+2\sum_{k=0}^{n-1} f\left(a+(2k+1)\frac{b-a}{2n}\right)+f(b)\right)\\
&= \frac{h}{4}\left(f(a)+2\sum_{k=1}^{n-1} f\left(a+k\frac{b-a}{n}\right)+f(b)\right)+\frac{h}{2}\sum_{k=0}^{n-1} f\left(a+(2k+1)\frac{b-a}{2n}\right)\\
&= \frac{1}{2}T_n+\frac{h}{2}\sum_{k=0}^{n-1} f\left(x_{k+\frac{1}{2}}\right)
\end{aligned}$$

We have the recursive trapezoid rule

$$T_{2n}=\frac{1}{2}T_n+\frac{h}{2}\sum_{k=0}^{n-1}f\left(x_{k+\frac{1}{2}}\right) \tag{4.5.1}$$

where $h=\dfrac{b-a}{n}$.

It is clear that if T_{2n} is to be computed, then we can take advantage of the work already done in the computation of T_n. Only the terms present in T_{2n} but not already present in T_n must be computed.

Example 4.5.1 Compute the integration $\int_0^1\frac{\sin x}{x}dx$ with the recursive trapezoid rule.

Solution $f(x)=\dfrac{\sin x}{x}$, $f(1)=0.841\,470\,9$, and we define $f(0)=1$, then

$$T_1=\frac{1}{2}(f(0)+f(1))=\frac{1}{2}(1+0.841\,470\,9)=0.920\,735\,5$$

$$f(1/2)=0.958\,851\,0$$

from the recursive trapezoid rule (4.5.1), we have

$$T_2=\frac{1}{2}T_1+\frac{1}{2}f\left(\frac{1}{2}\right)=0.939\,793\,3,$$

$$T_4=\frac{1}{2}T_2+\frac{1}{4}\left(f\left(\frac{1}{4}\right)+f\left(\frac{3}{4}\right)\right)=0.944\,513\,5$$

where

$$f\left(\frac{1}{4}\right)=0.989\,615\,8,\ f\left(\frac{3}{4}\right)=0.908\,851\,6$$

See Table 4-3.

Table 4-3 The results of recursive trapezoid rule

k	1	2	3	4	5
T_n	0.939 793 3	0.944 513 5	0.945 690 9	0.945 985 0	0.946 059 6
k	6	7	8	9	10
T_n	0.946 076 9	0.946 081 5	0.946 082 7	0.946 083 0	0.946 083 1

It indicates that in order to achieve 7 significant digits with the composite trapezoidal rule, it need 10 times bisection, i.e., there are 1 025 quadrature points, the quantity of computation is very large.

4.5.2 Romberg Algorithm（龙贝格算法）

From the representation of remainder of the composite trapezoidal rule, we obtain

$$\int_a^b f(x)\mathrm{d}x-T_n=-\frac{b-a}{12}h^2f''(\xi)$$

$$\int_a^b f(x)\mathrm{d}x-T_{2n}=-\frac{b-a}{12}\left(\frac{h}{2}\right)^2f''(\bar{\xi})$$

Assume $f''(\xi) \approx f''(\bar{\xi})$, then

$$\frac{I-T_{2n}}{I-T_n} \approx \frac{1}{4}$$

It implies that

$$I-T_{2n} \approx \frac{1}{3}(T_{2n}-T_n) \tag{4.5.2}$$

From this, we can estimate the error of T_{2n} with the results of T_n and T_{2n}. The method of estimating error with the results of computation is called **method of backward estimation** (事后估计法).

Denote

$$\overline{T_n} = T_{2n} + \frac{1}{3}(T_{2n}-T_n) = \frac{4}{3}T_{2n} - \frac{1}{3}T_n$$

It may be a better result of approximation, and what is it indeed?

In fact

$$\begin{aligned}
\frac{4}{3}T_{2n} - \frac{1}{3}T_n &= \frac{4}{3}\left(\frac{1}{2}T_n + \frac{b-a}{2n}\sum_{k=1}^{n} f\left(a+(2k-1)\frac{b-a}{2n}\right)\right) - \frac{1}{3}T_n \\
&= \frac{1}{3}T_n + \frac{2}{3}\cdot\frac{(b-a)}{n}\sum_{k=0}^{n-1} f\left(x_{k+\frac{1}{2}}\right) \\
&= \frac{1}{3}\cdot\frac{1}{2}\cdot\frac{b-a}{n}\left(f(a)+2\sum_{k=1}^{n-1} f(x_k)+f(b)\right) + \frac{4}{6}\cdot\frac{(b-a)}{n}\sum_{k=0}^{n-1} f\left(x_{k+\frac{1}{2}}\right) \\
&= \frac{1}{6}\cdot\frac{b-a}{n}\left(f(a)+4\sum_{k=0}^{n-1} f\left(x_{k+\frac{1}{2}}\right)+2\sum_{k=1}^{n-1} f(x_k)+f(b)\right) \\
&= S_n
\end{aligned}$$

that is

$$S_n = \frac{4}{3}T_{2n} - \frac{1}{3}T_n = \frac{4T_{2n}-T_n}{4-1} \tag{4.5.3}$$

Analogously , consider Simpson's rule we have

$$\frac{I-S_{2n}}{I-S_n} \approx \frac{1}{16}$$

it implies

$$I-S_{2n} \approx \frac{1}{15}(S_{2n}-S_n) \tag{4.5.4}$$

we can prove that

$$C_n = \frac{16}{15}S_{2n} - \frac{1}{15}S_n = \frac{4^2 S_{2n}-S_n}{4^2-1} \tag{4.5.5}$$

consider Cotes rule we have

$$I-C_{2n} \approx \frac{1}{63}(C_{2n}-C_n) \tag{4.5.6}$$

Denote

$$R_n = \frac{64}{63}C_{2n} - \frac{1}{63}C_n = \frac{4^3 C_{2n} - C_n}{4^3 - 1} \tag{4.5.7}$$

R_n is called **the sequence of Romberg** (龙贝格序列). See Table 4-4.

Example 4.5.2 Compute $\pi = \int_0^1 \frac{4}{1+x^2}\mathrm{d}x$ with the **recurrence formula** (递推公式), such that the error is less than 10^{-6}.

Solution

$$T_1 = \frac{1}{2}(f(0)+f(1)) = \frac{1}{2}(4+2) = 3$$

$$T_2 = \frac{1}{2}T_1 + \frac{1}{2}f\left(\frac{1}{2}\right) = 3.1$$

$$T_4 = \frac{1}{2}T_2 + \frac{1}{4}\left(f\left(\frac{1}{4}\right) + f\left(\frac{3}{4}\right)\right) = 3.131\,176\,47$$

analogously, we can compute T_8, $T_{16}, \cdots$. See Table 4-5.

Since

$$|T_{512} - T_{256}| < 3\times10^{-6}$$

therefore $T_{512} = 3.141\,592\,02$ is the approximate value satisfying the given accuracy.

Table 4-4 The sequence of Romberg

Trapezoidal	Simpson	Cotes	Romberg
T_1			
T_2	S_1		
T_4	S_2	C_1	
T_8	S_4	C_2	R_1
T_{16}	S_8	C_4	R_2
$\vdots$	$\vdots$	$\vdots$	$\vdots$

Table 4-5 The results of recursive trapezoid rule

n	T_n	n	T_n
1	3	32	3.141 429 89
2	3.1	64	3.141 551 96
4	3.131 176 47	128	3.141 582 48
8	3.1389 88 49	256	3.141 590 11
16	3.140 941 61	512	3.141 592 02

4.5.3 Richardson's Extrapolation（理查森外推法）

We use $T_m^{(k)}$ to represent each approximate value, k represents the number of bisection, m represents the sequence of the approximate value: m=0 represents the sequence of trapezoid, m=1 represents the sequence of Simpson, m=2 represents the sequence of Cotes. Then we have Table 4-6.

Then for $m = 1, 2, \cdots$, the Romberg quadratures are recursively defined by

$$\begin{cases} T_0^{(0)} = \dfrac{b-a}{2}(f(a)+f(b)) \\ T_0^{(l)} = \dfrac{1}{2}T_0^{(l-1)} + \dfrac{b-a}{2^l}\sum\limits_{i=1}^{2^{l-1}} f\left(a+(2i-1)\dfrac{b-a}{2^l}\right) \qquad (l=1,\ 2,\ 3,\cdots) \\ T_m^{(k)} = \dfrac{4^m T_{m-1}^{(k+1)} - T_{m-1}^{(k)}}{4^m - 1} \quad (k=0,\ 1,\ 2,\cdots;\ m=1,\ 2,\ 3,\cdots) \end{cases}$$

This interpretation of the Romberg quadrature as an extrapolation method in the sense of Richardson opens up the possibility of modifications using other than equidistant step sizes.

Example 4.5.3 Compute $I = \frac{2}{\sqrt{\pi}}\int_0^1 e^{-x}dx$ with Romberg algorithm, such that the error is less than 10^{-5}.

Solution Compute $I = \int_0^1 e^{-x}dx$ with Romberg algorithm, see Table 4-7.

Table 4-6 Richardson's extrapolation

k	T_0^k	T_1^{k-1}	T_2^{k-2}	T_3^{k-3}	T_4^{k-4}
0	$T_0^{(0)}$				
1	$T_0^{(1)}$	$T_1^{(0)}$			
2	$T_0^{(2)}$	$T_1^{(1)}$	$T_2^{(0)}$		
3	$T_0^{(3)}$	$T_1^{(2)}$	$T_2^{(1)}$	$T_3^{(0)}$	
4	$T_0^{(4)}$	$T_1^{(3)}$	$T_2^{(2)}$	$T_3^{(1)}$	$T_4^{(0)}$
⋮	⋮	⋮	⋮	⋮	⋮

Table 4-7 The sequence of Romberg

k	T_0^k	T_1^k	T_2^k	T_3^k
0	0.683 940	0.632 333	0.632 122	0.632 120
1	0.645 235	0.632 135	0.632 126	
2	0.635 410	0.632 121		
3	0.632 943			

$$\frac{2}{\sqrt{\pi}}\int_0^1 e^{-x}dx \approx \frac{2}{\sqrt{\pi}} \times 0.632120 \approx 0.71327\ T_4^{(0)}$$

The exact value of integration is 0.713 272.

4.6 Gaussian Quadrature Formula（高斯求积公式）

In Section 4.3 the Newton-Cotes formulas were derived by integrating interpolating polynomials. Since the error term in the interpolating polynomial of degree n involves the $(n+1)$st derivative of the function being approximated, a formula of this type is exact when approximating any polynomial of degree less than or equal to n.

All the Newton-Cotes formulas use values of the function at equally-spaced points. This restriction is convenient when the formulas are combined to form the composite rules we considered in Section 4.4, but it can significantly decrease the accuracy of the approximation. Consider, for example, the trapezoidal rule applied to determine the integrals of the functions show in Figure 4-5.

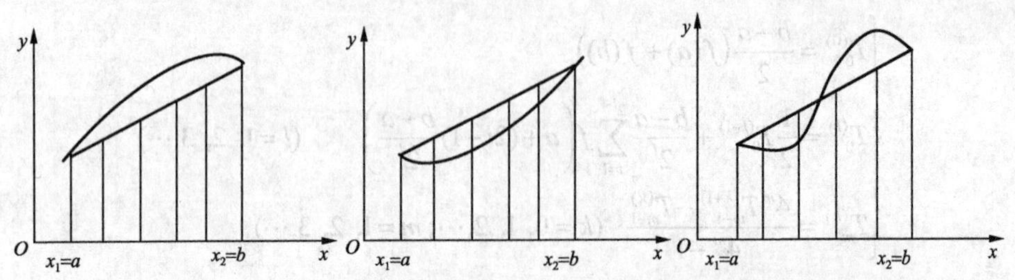

Figure 4-5 The trapezoidal rule applied to determine the integrals

The trapezoidal rule approximates the integral of the function by integrating the linear function that joins the endpoints of the graph of the function. But this is not likely the best line for approximating the integral. Lines such as those shown in Figure 4-6 would likely give much better approximations in most cases.

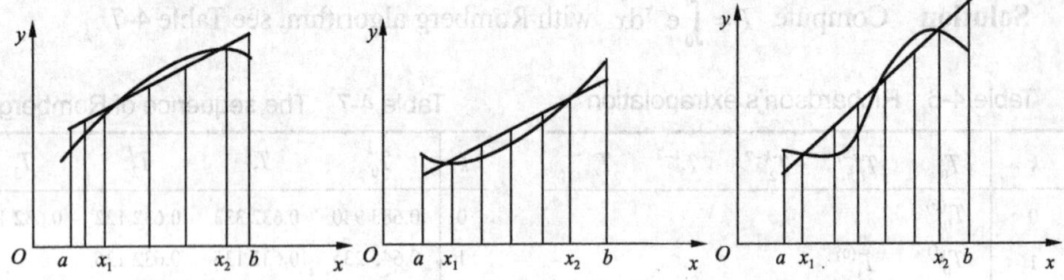

Figure 4-6 The trapezoidal rule applied to determine the integrals

Gaussian quadrature chooses the points for evaluation in an optimal, rather than **equally spaced** (等距). The nodes $x_1, x_2, \cdots, x_n$ in the interval $[a,b]$ and coefficients $c_1, c_2, \cdots, c_n$ are chosen to minimize the expected error obtained in the approximation

$$\int_a^b f(x)\mathrm{d}x \approx \sum_{k=1}^{n} c_k f(x_k)$$

To measure this accuracy, we assume that the best choice of these values is that which produces the exact result for the largest class of polynomials, that is, the choice gives the greatest degree of precision.

The coefficients $c_1, c_2, \cdots, c_n$ in the approximation formula are arbitrary, and the nodes $x_1, x_2, \cdots, x_n$ are restricted only by the fact that they must lie in $[a,b]$, the interval of integration. This gives us $2n$ parameters to choose. If the coefficients of a polynomial are considered parameters, the class of polynomials of degree at most $2n-1$ also contains $2n$ parameters. This, then, is the largest class of polynomials for which it is reasonable to expect the formula to be exact. With the proper choice of the values and constants, exactness on this set can be obtained.

To illustrate the procedure for choosing the approximate parameters, we will show how to select the coefficients and nodes when $n=2$ and the interval of integration is $[-1,1]$. We will then discuss the more general situation for an arbitrary choice of nodes and coefficients and show how the technique is modified when integrating over an arbitrary interval.

Suppose we want to determine c_1, c_2, x_1 and x_2 so that the integration formula

$$\int_{-1}^{1} f(x)\mathrm{d}x \approx c_1 f(x_1) + c_2 f(x_2)$$

gives the exact result whenever $f(x)$ is a polynomial of degree 2×2−1=3 or less, that is, when

$$f(x) = a_0 + a_1 x + a_2 x^2 + a_3 x^3$$

from some collection of constants, a_0, a_1, a_2 and a_3. Since

$$\int_a^b (a_0 + a_1 x + a_2 x^2 + a_3 x^3)\mathrm{d}x = a_0 \int_a^b 1\mathrm{d}x + a_1 \int_a^b x\mathrm{d}x + a_2 \int_a^b x^2\mathrm{d}x + a_3 \int_a^b x^3\mathrm{d}x$$

this is equivalent to showing that the formula gives exact results when $f(x)$ is $1, x, x^2$ and x^3. Therefore, we need c_1, c_2, x_1 and x_2, so that

$$c_1 + c_2 = \int_{-1}^{1} 1\mathrm{d}x = 2$$

$$c_1 x_1 + c_2 x_2 = \int_{-1}^{1} x\mathrm{d}x = 0$$

$$c_1 x_1^2 + c_2 x_2^2 = \int_{-1}^{1} x^2\mathrm{d}x = \frac{2}{3}$$

$$c_1 x_1^3 + c_2 x_2^3 = \int_{-1}^{1} x^3\mathrm{d}x = 0$$

This system of equations has the solution

$$c_1 = c_2 = 1,\ x_1 = -\frac{\sqrt{3}}{3},\ \text{and}\ x_2 = \frac{\sqrt{3}}{3}$$

we obtain the approximation formula

$$\int_{-1}^{1} f(x)\mathrm{d}x \approx f\left(-\frac{\sqrt{3}}{3}\right) + f\left(\frac{\sqrt{3}}{3}\right) \tag{4.6.1}$$

It is clear that this formula has degree of precision 3, that is, it produces the exact result for every polynomial of degree 3 or less.

This method could be used to determine the nodes and coefficients for formulas that give exact results for higher-degree polynomials, but an alternative method obtains them more easily. In section 3.6 we discussed collections of orthogonal polynomials, functions that have the property that a particular definite integral of the product of any two of them is 0. The set that is relevant to our problem is the set of Legendre polynomials, a collection $\{P_0(x), P_1(x), \cdots, P_n(x), \cdots\}$. The first few Legendre polynomials are

$$P_0(x) \equiv 1$$

$$P_1(x) = x - B_1 = x$$

$$P_2(x) = x^2 - \frac{1}{3}$$

$$P_3(x) = x^3 - \frac{3}{5}x$$

$$P_4(x) = x^4 - \frac{6}{7}x^2 + \frac{3}{35}$$
$$P_5(x) = x^5 - \frac{10}{9}x^3 + \frac{5}{21}x$$

The roots of these polynomials are distinct, lie in the interval (−1,1), have a symmetry with respect to the origin, and most importantly, are the correct choice for determining the parameters that solve our problem. The nodes $x_1,x_2,\cdots,x_n$ needed to produce an integral approximation formula that gives exact results for any polynomial of degree less than $2n$ are the roots of the nth-degree Legendre polynomial. This is established by the following result.

Theorem 4.6.1 Suppose that $x_1,x_2,\cdots,x_n$ are the roots of the nth-degree Legendre polynomial $P_n(x)$, and that for each $k=1,2,\cdots,n$, the numbers c_k are defined by

$$c_k = \int_{-1}^{1} \prod_{\substack{j=1 \\ j\neq k}}^{n} \frac{x - x_j}{x_k - x_j} \mathrm{d}x$$

If $P(x)$ is any polynomial of degree less than $2n$, then

$$\int_{-1}^{1} P(x)\mathrm{d}x = \sum_{k=1}^{n} c_k P(x_k)$$

Proof Consider the situation for a polynomial $P(x)$ of degree less than n. Rewrite $P(x)$ as an $(n-1)$st Lagrange polynomial with nodes at the roots of the nth Legendre polynomial $P_n(x)$. This representation of $P(x)$ is exact, since the error term involves the nth derivative of $P(x)$ and the nth derivative of $P(x)$ is 0. Hence,

$$\int_{-1}^{1} P(x)\mathrm{d}x = \int_{-1}^{1} \left(\sum_{k=1}^{n} \prod_{\substack{j=1 \\ j\neq k}}^{n} \frac{x - x_j}{x_k - x_j} P(x_k) \right) \mathrm{d}x$$
$$= \sum_{k=1}^{n} \left(\int_{-1}^{1} \prod_{\substack{j=1 \\ j\neq k}}^{n} \frac{x - x_j}{x_k - x_j} \mathrm{d}x \right) P(x_k) = \sum_{k=1}^{n} c_k P(x_k)$$

which verifies the result for polynomials of degree less than n.

If the polynomial $P(x)$ of degree at least n but less than $2n$ is divided by the nth Legendre polynomial $P_n(x)$, we get two polynomials $Q(x)$ and $R(x)$ of degree less than n:

$$P(x) = Q(x)P_n(x) + R(x)$$

Now, we **invoke** (要求) the unique power of the Legendre polynomials. First, the degree of the polynomial $Q(x)$ is less than n, so

$$\int_{-1}^{1} Q(x)P_n(x)\mathrm{d}x = 0$$

Next, since x_k is a root of $P_n(x)$ for each $k=1,2,\cdots,n$, we have

$$P(x_k) = Q(x_k)P_n(x_k) + R(x_k) = R(x_k)$$

Finally, since $R(x)$ is a polynomial of degree less than n, the opening argument implies that

$$\int_{-1}^{1} R(x)\mathrm{d}x = \sum_{k=1}^{n} c_k R(x_k)$$

Putting these facts together verifies that the formula is exact for the polynomial $P(x)$:

$$\int_{-1}^{1} P(x)\mathrm{d}x = \int_{-1}^{1} (Q(x)P_n(x) + R(x))\mathrm{d}x$$

$$= \int_{-1}^{1} R(x)\mathrm{d}x = \sum_{k=1}^{n} c_k R(x_k) = \sum_{k=1}^{n} c_k P(x_k)$$ □

The constants c_k needed for the quadrature rule can be generated from the equation in Theorem 4.6.1, but both these constants and roots of the Legendre polynomials are extensively tabulated. Table 4-8 lists these values for n=2,3,4, and 5.

Table 4-8 Gaussian quadrature coefficients

n	Roots $r_{n,k}$	Coefficients $c_{n,k}$
2	0.577 350 269 2	1.000 000 000 0
	−0.577 350 269 2	1.000 000 000 0
3	0.774 596 669 2	0.555 555 555 6
	0.000 000 000 0	0.888 888 888 9
	−0.774 596 669 2	0.555 555 555 6
4	0.861 136 311 6	0.347 854 845 1
	0.339 981 043 6	0.652 145 154 9
	−0.339 981 043 6	0.652 145 154 9
	−0.861 136 311 6	0.347 854 845 1
5	0.906 179 845 9	0.236 926 885 0
	0.538 469 310 1	0.478 628 670 5
	0.000 000 000 0	0.568 888 888 9
	−0.538 469 310 1	0.478 628 670 5
	−0.906 179 845 9	0.236 926 885 0

An integral $\int_a^b f(x)\mathrm{d}x$ over an arbitrary $[a,b]$ can be transformed into an integral over $[-1,1]$ by using the change of variables :

$$t = \frac{2x-a-b}{b-a} \Leftrightarrow x = \frac{1}{2}((b-a)t + a + b)$$

This permits Gaussian quadrature to be applied to any interval $[a,b]$, since

$$\int_a^b f(x)\mathrm{d}x = \int_{-1}^{1} f\left(\frac{(b-a)t+(b+a)}{2}\right)\frac{(b-a)}{2}\mathrm{d}t \tag{4.6.2}$$

Example 4.6.1 Consider the problem of finding approximations to $\int_1^{1.5} \mathrm{e}^{-x^2}\mathrm{d}x$.

Solution The Gaussian quadrature procedure applied to this problem requires that the integral first be transformed into a problem whose interval of integration is $[-1,1]$. Using Equation (4.6.2), we have

$$\int_1^{1.5} \mathrm{e}^{-x^2}\mathrm{d}x = \frac{1}{4}\int_{-1}^{1} \mathrm{e}^{(-(t+5)^2/16)}\mathrm{d}t$$

The values in Table 4-8 give the following Gaussian quadrature approximations for this problem:

n=2

$$\int_1^{1.5} \mathrm{e}^{-x^2}\mathrm{d}x \approx \frac{1}{4}\left(\mathrm{e}^{(-(0.577\,350\,269\,2+5)^2/16)} + \mathrm{e}^{(-(-0.577\,350\,269\,2+5)^2/16)}\right) = 0.109\,400\,3$$

n=3

$$\int_1^{1.5} \mathrm{e}^{-x^2}\mathrm{d}x \approx \frac{1}{4}\Big(0.555\,555\,555\,6\mathrm{e}^{-(0.774\,596\,669\,2+5)^2/16} + 0.888\,888\,888\,9\mathrm{e}^{-(5)^2/16} +$$

$$0.5\,555\,555\,556\mathrm{e}^{(-(-0.774\,596\,669\,2+5)^2/16)}\Big)$$

$$= 0.109\,364\,2$$

For further comparison, the values obtained using the Romberg procedure with $n=4$ are listed in Table 4-9.

Table 4-9 The values obtained using the Romberg procedure

Trapezoidal	Simpson	Cotes	Romberg
0.118 319 7			
0.111 562 7	0.109 310 4		
0.109 9114	0.109 361 0	0.109 364 3	
0.109 500 9	0.109 364 1	0.109 364 3	0.109 364 3

4.7 Multiple Integrals（多重积分）

The techniques discussed in the previous sections can be modified in a straightforward manner for use in the approximation of multiple integrals. Consider the double integral

$$\iint_R f(x,y)\mathrm{d}A$$

where $R=\{(x,y)\mid a\leqslant x\leqslant b,\ c\leqslant y\leqslant d\}$, for some constants a,b,c and d, is a rectangular region in the plane. We will employ the Composite Simpson's rule to illustrate the approximation technique, although any other composite formula could be used in its place.

To apply the Composite Simpson's rule, we divide the region R by partitioning both $[a,b]$ and $[c,d]$ into subintervals. To simplify the notation, we choose even integers n and m and partition $[a,b]$ and $[c,d]$ with the evenly spaced mesh points $x_0,x_1,\ldots,x_n$ and $y_0,y_1,\ldots,y_m$, respectively. These subdivisions determine step sizes $h=(b-a)/n$ and $k=(d-c)/m$. Writing the double integral as the **iterated integral** (累次积分)

$$\iint_R f(x,y)\mathrm{d}A=\int_a^b\left(\int_c^d f(x,y)\mathrm{d}y\right)\mathrm{d}x$$

we first use the Composite Simpson's rule to approximate

$$\int_c^d f(x,y)\mathrm{d}y$$

treating x as a constant. Let $y_j=c+jk,\ y_j=c+\left(j+\dfrac{1}{2}\right)k$, for each $j=0,1,\ldots,m$. Then

$$\int_c^d f(x,y)\mathrm{d}y=\frac{k}{6}[f(x,y_0)+4\sum_{j=0}^{m-1}f(x,y_{j+1/2})+2\sum_{j=1}^{m-1}f(x,y_j)+f(x,y_m)]-$$

$$\frac{(d-c)}{180}\left(\frac{k}{2}\right)^4\frac{\partial^4 f(x,\mu)}{\partial y^4}$$

for some μ in (c,d). Thus

$$\int_a^b\left(\int_c^d f(x,y)\mathrm{d}y\right)\mathrm{d}x=\frac{k}{6}\left[\int_a^b f(x,y_0)\mathrm{d}x+4\sum_{j=0}^{m-1}\int_a^b f(x,y_{j+1/2})\mathrm{d}x+ 2\sum_{j=1}^{m-1}\int_a^b f(x,y_j)\mathrm{d}x+\int_a^b f(x,y_m)\mathrm{d}x\right]- \frac{(d-c)}{180}\left(\frac{k}{2}\right)^4\int_a^b\frac{\partial^4 f(x,\mu)}{\partial y^4}\mathrm{d}x$$

The composite Simpson's rule is now employed on the integral in this equation. Let

$$x_i=a+ih,\quad x_i=a+\left(i+\frac{1}{2}\right)h,\quad \text{for each } i=0,1,\ldots,n$$

Then for each $j=0,1,\ldots,m$, we use the Composite Simpson's rule respectively to get the resulting approximation.

Example 4.7.1 The Composite Simpson's rule applied to approximate

$$\int_{1.4}^{2.0}\int_{1.0}^{1.5}\ln(x+2y)\mathrm{d}y\mathrm{d}x$$

with n=2 and m=1.

Solution h=0.3, k=0.5, then

$$\begin{aligned}
&\int_{1.4}^{2.0}\int_{1.0}^{1.5}\ln(x+2y)\mathrm{d}y\mathrm{d}x\\
&\approx\frac{k}{6}\left[\int_{1.4}^{2.0}\ln(x+2)\mathrm{d}x+4\int_{1.4}^{2.0}\ln(x+2.5)\mathrm{d}x+\int_{1.4}^{2.0}\ln(x+3)\mathrm{d}x\right]\\
&\approx\frac{0.5}{6}\times\frac{0.3}{6}[\ln 3.4+4(\ln 3.55+\ln 3.85)+2\ln 3.7+\ln 4]+\\
&\quad\frac{0.5}{6}\times\frac{1.2}{6}[\ln 3.9+4(\ln 4.05+\ln 4.35)+2\ln 4.2+\ln 4.5]+\\
&\quad\frac{0.5}{6}\times\frac{0.3}{6}[\ln 4.4+4(\ln 4.55+\ln 4.85)+2\ln 4.7+\ln 5]\\
&=0.429\ 552\ 44
\end{aligned}$$

The actual value of the integral to ten decimal places is 0.429 554 526 5.

The same techniques can be applied for the approximation of triple integrals as well as higher integrals for functions of more than three variables. The number of functional evaluations required for the approximation is the product of the number of functional evaluation required when the method is applied to each variable.

4.8 Numerical Differentiation（数值微分）

4.8.1 Numerical Differentiation

We illustrate these matters by examining a formula for numerical differentiation that emerges directly from the limit definition of $f'(x)$

$$f'(x) \approx \frac{1}{h}(f(x+h)-f(x)) \tag{4.8.1}$$

This formula is known as the forward-difference formula if $h>0$ and the backward-difference formula if $h<0$.

For a linear function, $f(x)=ax+b$, the approximate Formula (4.8.1) is exact, that is, it yields the correct value of $f'(x)$ for any nonzero value of h. The formula may be exact in other cases too, but only fortuitously. Let us therefore attempt to assess the error involved in this formula for numerical differentiation. The starting point is Taylor's theorem in this form:

$$f(x+h)=f(x)+hf'(x)+\frac{h^2}{2}f''(\xi) \tag{4.8.2}$$

Here ξ is a point in the open interval between x and $x+h$. For the validity of Equation (4.8.2), f and f' should be continuous on the closed interval between x and $x+h$, and f'' should exist on the corresponding open interval. A rearrangement of Equation (4.8.2) yields

$$f'(x)=\frac{1}{h}(f(x+h)-f(x))-\frac{h}{2}f''(\xi) \tag{4.8.3}$$

Equation (4.8.3) is more useful than Equation (4.8.1) because now on a large class of functions as described above, an error term is available along with the basic numerical formula. Notice that the error term in Equation (4.8.3) has two parts, a power of h and a factor involving some higher-order derivative of f. The latter gives us an indication of the class of functions to which the error estimate is applicable. The h-term in the error makes the entire expression converge to 0 as h approaches 0. The rapidity of this convergence will depend on the power of h. These remarks apply to many error estimates in numerical analysis. There will usually be a power of h and a factor telling us to what smoothness class the function must belong so that the estimate is valid.

Example 4.8.1 If Formula (4.8.1) is used to compute the derivative of $f(x)=\cos x$ at $x=\pi/4$ with $h=0.01$, what is the answer and how accurate is it?

Solution Using a calculator, we find

$$f'(x) \approx \frac{1}{h}(f(x+h)-f(x))=\frac{1}{0.01}(0.700\,000\,476-0.707\,106\,781)=-0.710\,630\,51$$

The error term in Equation (4.7.3) can be estimated like this

$$\left|\frac{h}{2}f''(\xi)\right|=0.005|\cos\xi|\leqslant 0.005$$

We can obtain a sharper bound by using the fact that $\pi/4<\xi<\pi/4+h$, so that $|\cos\xi|<0.707107$. This gives the bound 0.003 535 5. The actual error is

$$-\sin\frac{\pi}{4}+0.710\,630\,51=0.003\,523\,729$$

Example 4.8.2 Let $f(x)=\ln x$ and $x_0=1.8$. Solve $f'(x_0)$ and estimate its error.

Solution The formula

$$f'(x)=\frac{1}{h}(f(x+h)-f(x))-\frac{h}{2}f''(\xi)$$

is used to approximate $f'(1.8)$ with error

$$\left|\frac{h}{2}f''(\xi)\right|=\frac{|h|}{2\xi^2}\leqslant\frac{|h|}{2(1.8)^2}$$

where $1.8\leqslant\xi\leqslant1.8+h$.

The results in Table 4-10 are produced when h=0.1,0.01, and 0.001.

Since $f'(x)=1/x$, the exact value of $f'(1.8)$ is $0.55\bar{5}$, and the error bounds are quite close to the true approximation error.

Table 4-10 The computational results

h	$f(1.8+h)$	$\frac{f(1.8+h)-f(1.8)}{h}$	$\frac{\lvert h\rvert}{2(1.8)^2}$
0.1	0.641 853 89	0.540 672 2	0.015 432 1
0.01	0.593 326 85	0.554 018 0	0.001 543 2
0.001	0.588 342 07	0.555 401 3	0.000 154 3

4.8.2 Differentiation Polynomial Interpolation（多项式插值微分）

A general approach to numerical differentiation and integration can be based on polynomial interpolation. Suppose that we have n +1 values of a function f at points $x_0,x_1,\cdots,x_n$. A polynomial that interpolates f at the nodes x_i can be written in the language form in Section 2.3, Let us include the error term from Theorem 2.3.1, obtaining

$$f(x)=\sum_{k=0}^{n}f(x_k)l_k(x)+\frac{1}{(n+1)!}f^{(n+1)}(\xi_x)\omega(x) \tag{4.8.4}$$

Here we have written $\omega(x)=\prod_{k=0}^{n}(x-x_k)$. Taking derivatives in Equation (4.7.4), we have

$$f'(x)=\sum_{k=0}^{n}f(x_k)l_k'(x)+\frac{1}{(n+1)!}f^{(n+1)}(\xi_x)\omega'(x)+$$
$$\frac{1}{(n+1)!}\omega(x)\frac{\mathrm{d}}{\mathrm{d}x}f^{(n+1)}(\xi_x)$$

If x is one of the nodes, say $x=x_j$, then the preceding equation simplifies, because $\omega(x_j)=0$, and the result is

$$f'(x_j)=\sum_{k=0}^{n}f(x_k)l_k'(x_j)+\frac{1}{(n+1)!}f^{(n+1)}(\xi_j)\omega'(x_j)$$

Thus, in turn, can be simplified by computing $\omega'(x_\alpha)$. To do this, we note that

$$\omega'(x)=\sum_{i=0}^{n}\prod_{\substack{k=0\\k\neq i}}^{n}(x-x_k)$$

so

$$\omega'(x_j)=\prod_{\substack{k=0\\k\neq j}}^{n}(x_j-x_k)$$

The final differentiation formula with error term is

$$f'(x_j)=\sum_{k=0}^{n}f(x_k)l_k'(x_j)+\frac{1}{(n+1)!}f^{(n+1)}(\xi_j)\prod_{\substack{k=0\\k\neq j}}^{n}(x_j-x_k) \tag{4.8.5}$$

This formula is particularly well suited for nonequally spaced nodes. (4.8.5) is called an (n+1)-**point formula** to approximate $f'(x_j)$.

Given the explicit form of Equation (4.8.5) when n=2, the three cardinal functions for Lagrange interpolation in this case are

$$l_0(x)=\frac{(x-x_1)(x-x_2)}{(x_0-x_1)(x_0-x_2)}$$

$$l_1(x)=\frac{(x-x_0)(x-x_2)}{(x_1-x_0)(x_1-x_2)}$$

$$l_2(x)=\frac{(x-x_0)(x-x_1)}{(x_2-x_0)(x_2-x_1)}$$

their derivatives are

$$l_0'(x)=\frac{2x-x_1-x_2}{(x_0-x_1)(x_0-x_2)}$$

$$l_1'(x)=\frac{2x-x_0-x_2}{(x_1-x_0)(x_1-x_2)}$$

$$l_2'(x)=\frac{2x-x_0-x_1}{(x_2-x_0)(x_2-x_1)}$$

Evaluating at x_1, we obtain

$$l_0'(x_1)=\frac{x_1-x_2}{(x_0-x_1)(x_0-x_2)}$$

$$l_1'(x_1)=\frac{2x_1-x_0-x_2}{(x_1-x_0)(x_1-x_2)}$$

$$l_2'(x_1)=\frac{x_1-x_0}{(x_2-x_0)(x_2-x_1)}$$

Hence, from Equation (4.8.5), the numerical differentiation formula with its error term is thus

$$f'(x_j)=f(x_0)\left(\frac{2x_j-x_1-x_2}{(x_0-x_1)(x_0-x_2)}\right)+f(x_1)\left(\frac{2x_j-x_0-x_2}{(x_1-x_0)(x_1-x_2)}\right)+ \\ f(x_2)\left(\frac{2x_j-x_0-x_1}{(x_2-x_0)(x_2-x_1)}\right)+\frac{1}{6}f^{(3)}(\xi_j)\prod_{\substack{k=0\\k\neq j}}^{2}(x_j-x_k) \tag{4.8.6}$$

As an example for j=1, we have

$$f'(x_1)=f(x_0)\frac{x_1-x_2}{(x_0-x_1)(x_0-x_2)}+f(x_1)\frac{2x_1-x_0-x_2}{(x_1-x_0)(x_1-x_2)}+ \\ f(x_2)\frac{x_1-x_0}{(x_2-x_0)(x_2-x_1)}+\frac{1}{6}f'''(\xi_1)(x_1-x_0)(x_1-x_2)$$

What formula results if the nodes are equally spaced?

When $x_1=x_0+h$ and $x_2=x_0+2h$, for some $h\neq 0$.

We will assume equally spaced nodes throughout the remainder of this section.

Using Equation (4.8.6) with $x_j=x_0, x_1=x_0+h$, and $x_2=x_0+2h$ gives

$$f'(x_0)=\frac{1}{h}\left(-\frac{3}{2}f(x_0)+2f(x_1)-\frac{1}{2}f(x_2)\right)+\frac{h^2}{3}f^{(3)}(\xi_0)$$

Doing the same for $x_j=x_1$ gives

$$f'(x_1)=\frac{1}{h}\left(-\frac{1}{2}f(x_0)+\frac{1}{2}f(x_2)\right)-\frac{h^2}{6}f^{(3)}(\xi_1)$$

and for $x_j=x_2$

$$f'(x_2)=\frac{1}{h}\left(\frac{1}{2}f(x_0)-2f(x_1)+\frac{3}{2}f(x_2)\right)+\frac{h^2}{3}f^{(3)}(\xi_2)$$

Note that $x_1=x_0+h$, and $x_2=x_0+2h$, these formulas can be rewritten as

$$f'(x_0)=\frac{1}{h}\left(-\frac{3}{2}f(x_0)+2f(x_0+h)-\frac{1}{2}f(x_0+2h)\right)+\frac{h^2}{3}f^{(3)}(\xi_0)$$

$$f'(x_0+h)=\frac{1}{h}\left(-\frac{1}{2}f(x_0)+\frac{1}{2}f(x_0+2h)\right)-\frac{h^2}{6}f^{(3)}(\xi_1)$$

$$f'(x_0+2h)=\frac{1}{h}\left(\frac{1}{2}f(x_0)-2f(x_0+h)+\frac{3}{2}f(x_0+2h)\right)+\frac{h^2}{3}f^{(3)}(\xi_2)$$

We substitute x_0 for x_0+h in the middle equation to change this formula to an approximation for $f'(x_0)$. A similar change, we substitute x_0 for x_0+2h, in the last equation. We obtain three formulas for approximating $f'(x_0)$:

$$f'(x_0)=\frac{1}{2h}(-3f(x_0)+4f(x_0+h)-f(x_0+2h))+\frac{h^2}{3}f^{(3)}(\xi_0)$$

$$f'(x_0)=\frac{1}{2h}(-f(x_0-h)+f(x_0+h))-\frac{h^2}{6}f^{(3)}(\xi_1)$$

$$f'(x_0)=\frac{1}{2h}(f(x_0-2h)-4f(x_0-h)+3f(x_0))+\frac{h^2}{3}f^{(3)}(\xi_2)$$

Note that since the last of these equations can be obtained from the first by simply replacing h with $-h$, there are actually only two formulas:

$$f'(x_0)=\frac{1}{2h}(-3f(x_0)+4f(x_0+h)-f(x_0+2h))+\frac{h^2}{3}f^{(3)}(\xi_0) \tag{4.8.7}$$

$$f'(x_0)=\frac{1}{2h}(f(x_0+h)-f(x_0-h))-\frac{h^2}{6}f^{(3)}(\xi_1) \tag{4.8.8}$$

where ξ_0 lies between x_0 and x_0+2h, ξ_1 lies between x_0-h and x_0+h.

The methods presented in Equations (4.8.7) and (4.8.8) are called **three-point formulas.** Similarly, there are **five-point formulas** that involve evaluating the function at two more points. One is

$$f'(x_0)=\frac{1}{12h}(f(x_0-2h)-8f(x_0-h)+8f(x_0+h)-f(x_0+2h))+\frac{h^4}{30}f^{(5)}(\xi) \tag{4.8.9}$$

where ξ lies between x_0-2h and x_0+2h. The other five-point formula is useful for end-point approximations, particularly with regard to the clamped cubic spline interpolation. It is

$$f'(x_0)=\frac{1}{12h}(-25f(x_0)+48f(x_0+h)-36f(x_0+2h)+16f(x_0+3h)-3f(x_0+4h))+\frac{h^4}{5}f^{(5)}(\xi)$$

where ξ lies between x_0 and x_0+4h.

Example 4.8.3 Given values for $f(x)=xe^x$ in Table 4-11.

Table 4-11 Values for $f(x)=xe^x$

x	$f(x)$
1.8	10.889 365
1.9	12.703 199
2.0	14.778 112
2.1	17.148 957
2.2	19.855 030

Compute $f'(2.0)$ using the various three-point and five-point formulas.

Solution Three-point formula

Using (4.8.7) with

$h=0.1$

$$f'(2.0)\approx\frac{1}{0.2}(-3f(2.0)+4f(2.1)-f(2.2))=22.032\ 310$$

$$h=-0.1\quad f'(2.0)\approx\frac{1}{-0.2}(-3f(2.0)+4f(1.9)-f(1.8))=22.054\ 525$$

Using (4.8.8) with

$$h=0.1\quad f'(2.0)\approx\frac{1}{0.2}(f(2.1)-f(1.9))=22.228\ 790$$

$$h=0.2\quad f'(2.0)\approx\frac{1}{0.4}(f(2.2)-f(1.8))=22.414\ 163$$

The errors in the formulas are approximately

$$1.35\times10^{-1},\quad 1.13\times10^{-1},\quad -6.16\times10^{-2},\quad \text{and}\quad -2.47\times10^{-1}$$

respectively.

Five-point formula

Using (4.8.9) with $h=0.1$

$$f'(2.0)\approx\frac{1}{1.2}(f(1.8)-8f(1.9)+8f(2.1)-f(2.2))=22.166\ 996$$

The error is approximately 1.69×10^{-4}.

Expand a function f in a third Taylor polynomial about a point x_0 and evaluate at x_0+h and x_0-h. Then

$$f(x_0+h)=f(x_0)+hf'(x_0)+\frac{h^2}{2}f''(x_0)+\frac{h^3}{6}f'''(x_0)+\frac{h^4}{24}f^{(4)}(\xi_1)$$

and

$$f(x_0-h)=f(x_0)-hf'(x_0)+\frac{h^2}{2}f''(x_0)-\frac{h^3}{6}f'''(x_0)+\frac{h^4}{24}f^{(4)}(\xi_2)$$

where $x_0-h<\xi_2<x_0<\xi_1<x_0+h$.

Add these equations, we have

$$f(x_0+h)+f(x_0-h)=2f(x_0)+h^2f''(x_0)+\frac{h^4}{24}\left(f^{(4)}(\xi_1)+f^{(4)}(\xi_2)\right)$$

From this equation $f''(x_0)$ gives

$$f''(x_0)=\frac{1}{h^2}\left(f(x_0-h)-2f(x_0)+f(x_0+h)\right)-\frac{h^2}{24}\left(f^{(4)}(\xi_1)+f^{(4)}(\xi_2)\right) \tag{4.8.10}$$

Assume $f^{(4)}(x)$ is continuous on (x_0-h,x_0+h). Since $\frac{1}{2}\left(f^{(4)}(\xi_1)+f^{(4)}(\xi_2)\right)$ is between $f^{(4)}(\xi_1)$ and $f^{(4)}(\xi_2)$, the intermediate value theorem implies that a number ξ exists between ξ_1 and ξ_2, and therefore in (x_0-h,x_0+h), with

$$f^{(4)}(\xi)=\frac{1}{2}\left(f^{(4)}(\xi_1)+f^{(4)}(\xi_2)\right)$$

We rewrite Equation (4.8.10) as

$$f''(x_0)=\frac{1}{h^2}\left(f(x_0-h)-2f(x_0)+f(x_0+h)\right)-\frac{h^2}{12}f^{(4)}(\xi) \tag{4.8.11}$$

where $x_0-h<\xi<x_0+h$.

Example 4.8.4 For $f(x)=xe^x$ in Table 4-11, compute $f''(2.0)$.

Solution Using (4.8.11) with

$$h{=}0.1 \quad f''(2.0)\approx\frac{1}{0.01}\left(f(1.9)-2f(2.0)+f(2.1)\right)=29.593\,200$$

$$h{=}0.2 \quad f''(2.0)\approx\frac{1}{0.04}\left(f(1.8)-2f(2.0)+f(2.2)\right)=29.704\,275$$

The errors are approximately -3.70×10^{-2} and -1.48×10^{-1}, respectively.

4.8.3 Richardson's Extrapolation

We shall now indicate how to procedure known as **Richardson extrapolation** can be to more accuracy out of some numerical formulas. We assume that $f(x)$ is represented by its Taylor series

$$f(x+h)=\sum_{k=0}^{\infty}\frac{1}{k!}h^k f^{(k)}(x) \tag{4.8.12}$$

$$f(x-h)=\sum_{k=0}^{\infty}\frac{1}{k!}(-1)^k h^k f^{(k)}(x) \tag{4.8.13}$$

If the second equation is subtracted from the first, all terms with an even value of k will cancel, leaving

$$f(x+h)-f(x-h)=2hf'(x)+\frac{2}{3!}h^3f'''(x)+\frac{2}{5!}h^5f^{(5)}(x)+\cdots$$

A rearrangement yields

$$f'(x)=\frac{1}{2h}\big(f(x+h)-f(x-h)\big)-\left(\frac{1}{3!}h^2f^{(3)}(x)+\frac{1}{5!}h^4f^{(5)}(x)+\frac{1}{7!}h^6f^{(7)}(x)+\cdots\right)$$

This equation has the form

$$L=\varphi(h)+a_2h^2+a_4h^4+a_6h^6+\cdots \tag{4.8.14}$$

where L stands for $f'(x)$ and $\varphi(h)$ stands for the numerical differentiation Formula (4.8.8), that is

$$\varphi(h)=\frac{1}{2h}\big(f(x+h)-f(x-h)\big)$$

The numerical procedure is designed to estimate L. The function φ is such that $\varphi(h)$ can be evaluated for h>0, but it cannot be evaluated for h=0. Thus, we can only let h approach 0 in our attempt to determine L. For each h>0, the error is given by the series of terms $a_2h^2+a_4h^4+\cdots$. Assuming $a_2\neq 0$, we see that the first term, a_2h^2, is greater than the others when h is sufficiently small. We shall therefore look for a way of eliminating this dominant term, a_2h^2. Our analysis depends only on Equation (4.8.13) and will be applicable to other numerical processes.

Write out Equation (4.8.14) with h replaced by h/2.

$$L=\varphi(h/2)+a_2h^2/4+a_4h^4/16+a_6h^6/64+\cdots \tag{4.8.15}$$

The leading term in the error series, a_2h^2, can be eliminated between Equations (4.8.14) and (4.8.15) by multiplying the latter by 4 and subtracting the former from it.

Here is the work:

$$L=\varphi(h)+a_2h^2+a_4h^4+a_6h^6+\cdots$$

$$4L=4\varphi(h/2)+a_2h^2+a_4h^4/4+a_6h^6/16+\cdots$$

$$3L=4\varphi(h/2)-\varphi(h)-3a_4h^4/4-15a_6h^6/16+\cdots$$

Thus, we have

$$L=\frac{4}{3}\varphi(h/2)-\frac{1}{3}\varphi(h)-a_4h^4/4-5a_6h^6/16+\cdots \tag{4.8.16}$$

Equation (4.8.16) embodies the first step in Richardson extrapolation. It shows that a simple combination of $\varphi(h)$ and $\varphi(h/2)$ furnishes an estimate of L with accuracy $o(h^4)$.

Put

$$\psi = \frac{4}{3}\varphi(h/2) - \frac{1}{3}\varphi(h)$$

in Equation (4.8.16). Then

$$L = \psi(h) + b_4 h^4 + b_6 h^6 + \cdots$$
$$L = \psi(h/2) + b_4 h^4/16 + b_6 h^6/64 + \cdots$$

Now

$$L = \psi(h) + b_4 h^4 + b_6 h^6 + \cdots$$
$$16L = 16\psi(h/2) + b_4 h^4 + b_6 h^6/4 + \cdots$$
$$15L = 16\psi(h/2) - \psi(h) - 3b_6 h^6/4 - \cdots$$

Therefore, we have

$$L = \frac{16}{15}\psi(h/2) - \frac{1}{15}\psi(h) - b_6 h^6/20 - \cdots \quad (4.8.17)$$

Again, we can repeat this process by putting

$$\theta(h) = \frac{16}{15}\psi(h/2) - \frac{1}{15}\psi(h) \quad (4.8.18)$$

in Equation (4.8.18), so that

$$L = \theta(h) + c_6 h^6 + c_8 h^8 + \cdots$$

In a manner similar to that used above, it follows that

$$L = \frac{64}{63}\theta(h/2) - \frac{1}{63}\theta(h) - 3c_8 h^8/252 - \cdots$$

As a matter of fact, any number of steps can be carried out to obtain formulas of increasing precision. The complete algorithm, allowing for m steps of Richardson extrapolation algorithm, is given next:

(1) Select a convenient h (say h=1) and compute the m+1 numbers

$$D(n,0) = \varphi(h/2^n) \qquad (0 \leqslant n \leqslant m)$$

(2) Compute additional quantities by the formula

$$D(n,k) = \frac{4^k D(n,k-1) - D(n-1,k-1)}{4^k - 1} \quad (4.8.19)$$

where $k = 1,2,\cdots,m$ and $n = k, k+1,\cdots,m$.

The formulas given for $D(n,0)$ and $D(n,k)$ allow us to construct a triangular array:

$$\begin{array}{lllll} D(0,0) & & & & \\ D(1,0) & D(1,1) & & & \\ D(2,0) & D(2,1) & D(2,2) & & \\ \vdots & \vdots & \vdots & \ddots & \\ D(m,0) & D(m,1) & D(m,2) & \cdots & D(m,m) \end{array}$$

Example 4.8.5 Compute $f'(0.5)$ for $f(x) = x^2 e^{-x}$ using the complete Richardson extrapolation algorithm.

Solution

$$\varphi(h)=\frac{1}{2h}\left(\left(\frac{1}{2}+h\right)^2 \mathrm{e}^{-\left(\frac{1}{2}+h\right)}-\left(\frac{1}{2}-h\right)^2 \mathrm{e}^{-\left(\frac{1}{2}-h\right)}\right)$$

When $h=0.1,\ 0.05,\ 0.025$, from Richardson extrapolation algorithm, we obtain

$D(0,0)=0.451\,604\,908\,1$

$D(1,0)=0.454\,076\,169\,3$ $\quad D(1,1)=0.454\,899\,923\,1$

$D(2,0)=0.454\,692\,628\,8$ $\quad D(2,1)=0.454\,898\,115\,2$ $\quad D(2,2)=0.454\,897\,994$

4.9 Computer Experiments

4.9.1 Functions Needed in the Experiments by Mathematica

1．Input——键盘输入函数

格式 1：`变量=Input["提示信息"]`

功能：接受由键盘输入的信息（可以是常量、变量、函数或表达式）并赋值给变量.

格式 2：`f[x_]=Input["提示信息"]`

功能：自定义函数 $f(x)$，函数关系式由键盘输入.

2．Print——显示输出函数

格式：`Print[expr1,expr2,…,exprn]`

功能：依次显示表达式 expr1, expr2 , ⋯ , expr$n$ 的值.

说明：

① 若某个表达式为字符串，则该字符串两端要加定界符" "（引号）；

② 若在两个输出项之间需要有空出间隔，可以插入空格字符串输出项，如

`Print["系数 a=",a,"    ","系数 b=",b ]`

3．Abs——取绝对值函数

格式：`Abs [expr]`

4.9.2 Experiments by Mathematica

1．Newton-Cotes 求积公式

（1）算法.

① 输入被积函数 $f(x)$, 积分下限 a, 上限 b;

② 选择 Cotes 系数构造求积公式；

③ 用求积公式计算积分值.

（2）程序清单.

```
(*Newton-Cotes formula*)
Clear[a,b,x,n,s];
a=Input["a="];
b=Input["b="];
```

```
f[x_]=Input["f(x)="]
n=Input["n="]; (*输入节点个数*)
cotesc={{1/2,1/2},
      {1/6,4/6,1/6},
      {1/8,3/8,3/8,1/8},
      {7/90,16/45,2/15,16/45,7/90},
      {19/288,25/96,25/144,25/144,25/96,19/288},
      {41/840,9/35,9/280,34/105,9/280,9/35,41/840}};
h=(b-a)/(n-1);
x=Table[a+k*h,{k,0,n-1}];
s=(b-a)*Sum[cotesc[[n-1,k]]*f[x[[k]]],{k,1,n}];
Print["积分区间: [",a,",",b,"]"];
Print["节点个数=",n]
Print["Cotes 积分值=",s]
Print["定积分近似值=",N[s,8]]
```

（3）变量说明.

f[x_]：保存被积函数（由键盘输入）；

a：保存积分下限；

b：保存积分上限；

n：保存节点个数；

cotesc：保存 Cotes 系数；

h：保存节点步长；

x：保存节点 x_i；

s：保存积分值.

（4）计算实例.

Example 4.9.1 Solve approximate value to the integral $\int_2^5 xe^{-x/3}dx$ by using Newton-Cotes formula with four points.

操作步骤如下：

① 将光标定位在要执行的 Cell 中，按小键盘的【Enter】键；

② 在弹出的对话框中按提示分别输入

```
2, 5, x*Exp[-x/3], 4
```

③ 每次输入后单击 OK 命令按钮，得图 4-7 所示的输出结果。

```
Out[115]= e^(-x/3) x
积分区间: [2,5]
节点个数=4
Cotes积分值=3 (5/(8 e^(5/3)) + 3/(2 e^(4/3)) + 9/(8 e) + 1/(4 e^(2/3)))
定积分近似值=3.1669848
```

图 4-7 Newton-Cotes 求积公式计算结果

2. 复化梯形公式

（1）算法.

① 输入被积函数 $f(x)$，积分下限 a，上限 b，精度 eps；

② 置 $n \leftarrow 1$，计算 T_n；

③ 用递推公式计算 T_{2n}；

④ 判断误差：$|T_{2n} - T_n| < \varepsilon$?

若成立，输出积分近似值，停止；

否则，置 $T_n \leftarrow T_{2n}$， $n \leftarrow 2_n$，转到③.

（2）程序清单.

```
(*Composite Trapezoidal Rule*)
a=Input["a="];
b=Input["b="];
f[x_]=Input["f(x)="]
eps=Input["eps="];
Print["积分区间: [", a, ",", b, "]"];
n=1;
h=b-a;
t1=(f[a]+f[b])*h/2;
h=h/2;
t2=t1/2+h*f[a+h];
er=(t2-t1)//N;
While[Abs[er]>eps,
    Print["n=",2n, "  ", "定积分值为",N[t2,10],"  ","误差=",er];
    h=h/2;
    t1=t2;
    n=2n;
    t2=t1/2+h*Sum[f[a+i*h],{i,1,2n,2}];
    er=(t2-t1)//N
];
Print["n=",2n,"  ","定积分值为",N[t2,10],"  ","误差=",er]
```

（3）变量说明.

f[x_]：保存被积函数（由键盘输入）；

a：保存积分下限；

b：保存积分上限；

eps：保存积分精度要求（键盘输入）；

er：保存积分误差值；

s：保存积分近似值；

n：保存积分区间等分数.

（4）计算实例.

Example 4.9.2 Solve approximate value to the integral $\int_0^{\frac{\pi}{2}} e^x \sin x dx$ by using composite trapezoidal rule so that its error is less than 10^{-6}.

操作步骤如下：

① 将光标定位在要执行的Cell中，按小键盘的【Enter】键；

② 在弹出的对话框中按提示分别输入

```
0,Pi/2,Exp[x]*Sin[x],10^(-6)
```

③ 每次输入后单击 OK 命令按钮，得图 4-8 所示的输出结果.

Out[52]= e^x Sin[x]

积分区间: [0, $\frac{\pi}{2}$]

n	定积分值	误差
n=2	定积分值为3.107130865	误差=-0.671009
n=4	定积分值为2.954589285	误差=-0.152542
n=8	定积分值为2.917504808	误差=-0.0370845
n=16	定积分值为2.908300728	误差=-0.00920408
n=32	定积分值为2.906003919	误差=-0.00229681
n=64	定积分值为2.905429980	误差=-0.000573939
n=128	定积分值为2.905286512	误差=-0.000143468
n=256	定积分值为2.905250646	误差=-0.000035866
n=512	定积分值为2.905241679	误差=-8.96644×10^{-6}
n=1024	定积分值为2.905239438	误差=-2.24161×10^{-6}
n=2048	定积分值为2.905238877	误差=-5.60402×10^{-7}

图 4-8 复化梯形公式计算结果

3. 复化辛普森求积公式

（1）算法.

① 输入被积函数 $f(x)$，积分下限 a，上限 b，积分精度；

② 根据区间等分数 n，计算 S_n；

③ 输出积分值.

（2）程序清单.

```
(*  Composite  Simpson's Rule*)
Clear[a,b,x,n,f,s];
a=Input["a="];
b=Input["b="];
f[x_]=Input["f(x)="]
n=Input["n="];
Print["积分区间: [", a, ",", b, "]"];
    h=(b-a)/n;
    a1=a+h/2;
    s=h/6*(f[a]+f[b]+2Sum[f[a+k*h],{k,1,n-1}]+4Sum[f[a1+k*h],{k,0,n-1}]);
Print["n=",n,"   ","定积分值为",N[s,10]];
```

（3）变量说明.

f[x_]：保存被积函数（由键盘输入）；

a：保存积分下限；

b：保存积分上限；

n：保存区间等分数；

h：保存步长；

s：保存积分值.

（4）计算实例.

Example 4.9.3 Solve approximate value to the integral $\int_0^{\frac{\pi}{2}} e^x \sin x dx$ by using composite Simpson's rule, taking n=16.

① 将光标定位在要执行的 Cell 中，按小键盘的【Enter】键；

② 在弹出的对话框中按提示分别输入

```
0,  Pi/2,  Exp[x]*Sin[x],  16
```

③ 每次输入后单击 OK 命令按钮，得图 4-9 所示的输出结果。

Out[17]= e^x Sin[x]

积分区间: [0, $\frac{\pi}{2}$]

n=16   定积分值为2.905238316

图 4-9 复化辛普森公式计算结果

4. Romberg 求积方法

（1）算法.

① 输入被积函数 $f(x)$, 积分下

限 a, 上限 b, 精度 eps;

② 依次计算 T_1，T_2，T_4，T_8;

③ 依次计算:

t_1，t_2，t_3，t_4;

s_1，s_2，s_3;

c_1，c_2;

R_2;

做变换:

$$n \leftarrow 1, R_1 \leftarrow c_2, t_1 \leftarrow t_4, s_1 \leftarrow s_3, c_1 \leftarrow c_2$$

④ 判断: $|R_2 - R_1| < \varepsilon$?

若成立，输出积分值 R_2，停止;

否则，依次计算并做变换:

$1 \leftarrow R_2$;

$t_2 \leftarrow t_{16n}$;

$s_2=(4t_2-t_1)/3, t_1 \leftarrow t_2$;

$c_2=(16s_2-s_1)/15, s_1 \leftarrow s_2$;

$R_2=(64c_2-c_1)/63, c_1 \leftarrow c_2$

⑤ $n \leftarrow 2n$ 转到④.

（2）程序清单.

```
(*Romberg Algorithm*)
Clear[a,b,x,m,f];
a=Input["a="];b=Input["b="];
f[x_]=Input["f(x)="]
eps=Input["eps="];
m=1;h=b-a;t=Table[0,{4}];
t[[1]]=(f[a]+f[b])*h/2;
(*求 t2,t3,t4*)
Do[m=2m;
  h=h/2;
  t[[k+1]]=t[[k]]/2+h*Sum[f[a+i*h],{i,1,m,2}], {k,1,3}
]
s1=(4t[[2]]-t[[1]])/3;s2=(4t[[3]]-t[[2]])/3;s3=(4t[[4]]-t[[3]])/3;
c1=(16s2-s1)/15;c2=(16s3-s2)/15;
r1=c2;r2=(64c2-c1)/63;er=r2-r1;
t1=t[[4]];s1=s3;c1=c2;
k=1;
Print["积分区间: [",a,",",b,"]"];
Print["积分过程:"];
While[Abs[er]>eps,
   r1=r2;
   Print["r(",k,")=",N[r2,20],"     ","er=", N[er,10]];
```

```
    h=h/2;m=2m;
    t2=t1/2+h*Sum[f[a+i*h],{i,1,m,2}];
    s2=(4t2-t1)/3;c2=(16s2-s1)/15;r2=(64c2-c1)/63;
    t1=t2;s1=s2;er=r2-r1;k=k+1;c1=c2
];
Print["r(",k,")=",N[r2,20],"     ","er=",N[er,10]];
```

（3）变量说明.

f[x_]：保存被积函数（由键盘输入）；

a：保存积分下限；

b：保存积分上限；

eps：保存积分精度要求（键盘输入）；

er：保存积分误差值；

r(k)：保存积分近似值.

（4）计算实例.

Example 4.9.4 Solve approximate value to the integral $\int_0^1 \frac{4}{1+x^4}\mathrm{d}x$ by using Romberg algorithm so that its error is less than 10^{-20}.

① 将光标定位在要执行的 Cell 中，按小键盘的【Enter】键；

② 在弹出的对话框中按提示分别输入

```
0,1,4/(1+x^4),10^(-20)
```

③ 每次输入后单击 OK 命令按钮，得图4-10所示的输出结果。

```
Out[117]= 4/(1+x^4)
积分区间：[0,1]
积分过程：
r(1)=3.4679235940878319059      er=0.00003477187316
r(2)=3.4678920185484964638      er=-0.00003157553934
r(3)=3.4678919491210037334      er=-6.942749273×10^-8
r(4)=3.4678919493588465732      er=2.378428398×10^-10
r(5)=3.4678919493596412424      er=7.946691938×10^-13
r(6)=3.4678919493596441391      er=2.896699838×10^-15
r(7)=3.4678919493596441503      er=1.112229235×10^-17
r(8)=3.4678919493596441503      er=4.326046186×10^-20
r(9)=3.4678919493596441503      er=1.688051390×10^-22
```

图4-10 Romberg 算法计算结果

4.9.3 Experiments by Matlab

1. 复梯形公式

（1）函数语句.

```
function [n,t2]=compT(a,b,eps)
```

（2）参数说明.

在提示语句下输入相关参数。

fun：被积函数；

a：输入参数，积分下限；

b：输入参数，积分上限；

n：区间等分数，为偶数；

eps：需要达到的计算精度；

n：自然数，输出参数，达到精度所需分半次数；

t2：实变量，输出参数，积分近似值.

（3）compT.m 程序.

```
function [n,t2]=compT(a,b,eps)
% 利用复化梯形递推公式求 fun 在[a,b]的定积分;
h=b-a;                              %节点步长
t1=(fun(a)+fun(b))*h/2;             %梯形值 T0
h=h/2;
t2=t1/2+h*fun(a+h);                 %经 1 次分半复合梯形值
err=t2-t1;                          %误差
n=1;
while abs(err)>eps
   sum=0;
   h=h/2;
   t1=t2;
   n=n+1;
   k=1;
   while k<2^n
     sum=sum+fun(a+k*h);
     k=k+2;
   end
   t2=t1/2+h*sum
   err=t2-t1;
end
```

fun.m 程序

```
function y=fun(x)
y=被积函数表达式;
```

（4）计算实例.

Example 4.9.5 Solve approximate value to the integral $\int_0^{4.3} e^{-x^2} dx$ by using Romberg algorithm so that its error is less than 10^{-6}.

建立 m 文件:

```
function y=fun(x)
y=exp(-x^2);
```

在命令窗口输入:

```
>> a=0; b=4.3; eps=10^-6;
>> [n,t2]=compT(a,b,eps)
```

2. 复化辛普森公式

（1）函数语句.

```
function y=comsimpson(fun,a,b,n)
```

（2）参数说明.

fun：被积函数;

a：输入参数，积分下限;

b：输入参数，积分上限;

n：区间等分数，为偶数.

（3）comsimpson.m 程序.

```
function y=comsimpson(fun,a,b,n)
z1=fun(a)+fun(b);m=n/2;
h=(b-a)/(2*m);x=a;
z2=0;z3=0;x2=0;x3=0;
for k=2:2:2*(m-1)
    x2=x+k*h;z2=z2+2*fun(x2);
end
for k=1:2:2*(m-1)
    x3=x+k*h;z3=z3+4* fun(x3);
end
y=(z1+z2+z3)*h/3;
```

（4）计算实例.

Example 4.9.6 Compute $\frac{1}{\sqrt{2\pi}}\int_0^1 e^{\frac{x^2}{2}}dx$ by taking n=20 000 .

建立 m 文件 fun.m：

```
function y=fun(x)
y=exp(x^2/2)/sqrt(2*pi)
```

在命令窗口输入：

```
>> a=0;b=1;n=20000;
>> y=comsimpson(fun,a,b,n)
```

3. Rormberg 求积方法

（1）函数语句.

```
[J,err]=romberg
```

（2）参数说明.

在提示语句下输入相关参数。

fun：符号函数，输入参数，被积函数；

a：实向量，输入参数，积分区间；

delta：实变量，输入参数，需要达到的计算精度；

m：实变量，输入参数，最高修正次数；

J：实变量，输出参数，积分近似值；

Err：实变量，输出参数，误差.

（3）romberg.m 程序.

```
function [J,err]=romberg
fun=input('请输入被积函数(以字符串的形式，用引号''括起来，自变量符号用 x)：');
a=input('请输入积分区间(以[a,b]形式)：');
delta=input('请输入误差限：');
m=input('请输入修正次数：');
if length(delta)<1
    delta=5e-6;
end
```

```
if length(m)<1
   m=100;
end
h1=a(2)-a(1);
x1=a(1):h1:a(2);
y1=subs(fun,'x',x1);
I(1,1)=sum(y1)*h1/2;
fprintf('\nromberg算法得到的数据如下: ')
fprintf('\n%6d %14.10f ',0,I(1,1));
err=1;
for i=1:m
   if err>delta
     h2=h1/2;
     x2=x1(1:2^(i-1))+h2;
     y2=subs(fun,'x',x2);
     I(i+1,1)=I(i,1)/2+sum(y2)*h2;
     fprintf('\n%6d %14.10f',i,I(i+1,1));
     for j=2:i+1
       I(i+1,j)=(4^(j-1)*I(i+1,j-1)-I(i,j-1))/(4^(j-1)-1);
       fprintf('%14.10f',I(i+1,j));
     end
     err=abs(I(i+1,i+1)-I(i,i));
     h1=h2;
     x1(1:2:2^(i)+1)=x1;
     x1(2:2:2^(i))=x2;
   else
     J=I(i,i);
     return
   end
end
J=I(m+1,m+1);
fprintf('romberg算法进行到%d步还没有达到给定的误差要求\n',m);
```

（4）计算实例.

Example 4.9.7 Solve approximate value to the integral $\int_0^{4.3} e^{-x^2} dx$ by using Romberg algorithm so that its error is less than 10^{-6}.

在命令窗口输入：

```
[J,err]=romberg
请输入被积函数(以字符串的形式，用引号'括起来，自变量符号用x): 'exp(-x^2)'
请输入积分区间(以[a,b]形式): [0,4.3]
请输入误差限: 10^(-6)
请输入修正次数:
```

Romberg 算法得到的数据如下：

```
0   2.1500000201
1   1.0961306289  0.7448408319
2   0.8865732480  0.8167207876  0.8215127847
```

```
   3   0.8862269229  0.8861114812  0.8907375275  0.8918363329
   4   0.8862269239  0.8862269243  0.8862346205  0.8861631458  0.8861408980
   5   0.8862269243  0.8862269244  0.8862269244  0.8862268022  0.8862270519
       0.8862271361
   6   0.8862269244  0.8862269244  0.8862269244  0.8862269244  0.8862269249
0.8862269247  0.8862269247
   J =
       0.8862
   err =
       2.1138e-007
```

Exercises 4

Questions

1. Find the parameters in the following quadrature formulas, such that the quadrature formulas have the highest possible degrees of precision.

(1) $\int_{-h}^{h} f(x)\mathrm{d}x \approx a_{-1}f(-h)+a_0 f(0)+a_1 f(h)$;

(2) $\int_{-1}^{1} f(x)\mathrm{d}x \approx (f(-1)+2f(x_1)+3f(x_2))/3$;

(3) $\int_{0}^{h} f(x)\mathrm{d}x \approx h(f(0)+f(h))/2+ah^2(f'(0)-f'(h))$;

(4) $\int_{-1}^{1} f(x)\mathrm{d}x \approx af(-1)+bf(1)+cf'(-1)+df'(1)$.

2. Given points $x_0=\frac{1}{4}, x_1=\frac{3}{4}$, write the interpolatory quadrature formula for computing $\int_0^1 f(x)\mathrm{d}x$ and its error.

3. For $\int_0^3 f(x)\mathrm{d}x$ structure a quadrature formula whose degree of accuracy is three at least.

4. Compute the following integrals with trapezoidal rule and Simpson's rule:

(1) $\int_0^1 \frac{x}{4+x^2}\mathrm{d}x, \quad n=8$; (2) $\int_1^9 \sqrt{x}\mathrm{d}x, \quad n=4$.

5. Compute the approximate value of integral $\int_1^2 \mathrm{e}^{\frac{1}{x}}\mathrm{d}x$ with Romberg's algorithm such that $\left|T_m^{(0)}-T_{m-1}^{(0)}\right|<10^{-5}$.

6. Deduce the following three rectangular quadrature formulae.

$$\int_a^b f(x)\mathrm{d}x=(b-a)f(a)+\frac{1}{2}f'(\eta)(b-a)^2$$

$$\int_a^b f(x)\mathrm{d}x=(b-a)f(b)-\frac{1}{2}f'(\eta)(b-a)^2$$

$$\int_a^b f(x)\mathrm{d}x=(b-a)f\left(\frac{a+b}{2}\right)+\frac{f''(\eta)}{24}(b-a)^3$$

7. Fill blanks

(1) Given the interpolatory quadrature formula

$$\sum_{k=0}^{n} A_k f(x_k) \approx \int_a^b f(x)\mathrm{d}x \quad \text{then} \quad \sum_{k=0}^{n} A_k = (\quad).$$

(2) Let $f(x) = x^5 + 2x^4 + 3x^2 + 4$, and $\sum_{k=0}^{2} A_k f(x_k) \approx \int_a^b f(x)\mathrm{d}x$ be Gaussion quadrature formula, then $\int_a^b f(x)\mathrm{d}x - \sum_{k=0}^{2} A_k f(x_k) = (\quad)$.

8. Approximate the following integrals using Gaussian quadrature with n=2, and compare your results to the exact values of the integrals.

(1) $\int_0^1 x^2 \mathrm{e}^{-x}\mathrm{d}x$; (2) $\int_0^{\pi/4} x^2 \sin x\mathrm{d}x$;

(3) $\int_1^{1.6} \frac{2x}{x^2-4}\mathrm{d}x$; (4) $\int_0^{0.35} \frac{2}{x^2-4}\mathrm{e}^{-x}\mathrm{d}x$.

9. Determine each missing entry in the following tables by using the forward-difference formulas and backward-difference formulas.

(1)

x	$f(x)$	$f'(x)$
0.0	0.000 00	
0.2	0.741 40	
0.4	1.371 8	

(2)

x	$f(x)$	$f'(x)$
0.5	0.479 4	
0.6	0.564 6	
0.7	0.644 2	

10. Consider the following table of data:

x	0.2	0.4	0.6	0.8	1.0
$f(x)$	0.979 865 2	0.917 771 0	0.808 038	0.638 609 3	0.384 373 5

(1) Use all the approximate formulas given in this section to approximate $f'(0.4)$ and $f''(0.4)$.

(2) Use all the approximate formulas given in this section to approximate $f'(0.6)$ and $f''(0.6)$.

11. Use Simpson's double integral with $n = m = 4$ to approximate the following double integrals, and compare the results to the exact answers.

(1) $\int_{2.1}^{2.5}\int_{1.2}^{1.4} xy^2\mathrm{d}y\mathrm{d}x$; (2) $\int_0^{0.5}\int_0^{0.5} \mathrm{e}^{y-x}\mathrm{d}y\mathrm{d}x$.

12. Given function $f(x) = \frac{1}{(1+x)^2}$, use three-point formula to approximate $f'(1.0)$, $f'(1.1)$, , $f'(1.2)$, and estimate the errors. The functional values are given in the following table:

x	1.0	1.1	1.2
$f(x)$	0.2500	0.2268	0.2066

13. Use the following methods to calculate the integral $\int_1^3 \frac{dx}{x}$, and compare the results.

(1) Romberg Algorithm;

(2) three-point and five-point formulas.

Computer Questions

1. Program the algorithm in the text in which Richardson extrapolation is used repeatedly to estimate $f'(x)$. Test your program on the following:

(1) $\ln x$ at $x=3$;

(2) $\tan x$ at $x = \arcsin 0.8$.

2. Carry out the following exercises:

Obtain the numerical value of $\int_0^1 \sqrt{1+\sin^3 x}\,dx$.

3. Use Romberg algorithm to carry out the following definite integrals:

(1) $\int_0^1 \frac{\sin x}{x} dx$; (2) $\int_{-1}^1 \frac{\cos x - e^x}{\sin x} dx$; (3) $\int_1^\infty (xe^x)^{-1} dx$.

5 Solution of Nonlinear Equations

提 要

在实际中，许多问题都归结为求解方程 $f(x)=0$. 本章主要介绍非线性方程 $f(x)=0$ 的数值解法，包括二分法、迭代法、加速收敛法、Newton 法、弦截法与抛物线法等.

词 汇

chord	弦	slope	斜率
simple root	单根	multiple root	重根
midpoint	中点	monotonically	单调地
bisection method	二分法	fixed point	不动点
tangent method	切线法	descent method	下山法
degree of convergence	收敛阶	acceleration method	加速法
secant method	割线法	intermediate-value theorem	介值定理
interval halving	区间分半	successive approximation	逐次近似
asymptotic relation	渐近关系	local convergence	局部收敛
exponential function	指数函数	parallel chord method	平行弦法
superlinear convergence	超线性收敛	reduced Newton method	简化牛顿法

5.1 Introduction

This chapter is devoted to the problem of root-finding. The problem occurs frequently in scientific work. This process involves solving a nonlinear equation or a system of nonlinear equations-namely, finding x such that $f(x)=0$ for a given function f, or finding $x=(x_1,x_2,\cdots,x_n)^{\mathrm{T}}$ so that $F(x)=0$. A root of this equation is also called a zero of the function f.

Before we examine numerical techniques for finding roots of nonlinear equations, it is instructive to consider some examples where nonlinear equations arise.

Tunnel Diode Circuit

Practical electrical circuits (电路) typically include a combination of linear and nonlinear circuit elements. One of the most fundamental nonlinear circuit elements is the diode which is essentially a switch which allows electrical current to flow in one direction, but not the other. The **voltage-current** (电压和电流) relationship of a tunnel diode is shown in Figure 5-1. Since the curve is not a straight line through the origin, it is clear from inspection that this is a nonlinear device.

The **nonmonotonic** (非单调) relationship between voltage x and current y of a tunnel diode can be modeled mathematically by fitting a curve $g(x)$ to the measured data supplied by the manufacture using the curve fitting techniques from Chapter 2.

$$y=g(x) \tag{5.1.1}$$

Let us consider the circuit in Figure 5-2, which employs the diode D in series with a resistor with resistance R and an applied voltage source E. The current in this circuit y can be determined once the voltage drop across the diode, x, is known. Applying Kirchhoff's voltage law, and using the voltage-current relationship of the diode, the diode voltage x must satisfy the following nonlinear equation:

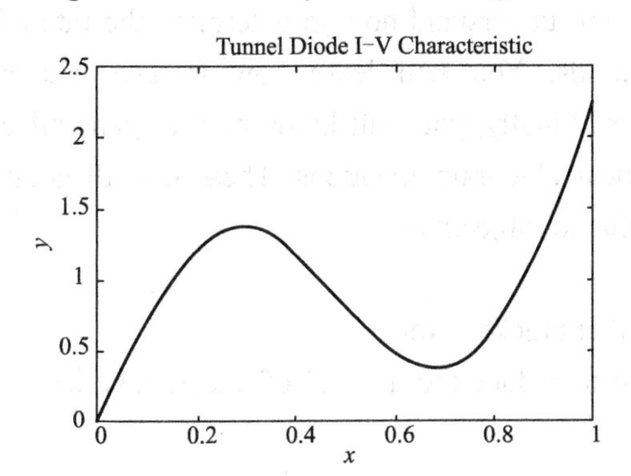

Figure 5-1 Tunnel diode voltage-current characteristic

Figure 5-2 A tunnel diode circuit

$$Rg(x)+x-E=0 \tag{5.1.2}$$

It is clear that Equation (5.1.2) is a special case of $f(x)=0$. It is of interest to note that a good approximation to the diode voltage can be obtained graphically in this case. First note that (5.1.2) can be rewritten as

$$g(x)=\frac{E-x}{R} \tag{5.1.3}$$

If the two terms, $g(x)$ and $(E-x)/R$, are plotted on the same graph as shown in Figure 5-3, then **intersections** (交点) of the two curves are roots. This graphical procedure can be used to get a good initial estimate of the solution. Iterative techniques then can be used to **refine** (加细，改善) this estimate. Observe from Figure 5-3 that when $E=1.5$ and $R=1$, there are three roots.

In the theory of **diffraction** (分解,绕射) of light, we need to find the roots of the equation

$$x - \tan x = 0$$

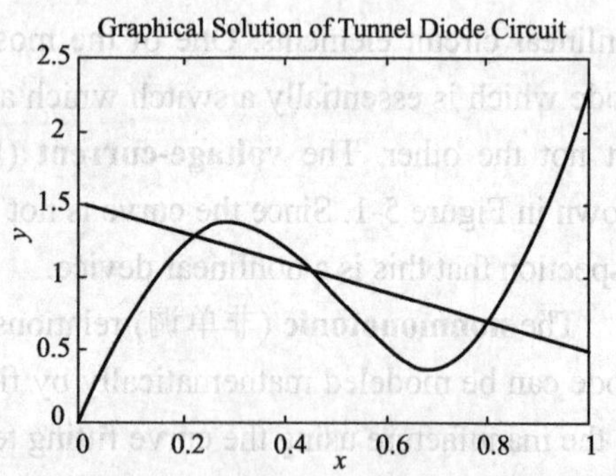

Figure 5-3 Graphical solution of tunnel diode circuit

In the calculation of **planetary orbits** (行星轨道), we need the roots of Kepler's equation

$$x - a\sin x = b$$

for various values of a and b.

Since locating the zeros of functions has been an active area of study for several hundred years, numerous methods have been developed. In this chapter, we begin with three simple methods that are quite useful: the bisection method, Newton's method, and the secant method. Then we explore the general theory of fixed-point methods.

When you finish this chapter, you will be able to find roots of nonlinear algebraic equations. You will know how to bracket a root and then narrow down the interval of uncertainty using the bisection method. You will understand how to determine the rate of convergence of different root finding methods. You will learn how to use Aitken extrapolation to improve the convergence rate. Finally, you will know how to generalize Newton's method to solve systems of nonlinear algebraic equations. These overall goals will be achieved by mastering the following chapter objectives.

Objectives

- Know how to identify a pair of points that bracket a root.
- Be able to apply the bisection method to reduce the interval of uncertainty within which a root is known to lie.
- Know how to specify the rate of convergence of different root finding methods.
- Be able to apply Aitken extrapolation to improve the convergence rate.
- Know how to find roots using the secant, Muller, and Newton's methods.
- Know how to solve nonlinear algebraic systems of equations iteratively using Newton's method.

5.2 Basic Theories

We discuss the solution of a single nonlinear equation with one unknown

$$f(x)=0$$

If $f(x)$ can be written in the following form

$$f(x) = (x - x^*)^m g(x)$$

where m is a positive integer and $g(x^*) \neq 0$, then x^* is called m **multiple root** (m 重根),or x^* is m **multiple zero** of function $f(x)$. when m=1, x^* is called simple root. If x^* is

m multiple zero of function $f(x)$, and $g(x)$ is sufficiently smooth, then

$$f(x^*)=f'(x^*)=\cdots=f^{(m-1)}(x^*)=0,\quad f^{(m)}(x^*)\neq 0$$

If f is a continuous function on the interval $[a,b]$ and if $f(a)f(b)<0$, then f must have a zero in (a,b). Since $f(a)f(b)<0$, the function f changes sign on the interval $[a,b]$ and, therefore, it has at least one zero in the interval. This is a consequence of the **Intermediate-Value Theorem** (介值定理). Usually, we can obtain the interval of the roots by **successive search method** (逐次搜索).

Example 5.2.1 Find the intervals of the roots for the following equation

$$f(x)=x^3-11.1x^2+38.8x-41.77=0$$

Solution By successive search method for $f(x)=0$ with step size $h=1$, we have Table 5-1.

Therefore the intervals containing roots are [2,3], [3,4] and [5,6]. See Figure 5-4.

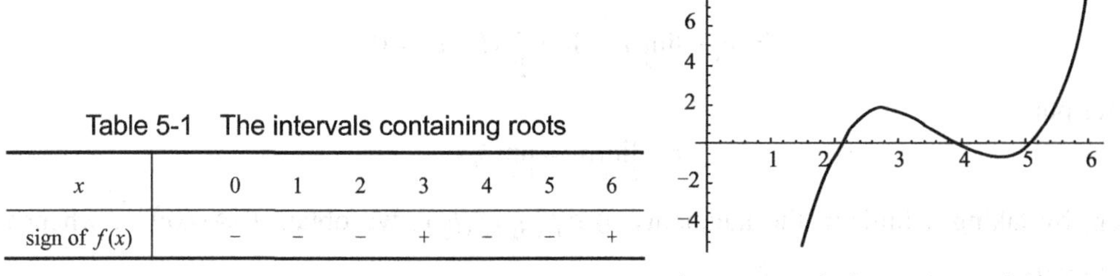

Table 5-1 The intervals containing roots

x	0	1	2	3	4	5	6
sign of $f(x)$	−	−	−	+	−	−	+

Figure 5-4 The intervals containing roots

5.3 Bisection Method（二分法）

The bisection method exploits this idea in the following way. Denote $[a,b]=[a_0,b_0]$. If $f(a)f(b)<0$, then we compute $x_0=(a_0+b_0)/2$ and test whether $f(a_0)f(x_0)<0$. If this is true, then f has a zero in $[a_0,x_0]$. So we rename $[a_0,x_0]=[a_1,b_1]$ and start again with the new interval $[a_1,b_1]$, which is half as large as the original interval. If $f(a_0)f(x_0)>0$, then $f(x_0)f(b_0)<0$, and in this case we rename $[x_0,b_0]=[a_1,b_1]$. In either case, a new interval containing a zero of f has been produced, and the process can be repeated. Figure 5-5 and Figure 5-6 show the two cases, assuming $f(a)>0>f(b)$. These figures make it clear why the bisection method finds one zero but not all the zeros in the interval $[a,b]$, of course, if $f(a_k)f(x_k)=0$, then $f(x_k)=0$ and a zero has been found. However, it is quite unlikely that $f(x_k)$ is exactly 0 in the computer because of round-off errors. Thus, the **stopping criterion** (迭代) should not be whether $f(x_k)=0$. A reasonable tolerance must be allowed. The bisection method is also known as **the method of interval halving** (区间分半法).

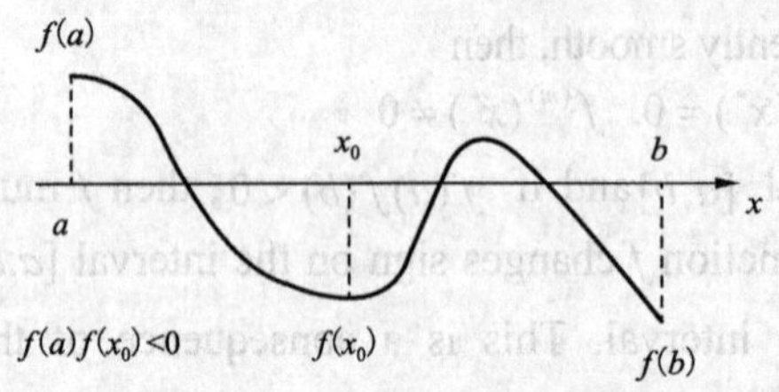

Figure 5-5 Bisection method selects left subinterval

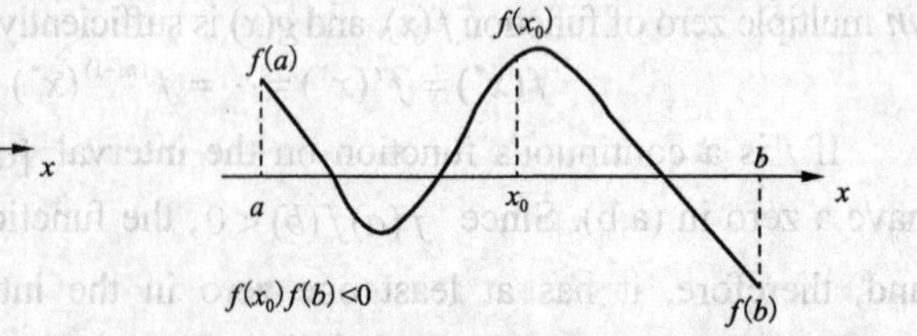

Figure 5-6 Bisection method selects right subinterval

To analyze the bisection method, let us denote the successive intervals that arise in the process by $[a_0,b_0]$, $[a_1,b_1]$, and so on. We have

$$[a_0,b_0]\supset[a_1,b_1]\supset\cdots\supset[a_k,b_k]\supset\cdots$$

They satisfy

$$b_k-a_k=\frac{1}{2}(b_{k-1}-a_{k-1})=\frac{1}{2^k}(b-a)\quad(k\geqslant 1) \tag{5.3.1}$$

Thus

$$\lim_{k\to\infty}b_k-\lim_{k\to\infty}a_k=\lim_{k\to\infty}\frac{1}{2^k}(b-a)=0$$

If we put

$$r=\lim_{k\to\infty}a_k=\lim_{k\to\infty}b_k$$

then, by taking a limit in the inequality $0\geqslant f(a_k)f(b_k)$, we obtain $0\geqslant(f(r))^2$, whence $f(r)=0$.

Suppose that, at a certain stage in the process, the interval $[a_k, b_k]$ has just been defined, if the process is now stopped, the root is certain to lie in this interval. The best estimate of the root at this stage is not a_k or b_k but the midpoint of the interval:

$$x_k=(a_k+b_k)/2$$

The error is then bounded as follows:

$$|x_k-x^*|\leqslant\frac{1}{2}(b_k-a_k)=\frac{1}{2^{k+1}}(b-a)$$

If

$$|x_k-x^*|\leqslant\frac{1}{2}(b_k-a_k)=\frac{1}{2^{k+1}}(b-a)<\varepsilon$$

then

$$k>\frac{\ln(b-a)-\ln\varepsilon}{\ln 2}-1 \tag{5.3.2}$$

where ε is the given accuracy.

Example 5.3.1 Find a real root of the equation $f(x)=x^3-x-1=0$ in [1.0, 1.5] such that the error is less than $\frac{1}{2}\times10^{-2}$.

Solution Here a=1.0, b=1.5, and $f(a)<0, f(b)>0$. Take $x_0=1.25$, since $f(x_0)<0$, therefore, $f(x_0)\cdot f(b)<0$, so we let $a_1=x_0=1.25, b_1=b=1.5$, and obtain a new interval containing a root $[a_1,b_1]$.

Repeated the process. Now we estimate the number of bisection, from the Inequality (5.3.2), we have

$$k>\frac{\ln(1.5-1.0)-\ln 0.005}{\ln 2}-1>5$$

hence, it needs 6 steps to reach the given accuracy. See Table 5-2.

When the bisection algorithm was run on a machine by using mathematica, the following output was produced and the graph is given(see Figure 5-7).

$a=1$	$b=1.5$	$f(x)=-1-x+x^3$
$n=1$	$x=1.25$	eps = 0.25
$n=2$	$x=1.375$	eps = 0.125
$n=3$	$x=1.312\,5$	eps = 0.062 5
$n=4$	$x=1.343\,75$	eps = 0.031 25
$n=5$	$x=1.328\,13$	eps = 0.015 625
$n=6$	$x=1.320\,31$	eps = 0.007 812 5
$n=7$	$x=1.324\,22$	eps = 0.003 906 25

Table 5-2 The results by using bisection method

k	a_k	b_k	x_k	Sign of $f(x_k)$
0	1.0	1.5	1.25	−
1	1.25	1.5	1.375	+
2	1.25	1.375	1.312 5	−
3	1.312 5	1.375	1.343 8	+
4	1.312 5	1.343 8	1.328 1	+
5	1.312 5	1.328 1	1.320 3	−
6	1.320 3	1.328 1	1.324 2	−

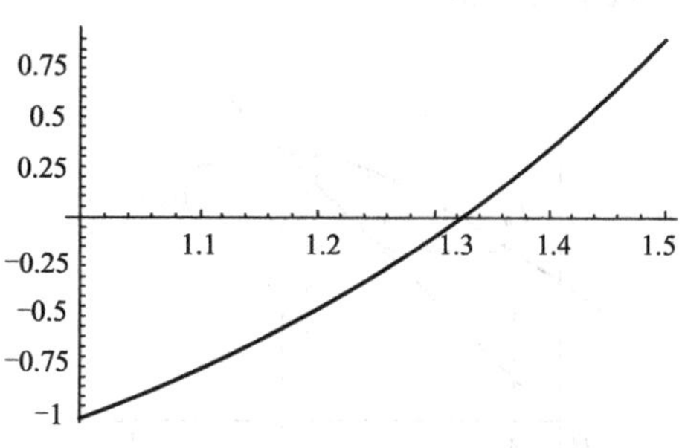

Figure 5-7 The graph of the root

Example 5.3.2 Assume that the bisection method is started with the interval [50, 63]. How many steps should be taken to compute a root with relative accuracy of one part in 10^{-12}?

Solution The stated requirement on relative accuracy means that

$$\frac{|x_k-x^*|}{|x^*|}\leqslant 10^{-12}$$

We know that $x^*\geqslant 50$, and thus it suffices to secure the inequality

$$\frac{|x_k-x^*|}{50}\leqslant 10^{-12}$$

We infer that the following condition is sufficient:

$$\frac{1}{2^{(k+1)}}\times\frac{63-50}{50}\leqslant 10^{-12}$$

Solving this for k, we conclude that $k\geqslant 37$.

5.4 Iterative Method and Its Convergence（迭代法及其收敛性）

5.4.1 Fixed Point and Iteration

Write the equation $f(x)=0$ in the equivalent form

$$x=\varphi(x) \tag{5.4.1}$$

If $f(x^*)=0$, then $x^*=\varphi(x^*)$, x^* is said to be a **fixed point** (不动点) of the function $\varphi(x)$. The zero of the function $f(x)$ is equivalent to the fixed point of the function $\varphi(x)$. Chose an initial value x_0 and structure a successive approximation

$$x_{k+1}=\varphi(x_k)\ (k=0,1,\cdots) \tag{5.4.2}$$

$\varphi(x)$ is called **iterated function.** If for arbitrary $x_0\in[a,b]$, the limit of sequence $\{x_k\}$ exists, i.e.,

$$\lim_{k\to\infty}x_k=x^*$$

then the iteration in (5.4.2) is convergent, and x^* is a fixed point of $\varphi(x)$.

This method is also called iterative method of fixed point.

See Figure 5-8.

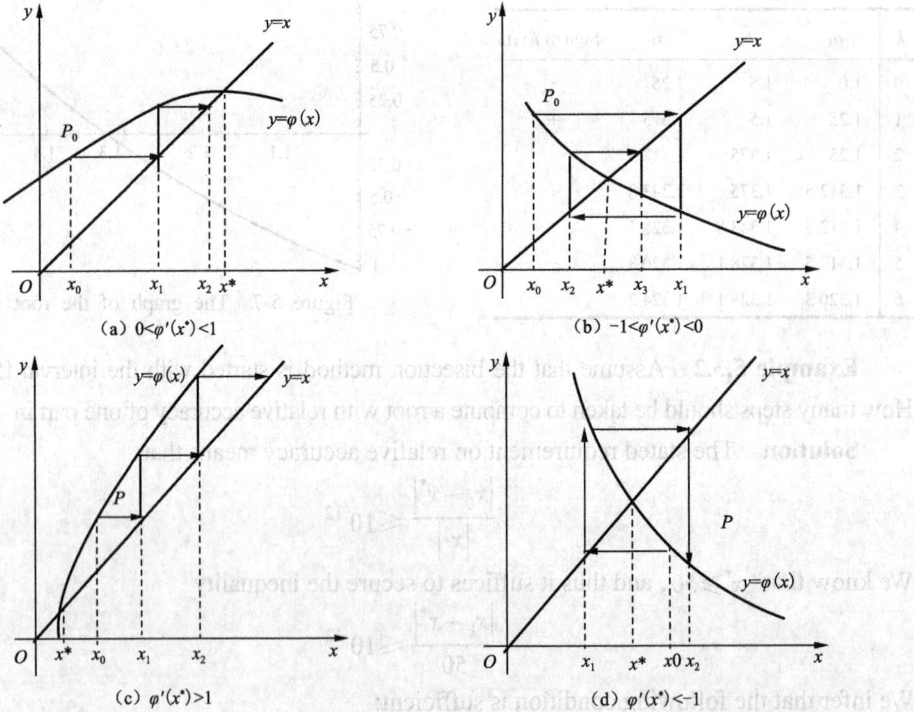

Figure 5-8 Iterative method

These figures illustrate graphically the successive approximations for functions φ with positive and negative slope, respectively. The sequence $\{x_k\}$ converges to the fixed point **monotonically** (单调) if the absolute value of the slope of function φ is less than one in a neighborhood of the fixed point. Otherwise, if the absolute value of the slope of function φ is greater than one, it can be seen that the corresponding iteration will move away from the fixed point, i.e., the sequence diverges (发散) .

5.4.2 Global Convergence (全局收敛)

Theorem 5.4.1 Let $\varphi(x)\in \mathbf{C}[a,b]$ satisfy the conditions

(1) $a\leqslant\varphi(x)\leqslant b$ holds for any $x\in[a,b]$.

(2) There exists a constant L<1, such that

$$|\varphi(x)-\varphi(y)|\leqslant L|x-y| \tag{5.4.3}$$

holds for any $x,y\in[a,b]$.

Then $\varphi(x)$ has a unique fixed point x^* on the closed interval $[a,b]$.

Proof The existence of fixed point.

If $\varphi(a)=a$ or $\varphi(b)=b$, then $\varphi(x)$ has a fixed point on $[a,b]$.

Since $a\leqslant\varphi(x)\leqslant b$, let $\varphi(a)>a$ and $\varphi(b)<b$, define the function by

$$f(x)=\varphi(x)-x$$

Obviously $f(x)\in\mathbf{C}[a,b]$, $f(a)=\varphi(a)-a>0$, and $f(b)=\varphi(b)-b<0$. From the properties of continuous functions, we have $x^*\in(a,b)$ such that $f(x^*)=0$, i.e., $x^*=\varphi(x^*)$, x^* is a fixed point of $\varphi(x)$.

The uniqueness.

Let x_1^* and x_2^* be fixed points of $\varphi(x)$. According to (5.4.3) we obtain

$$|x_1^*-x_2^*|=|\varphi(x_1^*)-\varphi(x_2^*)|\leqslant L|x_1^*-x_2^*|<|x_1^*-x_2^*|$$

Therefore $x_1^*=x_2^*$. □

Corollary Let $\varphi(x)\in\mathbf{C}[a,b]$ satisfy the conditions

(1) $a\leqslant\varphi(x)\leqslant b$ holds for any $x\in[a,b]$.

(2) $q=\sup\limits_{x\in[a,b]}|\varphi'(x)|<1$.

Then $\varphi(x)$ has a unique fixed point x^* on the closed interval $[a,b]$.

Proof By the **mean value theorem** (中值定理), for $x,y\in[a,b]$, we have that

$$\varphi(x)-\varphi(y)=\varphi'(\xi)(x-y)$$

for some intermediate point $\xi\in(x,y)$. Hence

$$|\varphi(x)-\varphi(y)|\leqslant\sup_{\xi\in(a,b)}|\varphi'(\xi)||x-y|=q|x-y|$$

The **assertion** (结论) follows from Theorem 5.4.1. □

Theorem 5.4.2 Let $\varphi(x)\in\mathbf{C}[a,b]$ satisfy the two conditions in Theorem 5.4.1, then the successive approximations

$$x_{k+1}=\varphi(x_k)\quad(k=0,1,\cdots)$$

with arbitrary $x_0\in[a,b]$ converge to the fixed point x^* of the function $\varphi(x)$, and the error is determined by

$$|x_k-x^*|\leqslant\frac{L^k}{1-L}|x_1-x_0|\tag{5.4.4}$$

Proof Let $x^*\in(a,b)$ is the unique fixed point of $\varphi(x)$, then from the condition (2) in the Theorem 5.4.1, we have

$$\begin{aligned}|x^*-x_k|&=|\varphi(x^*)-\varphi(x_{k-1})|\\&\leqslant L|x^*-x_{k-1}|\leqslant\cdots\leqslant L^k|x^*-x_0|\to0\quad(\text{when }k\to\infty)\end{aligned}$$

That is, the sequence $\{x_k\}$ converges to x^*.

$$|x_{k+1}-x_k|=|\varphi(x_k)-\varphi(x_{k-1})|\leqslant L|x_k-x_{k-1}|$$

Recursively, we obtain

$$|x_{k+1}-x_k|\leqslant L^k|x_1-x_0|$$

For arbitrary positive integer p we have

$$\begin{aligned}|x_{k+p}-x_k|&\leqslant|x_{k+p}-x_{k+p-1}|+|x_{k+p-1}-x_{k+p-2}|+\cdots+|x_{k+1}-x_k|\\&\leqslant(L^{k+p-1}+L^{k+p-2}+\cdots+L^k)|x_1-x_0|\\&=\frac{L^k(1-L^p)}{1-L}|x_1-x_0|\end{aligned}$$

Let $p\to\infty$ and notice that $\lim\limits_{p\to\infty}x_{k+p}=x^*$, we have (5.4.4). □

The following examples illustrate that in some cases we have global convergence, where the successive approximations converge for each starting point in the **domain of definition** (定义域) of the function φ.

Example 5.4.1 For computing the square root of a positive real number a by an iterative method we consider the function $\varphi:(0,\infty)\to(0,\infty)$ given by

$$\varphi(x)=\frac{1}{2}\left(x+\frac{a}{x}\right)$$

By solving the quadratic equation $\varphi(x)=x$, it can been seen that φ has the fixed point $x=\sqrt{a}$. By the **arithmetic geometric mean inequality** (算术几何平均不等式) we have that $\varphi(x)>\sqrt{a}$ for $x>0$, i.e., φ maps the open interval $(0,\infty)$ into $[\sqrt{a},\infty)$, and therefore it **maps** (映射) the closed interval $[\sqrt{a},\infty)$ into itself. From

$$\varphi'(x)=\frac{1}{2}\left(1-\frac{a}{x^2}\right)$$

it follows that

$$L=\sup_{\sqrt{a}\leqslant x<\infty}|\varphi'(x)|=\frac{1}{2}$$

Hence $\varphi:\left[\sqrt{a},\infty\right)\to\left[\sqrt{a},\infty\right)$ is a **contraction** (压缩). Therefore, the successive approximations

$$x_{k+1}=\frac{1}{2}\left(x_k+\frac{a}{x_k}\right),\quad k=0,1,\cdots$$

converge to the square root $\sqrt{a}$ for each $x_0>0$, and we have the **posteriori error estimate** (后验误差估计)

$$\left|\sqrt{a}-x_k\right|\leqslant\left|x_k-x_{k-1}\right|$$

Figure 5-9 illustrates the convergence. The numerical results again are for a=2.

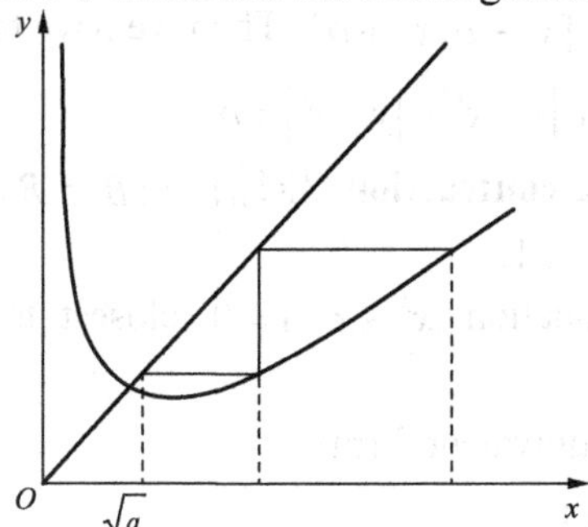

k	x_k
0	5.000 000 00
1	2.700 000 00
2	1.720 370 37
3	1.441 455 37
4	1.414 470 98
5	1.414 213 59
6	1.414 213 56

Figure 5-9 Square root by iteration

Example 5.4.2 The function $h:(0,1)\to(-\infty,\infty)$ given by $h(x)=x+\ln x$ is strictly monotonically increasing with limits $\lim\limits_{x\to 0}h(x)=-\infty$ and $\lim\limits_{x\to 1}h(x)=1$. Therefore , the function $\varphi(x)=-\ln x$ has an unique fixed point x. Since this fixed point must satisfy $0<x<1$, the derivative

$$\left|\varphi'(x)\right|=\frac{1}{x}>1$$

implies that φ is not contracting in a neighborhood of the fixed point. However, we can still design a convergent scheme because $x=-\ln x$ is equivalent to $e^{-x}=x$. We consider the inverse function

$$g(x)=e^{-x}$$

of φ, which has derivative $\left|g'(x)\right|=e^{-x}<1$ at the fixed point. Obviously, for each $0<a<1/e$ the **exponential function** (指数函数) g maps the interval $[a,1]$ into itself. Since

$$L=\sup_{a\leqslant x\leqslant 1}\left|g'(x)\right|=e^{-a}<1$$

by Corollary 5.4.1 it follows that for arbitrary $x_0>0$ the successive approximations $x_{k+1}=e^{-x_k}$ converge to the unique solution of $x=e^{-x}$.

5.4.3 Local Convergence (局部收敛)

Definition 5.4.1 Let $\varphi(x)$ have a fixed point x^*, if there exists a neighborhood of x^*: $R:\left|x-x^*\right|\leqslant\delta$ such that for arbitrary $x_0\in\mathbf{R}$, the iteration $x_{k+1}=\varphi(x_k)$ yields a sequence $\{x_k\}\subset\mathbf{R}$ and the sequence converges to x^*, then this iteration is called **local**

convergence.

The following theorem states that for a fixed point x^* with $|\varphi'(x^*)|<1$ we always can find starting points x_0 ensuring convergence of the successive approximations.

Theorem 5.4.3 Let x^* be a fixed point of a continuously differentiable function φ such that $|\varphi'(x^*)|<1$. Then the method of successive approximations $x_{k+1}=\varphi(x_k)$ is locally convergent.

Proof Since $\varphi'(x)$ is continuous and $|\varphi'(x^*)|<1$, there exist constants $0<q<1$ and $\delta>0$ such that $|\varphi'(y)|\leqslant q$ for all $y\in B=[x^*-\delta,x^*+\delta]$. Then we have that

$$|\varphi(y)-x^*|=|\varphi(y)-\varphi(x^*)|\leqslant q|y-x^*|\leqslant|y-x^*|<\delta$$

for all $y\in B$, i.e., φ maps B into itself and is a **contraction** (压缩) $\varphi:B\to B$. Now the statement of the theorem follows from Theorem 5.4.1. □

Example 5.4.3 Find the root of the equation $x^3-x-1=0$ closest to $x_0=1.5$ using iterative method.

Solution Write the given equation in the equivalent form:

$$x=\sqrt[3]{x+1}$$

Construct iteration scheme

$$x_{k+1}=\sqrt[3]{x_k+1}$$

where

$$\varphi(x)=\sqrt[3]{x+1}\quad\text{and}\quad\varphi'(x)=\frac{1}{3\cdot\sqrt[3]{(x+1)^2}}$$

In a neighborhood of $x_0=1.5$, we have $\varphi'(x)<1$, therefore, the iteration converges.

If the iterative scheme is $x_{k+1}=x_k^3-1$, then $\varphi'(x)=3x^2$, and $\varphi'(x)>1$ holds in a neighborhood of $x_0=1.5$, then the iteration diverges.

Example 5.4.4 Find the approximate value of $x^*=\sqrt{3}$ by different ways.

Solution $x^*=\sqrt{3}$ is the root of $x^2-3=0$. There are many ways to change the equation to the fixed-point form $x=\varphi(x)$ using simple algebraic manipulation.

(1) $x=x^2+x-3$.

The iteration scheme

$$x_{k+1}=x_k^2+x_k-3$$

$$\varphi(x)=x^2+x-3,\quad\varphi'(x)=2x+1,\quad\varphi'(x^*)=\varphi'(\sqrt{3})=2\sqrt{3}+1>1$$

So the iteration does not converge.

(2) $x=\dfrac{3}{x}$.

The iteration scheme

$$x_{k+1}=\frac{3}{x_k}$$

$$\varphi(x)=\frac{3}{x},\quad\varphi'(x)=-\frac{3}{x^2},\quad\varphi'(x^*)=-1.$$

Infact $x_1=\frac{3}{x_0}$, $x_2=\frac{3}{x_1}=x_0$, $x_3=\frac{3}{x_2}=x_1,\cdots$. So the iteration does not converge.

(3) $x=x-\frac{1}{4}(x^2-3)$.

The iteration scheme
$$x_{k+1}=x_k-\frac{1}{4}(x_k^2-3)$$

$$\varphi(x)=x-\frac{1}{4}(x^2-3),\quad \varphi'(x)=1-\frac{1}{2}x,\quad \varphi'(x^*)=\varphi'(\sqrt{3})=1-\frac{\sqrt{3}}{2}\approx 0.134<1$$

So the iteration converges to $x^*=\sqrt{3}$.

(4) $x=\frac{1}{2}\left(x+\frac{3}{x}\right)$.

The iteration scheme
$$x_{k+1}=\frac{1}{2}\left(x_k+\frac{3}{x_k}\right)$$

$$\varphi(x)=\frac{1}{2}\left(x+\frac{3}{x}\right),\quad \varphi'(x)=\frac{1}{2}\left(1-\frac{3}{x^2}\right),\quad \varphi'(x^*)=\varphi'(\sqrt{3})=0$$

So the iteration converges to $x^*=\sqrt{3}$.

5.4.4 Order of Convergence

Definition 5.4.2 Let the successive approximations
$$x_{k+1}=\varphi(x_k),\quad k=0,1,\cdots$$
converge to the root x^* of function $x=\varphi(x)$. If positive constants $p\geqslant 1$ and c exist with
$$\lim_{k\to\infty}\frac{|x_{k+1}-x^*|}{|x_k-x^*|^p}=c$$
then the iteration converges to x^* of order p, **with asymptotic error constant** (渐近误差常数) c.

In generally, a sequence with a higher order of convergence converges more rapidly than a sequence with a lower order. The asymptotic constant affects the speed of convergence but is not as important as the order. Three cases of order are given special attention.

(1) If $p=1$ and $0<c<1$, the iteration is **linearly convergent.**

(2) If $p>1$, the iteration is **superlinearly convergent** (超线性收敛).

(3) If $p=2$, the iteration is **quadratically convergent** .

Theorem 5.4.4 For the iteration $x_{k+1}=\varphi(x_k)$, if $\varphi^{(p)}(x)$ is continuously differential in a neighborhood of the given root x^*, and
$$\varphi'(x^*)=\varphi''(x^*)=\cdots\varphi^{(p-1)}(x^*)=0$$
$$\varphi^{(p)}(x^*)\neq 0 \tag{5.4.5}$$
Then this iteration is convergent with order p in a neighborhood of x^*.

Proof Since $\varphi'(x^*)=0$, from Theorem 5.4.3 we obtain that the iteration

$x_{k+1}=\varphi(x_k)$ is locally convergent.

From the Taylor expansion of $\varphi(x_k)$ at the root x^*, and using the condition (5.4.5), we have

$$\varphi(x_k)=\varphi(x^*)+\frac{\varphi^{(p)}(\xi)}{p!}(x_k-x^*)^p,\quad \xi \text{ is between } x_k \text{ and } x^*$$

Note that $x_{k+1}=\varphi(x_k)$, $\varphi(x^*)=x^*$, we obtain

$$x_{k+1}-x^*=\frac{\varphi^{(p)}(\xi)}{p!}(x_k-x^*)^p$$

Therefore

$$\lim_{k\to\infty}\frac{|x_{k+1}-x^*|}{|x_k-x^*|^p}=\left|\frac{\varphi^{(p)}(x^*)}{p!}\right| \tag{5.4.6}$$

It illustrates that the iteration is convergent with order p. □

Example 5.4.5 The iteration $x_{k+1}=\frac{2}{3}x_k+\frac{1}{x_k^2}$ converges to $x^*=\sqrt[3]{3}$, determine its order of convergence.

Solution Where $\varphi(x)=\frac{2}{3}x+\frac{1}{x^2}$

Since

$$\varphi'(x)=\frac{2}{3}-\frac{2}{x^3},\quad \varphi'(x^*)=\varphi'(\sqrt[3]{3})=\frac{2}{3}-\frac{2}{x^3}=0$$

$$\varphi''(x)=\frac{6}{x^4},\quad \varphi''(x^*)=\frac{6}{3\sqrt[3]{3}}=\frac{2}{\sqrt[3]{3}}\neq 0$$

Therefore this iteration is quadratically convergent.

5.5 Accelerating Convergence（加速收敛法）

We now consider a technique called **Aitken's Δ^2 method** (埃特金方法) that can be used to accelerate the convergence of a sequence that is linearly convergent, regardless of its origin or application.

Let x_{k+1} and x_{k+2} approximate to the root x^* of function $x=\varphi(x)$, then from the mean value theorem of differentiation

$$x_{k+1}-x^*=\varphi(x_k)-\varphi(x^*)=\varphi'(\xi_1)(x_k-x^*)$$

$$x_{k+2}-x^*=\varphi(x_{k+1})-\varphi(x^*)=\varphi'(\xi_2)(x_{k+1}-x^*)$$

We suppose $\varphi'(x)\approx L$ for all x, then

$$\frac{x_{k+1}-x^*}{x_{k+2}-x^*}\approx\frac{x_k-x^*}{x_{k+1}-x^*}$$

Solving for x^* gives

$$x^*\approx\frac{x_kx_{k+2}-x_{k+1}^2}{x_{k+2}-2x_{k+1}+x_k}$$

Adding and subtracting the terms x_k^2 and $2x_k x_{k+1}$ in the **numerator** (分子) and grouping terms appropriately gives

$$x^* \approx \frac{(x_k^2 - 2x_k x_{k+1} + x_k x_{k+2}) - (x_k^2 - 2x_k x_{k+1} + x_{k+1}^2)}{x_{k+2} - 2x_{k+1} + x_k} = x_k - \frac{(x_{k+1} - x_k)^2}{x_{k+2} - 2x_{k+1} + x_k}$$

Aitken's Δ^2 method is based on the assumption that the sequence $\{\overline{x}_k\}_{k=0}^{\infty}$ defined by

$$\overline{x}_{k+1} = x_k - \frac{(x_{k+1} - x_k)^2}{x_k - 2x_{k+1} + x_{k+2}} = x_k - (\Delta x_k)^2 / \Delta^2 x_k \quad (k = 0,1,\cdots) \tag{5.5.1}$$

converges more rapidly to x^* than the original sequence $\{x_k\}_{k=0}^{\infty}$ does.

Example 5.5.1 The sequence $\{x_k\}_{k=1}^{\infty}$, where $x_k = \cos(1/k)$ converges linearly to $x^* = 1$. The first few terms of the sequences $\{x_k\}_{k=1}^{\infty}$ and $\{\overline{x}_k\}_{k=1}^{\infty}$ are given in Table 5-3. It certainly appears that $\{\overline{x}_k\}_{k=1}^{\infty}$ converges more rapidly to $x^* = 1$ than does $\{x_k\}_{k=1}^{\infty}$.

We can prove that

$$\lim_{k \to \infty} \frac{\overline{x}_{k+1} - x^*}{x_k - x^*} = 0$$

This illustrates that the sequence $\{\overline{x}_k\}_{k=0}^{\infty}$ converges to x^* more rapidly than does the original sequence $\{x_k\}_{k=0}^{\infty}$.

Table 5-3 Accelerating convergence

k	x_k	$\bar{x}_k$
1	0.540 30	0.961 78
2	0.877 58	0.982 13
3	0.944 96	0.989 79
4	0.968 91	0.993 42
5	0.980 07	0.995 41
6	0.986 14	
7	0.989 81	

By applying a modification of Aitken's Δ^2 method to a linearly convergent sequence obtained from fixed-point iteration, we can accelerate the convergence to quadratic. This procedure is known as **Steffensen's method** (斯蒂芬森法) and differs slightly from application Aitken's Δ^2 method directly to the linearly convergent fixed-point iteration sequence.

Denote

$$y_k = \varphi(x_k), \quad z_k = \varphi(y_k)$$

then Steffensen's method is defined by

$$x_{k+1} = x_k - \frac{(y_k - x_k)^2}{z_k - 2y_k + x_k} \quad (k = 0,1,\cdots) \tag{5.5.2}$$

Example 5.5.2 Find the root of $x^3 - x - 1 = 0$ closest to $x_0 = 1.5$ using Steffensen's method.

Solution The iterative scheme

$$x_{k+1} = x_k^3 - 1$$

is not convergent . We now compute by using Equation (5.5.2), take

$$\varphi(x) = x^3 - 1$$

then

$$y_k = \varphi(x_k) = x_k^3 - 1, \quad z_k = \varphi(y_k) = y_k^3 - 1 = (x_k^3 - 1)^3 - 1$$

Steffensen's method with $x_0 = 1.5$ gives the values in Table 5-4.

Example 5.5.3 Solve $x^3 + 4x^2 - 10 = 0$ using Steffensen's method.

Solution The fixed-point method

$$\varphi(x) = \left(\frac{10}{x+4}\right)^{1/2}$$

Steffensen's method with $x_0 = 1.5$ gives the values in Table 5-5.

Table 5-4 Results by using Steffensen's method

k	x_k	y_k	z_k
0	1.5	2.375 00	12.39 65
1	1.416 29	1.840 92	5.238 88
2	1.355 65	1.491 40	2.317 28
3	1.328 95	1.341 70	1.444 35
4	1.324 80	1.325 18	1.327 14
5	1.324 72		

Table 5-5 Results by using Steffensen's method

k	x_k	y_k	z_k
0	1.5	1.348 399 725	1.367 376 372
1	1.365 265 224	1.365 225 534	1.365 230 583
2	1.365 230 013		

5.6 Newton's Method

Newton's method is a general procedure that can be applied in many **diverse** (不同的) situations. When specialized to the problem of locating a zero of a real-valued function of a real variable, it is often called the **Newton-Raphson iteration**. In general, Newton's method is faster than the bisection and the secant methods since its convergence is quadratic rather than linear or superlinear. Once the quadratic convergence becomes effective, that is, the values of Newton's method sequence are sufficiently close to the root, the convergence is so rapid that only a few more values are needed.

5.6.1 Newton's Method and Its Convergence

Let x^* be a zero of f and let x_k be an approximation to x^* (we assume that $f'(x_k) \neq 0$). In a neighborhood of x_k, by Taylor's formula we have that

$$f(x) \approx f(x_k) + f'(x_k)(x - x_k) = g(x) \tag{5.6.1}$$

therefore, we may consider the zero of the function g as a new approximation to the zero of f and denote it by x_{k+1}. From the linear equation

$$f(x_k) + f'(x_k)(x - x_k) = 0$$

We immediately obtain

$$x_{k+1} = x_k - \frac{f(x_k)}{f'(x_k)} \quad (k = 0, 1, \cdots) \tag{5.6.2}$$

This is called **Newton's method** or **Tangent method** (切线法).

Geometrically, the linear function g describes the tangent line to the graph of the function f at the point x_k (see Figure 5-10).

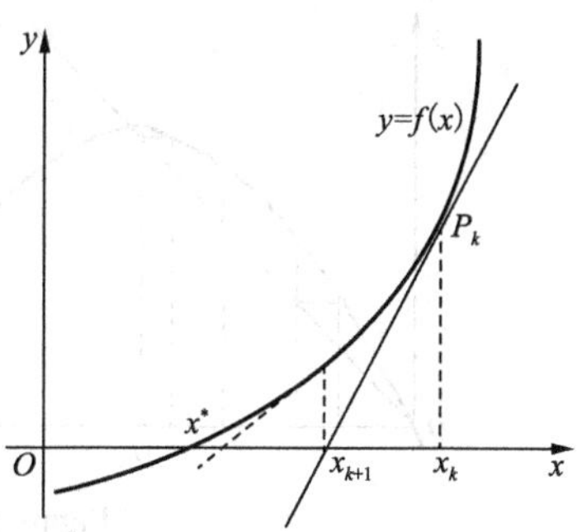

Figure 5-10 Newton's method

For (5.6.2), the iterated function

$$\varphi(x) = x - \frac{f(x)}{f'(x)}$$

since

$$\varphi'(x) = \frac{f(x)f''(x)}{(f'(x))^2}$$

Assume x^* is a simple root of f, i.e., $f(x^*)=0,\ f'(x^*)\neq 0$, then we obtain that $\varphi'(x^*)=0$, hence, Newton's method converges at least quadratically in a neighborhood of x^*. And since

$$\varphi''(x^*) = \frac{f''(x^*)}{f'(x^*)}$$

From (5.4.6) we have

$$\lim_{k\to\infty}\frac{|x_{k+1}-x^*|}{|x_k-x^*|^2} = \left|\frac{f''(x^*)}{2f'(x^*)}\right|$$

Example 5.6.1 Solve the equation $xe^x-1=0$ by Newton's method.

Solution For the given equation Newton's formula is

$$x_{k+1} = x_k - \frac{x_k - e^{-x_k}}{1+x_k}$$

Newton's formula with $x_0=0.5$ gives the values in Table 5-6.

Table 5-6 Results by using Newton's formula

k	x_k
0	0.5
1	0.571 02
2	0.567 16
3	0.567 14

Example 5.6.2 For the equation

$$f(x) = a - \frac{1}{x} = 0$$

where $a>0$, the Newton iteration is given by

$$x_{k+1} = 2x_k - ax_k^2,\quad k=0,1,\cdots$$

Let $\varphi(x)=2x-ax^2$, $\varphi'(x)=2-2ax$, $\varphi'(1/a)=0$. Hence, the successive approximations converge to the fixed point $x=1/a$, for arbitrary chosen $x_0\in(0,\ 2/a)$. Figure 5-11 illustrates the convergence. The numerical results are for $a=2$ and two different starting points, $x_0=0.3$ and $x_0=0.4$.

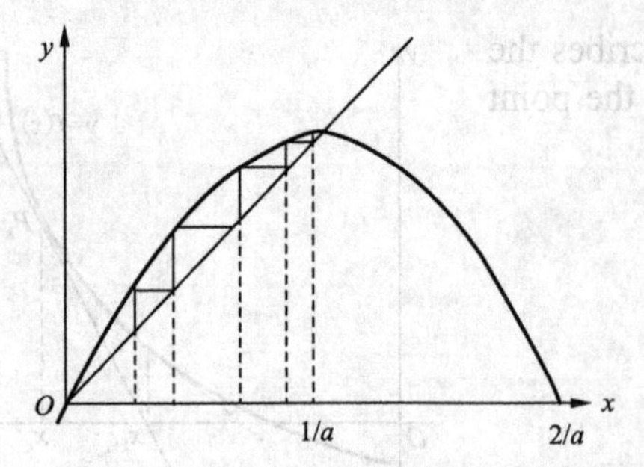

k	x_k	x_k
0	0.300 000 00	0.400 000 00
1	0.420 000 00	0.480 000 00
2	0.487 200 00	0.499 200 00
3	0.499 672 32	0.499 998 72

Figure 5-11　Division by Newton iteration

Example 5.6.3　For the function

$$f(x)=x^2-a$$

where $a>0$, the Newton iteration is given by

$$x_{k+1}=\frac{1}{2}\left(x_k+\frac{a}{x_k}\right)$$

it follows that

$$\varphi'(x)=\frac{1}{2}\left(1-\frac{a}{x^2}\right),\qquad L=\sup_{\sqrt{a}\leqslant x<\infty}|\varphi'(x)|=\frac{1}{2}$$

Therefore, the successive approximations

$$x_{k+1}=\frac{1}{2}\left(x_k+\frac{a}{x_k}\right),\quad k=0,1,\cdots$$

converge to the square root $\sqrt{a}$ for each $x_0>0$.

5.6.2 Reduced Newton Method and Newton's Descent Method（简化牛顿法与牛顿下山法）

Recall that the Newton iteration is defined by the equation

$$x_{k+1}=x_k-\frac{f(x_k)}{f'(x_k)}$$

The advantage of Newton's method is its rapid convergence(see Figure 5-12). One of the drawbacks of Newton's method is that it involves the derivative of the function whose zero is sought. To overcome this disadvantage, we use the following methods.

(1) **Reduced Newton method or parallel chord method**（简化牛顿法或平行弦法）

Let the iteration formula be

$$x_{k+1}=x_k-Cf(x_k),\quad C\neq 0,\quad k=0,1,\cdots$$

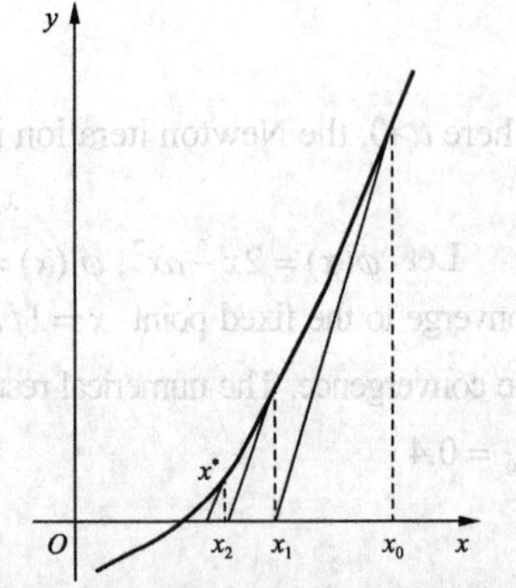

Figure 5-12　Reduced Newton method

The iterated function

$$\varphi(x) = x - Cf(x) \tag{5.6.3}$$

If $|\varphi'(x)| = |1 - Cf'(x)| < 1$, i.e., $0 < Cf'(x) < 2$ holds at the neighborhood of x^*, then the iteration method (5.6.3) is local convergent.

Take $C = \dfrac{1}{f'(x_0)}$ in (5.6.3), then it is called **Reduced Newton method.**

It is only linear convergence. Geometrically, take the **intersection** (交点) of **parallel chord** (平行弦) with x-axis as the approximate value of x^*.

(2) Newton's descent method

To prevent the divergence of the iteration, we assume that function f is monotonic, i.e.,

$$|f(x_{k+1})| < |f(x_k)|$$

It is called **descent method.**

We combine Newton's method and descent method. Let

$$\bar{x}_{k+1} = x_k - \frac{f(x_k)}{f'(x_k)}$$

$$x_{k+1} = \lambda \bar{x}_{k+1} + (1-\lambda)x_k \tag{5.6.4}$$

Where $\lambda(0<\lambda\leqslant 1)$ is called **descent factor** $(\lambda = 1, \frac{1}{2}, \frac{1}{2^2}, \cdots)$, the (5.6.4) is of the form

$$x_{k+1} = x_k - \lambda \frac{f(x_k)}{f'(x_k)} \quad (k = 0,1,\cdots)$$

It is called **Newton's descent method.**

5.6.3 The Case of Multiple Roots (重根情况)

Let x^* be a zero of $f(x)$ of multiplicity m, and $f(x) = (x - x^*)^m g(x)$, the integer $m \geqslant 2$, $g(x^*) \neq 0$, then

$$f(x^*) = f'(x^*) = \cdots = f^{(m-1)}(x^*) = 0, \quad f^{(m)}(x^*) \neq 0$$

If $f'(x_k) \neq 0$, Newton's method is applied to $f(x)$ to give iterative function

$$\varphi(x) = x - \frac{f(x)}{f'(x)}$$

We can prove $\varphi'(x^*) = 1 - \dfrac{1}{m} \neq 0$, and $|\varphi'(x^*)| < 1$, therefore, Newton's method is only linear convergence for solving the multiple roots. If we take

$$\varphi(x) = x - m\frac{f(x)}{f'(x)}$$

We can prove $\varphi'(x^*) = 0$. The iteration defined by

$$x_{k+1} = x_k - m\frac{f(x_k)}{f'(x_k)} \quad (k = 0,1,\cdots) \tag{5.6.5}$$

is **quadratically convergent** (二阶收敛). However, it needs to know m.

We let $\mu(x) = f(x)/f'(x)$, if x^* is multiplicity m roots of the equation $f(x) = 0$, then

$$\mu(x) = \frac{(x - x^*)g(x)}{mg(x) + (x - x^*)g'(x)} = (x - x^*)\frac{g(x)}{mg(x) + (x - x^*)g'(x)}$$

However, $g(x^*) \neq 0$, so

$$\mu'(x^*)\frac{g(x^*)}{mg(x^*) + (x^* - x^*)g'(x^*)} = \frac{1}{m} \neq 0$$

Hence x^* is a simple root of equation $\mu(x) = 0$. Newton' s method can then be applied to $\mu(x)$ to give

$$\varphi(x) = x - \frac{\mu(x)}{\mu'(x)} = x - \frac{f(x)f'(x)}{(f'(x))^2 - f(x)f''(x)} \quad (k = 0,1,\cdots)$$

If $\varphi(x)$ has the required continuity conditions, functional iteration applied to $\varphi(x)$ will be quadratically convergent regardless of the multiplicity of the zero of $f(x)$. We have the iterative scheme

$$x_{k+1} = x_k - \frac{f(x_k)f'(x_k)}{(f'(x_k))^2 - f(x_k)f''(x_k)} \quad (k = 0,1,\cdots) \tag{5.6.6}$$

Example 5.6.4 Given $x^* = \sqrt{2}$ is multiplicity two roots of equation $x^4 - 4x^2 + 4 = 0$, find the roots by the above three methods respectively.

Solution (1) Newton's method

$$x_{k+1} = x_k - \frac{x_k^2 - 2}{4x_k}$$

(2) By using (5.6.5)

$$x_{k+1} = x_k - \frac{x_k^2 - 2}{2x_k}$$

(3) By using (5.6.6)

$$x_{k+1} = x_k - \frac{x_k(x_k^2 - 2)}{(x_k^2 + 2)}$$

With $x_0 = 1.5$, the iterates for (1), (2) and (3) are shown in Table 5-7. The results illustrate the rapid convergence of both methods (2) and (3).

Table 5-7 The comparison of convergence

k	x_k	method (1)	method (2)	method (3)
1	x_1	1.458 333 333	1.416 666 667	1.411 764 706
2	x_2	1.436 607 143	1.414 215 686	1.414 211 438
3	x_3	1.425 497 619	1.414 213 562	1.414 213 562

5.7 Secant Method and Muller Method（割线法与抛物线法）

5.7.1 Secant Method

Let x_k and x_{k-1} be approximate roots of the equation $f(x) = 0$, we structure a linear interpolation polynomial $p_1(x)$ by using $f(x_k)$ and $f(x_{k-1})$, and take the root x_{k+1} of $p_1(x)=0$ as the new approximate root of $f(x)=0$.

Since

$$p_1(x)=f(x_k)+\frac{f(x_k)-f(x_{k-1})}{x_k-x_{k-1}}(x-x_k) \tag{5.7.1}$$

Therefore

$$x_{k+1}=x_k-\frac{f(x_k)}{f(x_k)-f(x_{k-1})}(x_k-x_{k-1}) \tag{5.7.2}$$

Formula (5.7.2) can be obtained from Newton's method if $f'(x_k)$ is replaced by the difference quotient

$$\frac{f(x_k)-f(x_{k-1})}{x_k-x_{k-1}}$$

See Figure 5-13. Geometrically, given points $P_{k-1}(x_{k-1}, f(x_{k-1}))$ and $P_k(x_k, f(x_k))$, the **slope** (斜率) of the **chord** (弦) $\overline{P_kP_{k-1}}$ equals the difference quotient

$$\frac{f(x_k)-f(x_{k-1})}{x_k-x_{k-1}}$$

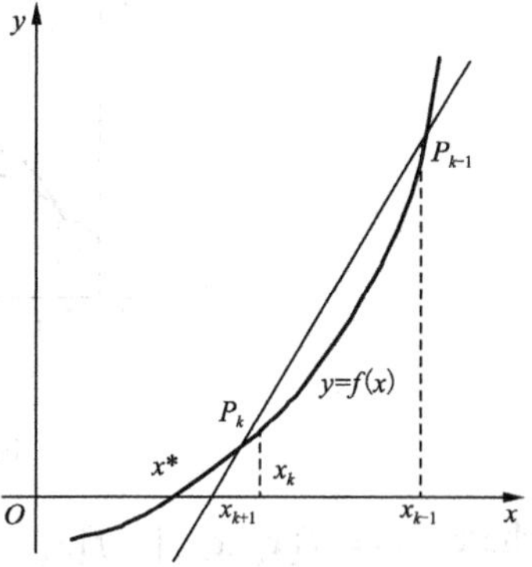

Figure 5-13 The secant method

its equation is of the form $y=f(x_k)+\dfrac{f(x_k)-f(x_{k-1})}{x_k-x_{k-1}}(x-x_k)$

Let y=0, we have the intersection x_{k+1} of the chord $\overline{P_kP_{k-1}}$ and x-axis (轴), i.e., the Formula (5.7.2).

Theorem 5.7.1 Suppose $f(x)$ is twice continuously differentiable in a neighborhood $\Delta:|x-x^*|\leqslant\delta$ at the root x^*, and $f'(x)\neq 0$ for arbitrary $x\in\Delta$, and the initial value $x_0,x_1\in\Delta$, then the secant method will converge to the root x^* with order $p=\dfrac{1+\sqrt{5}}{2}\approx 1.618$. Where p is the positive root of the equation

$$\lambda^2-\lambda-1=0$$

Example 5.7.1 Solve the following equation by the secant method.

$$f(x)=x^3-3x^2-x+9$$

Solution Take x_0=−2, x_1=−1

$$\begin{aligned}x_{k+1}&=x_k-\frac{f(x_k)}{f(x_k)-f(x_{k-1})}(x_k-x_{k-1})\\&=x_k-\frac{x_k^3-3x_k^2-x_k+9}{(x_k^3-3x_k^2-x_k+9)-(x_{k-1}^3-3x_{k-1}^2-x_{k-1}+9)}(x_k-x_{k-1})\end{aligned}$$

5.7.2 Muller Method

Given three approximate roots x_k, x_{k-1}, x_{k-2} of the equation $f(x)=0$, we structure quadratic interpolation polynomial $P_2(x)$, and chose one zero x_{k+1} of $P_2(x)$ as new approximate root, this iteration is called **Muller method**.

See Figure 5-14. The interpolation polynomial

$$P_2(x)=f(x_k)+f[x_k,x_{k-1}](x-x_k)+f[x_k,x_{k-1},x_{k-2}](x-x_k)(x-x_{k-1})$$

has two zeros

$$x_{k+1}=x_k-\frac{2f(x_k)}{\omega\pm\sqrt{\omega^2-4f(x_k)f[x_k,x_{k-1},x_{k-2}]}}$$

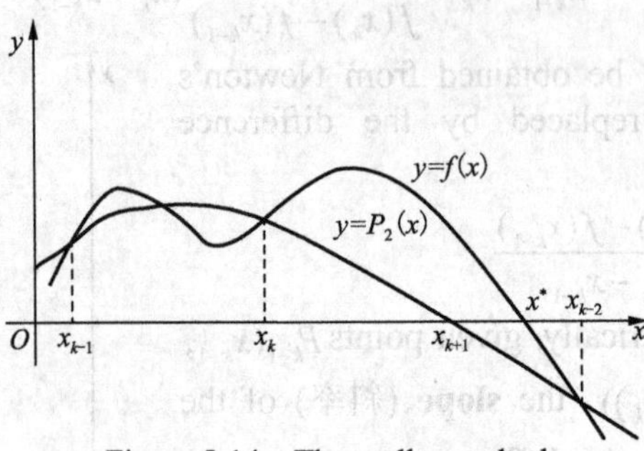

Figure 5-14 The muller method

where $\omega=f[x_k,x_{k-1}]+f[x_k,x_{k-1},x_{k-2}](x_k-x_{k-1})$.

Muller method is faster than secant method. In fact, in some conditions we can prove, for the Muller method, the error of iteration is as follows

$$\frac{|e_{k+1}|}{|e_k|^{1.840}}\to\left|\frac{f'''(x^*)}{6f'(x^*)}\right|^{0.42}$$

Obviously, the Muller method is superlinear convergence with order p=1.840(i.e., the root of $\lambda^3-\lambda^2-\lambda-1=0$).

Since $(1+\sqrt{5})/2\approx1.618<2$, the rapidity of convergence of the secant method is not as good as Newton's method but is better than the bisection method. However, each step of the secant method requires only one new function evaluation, whereas each step of the Newton algorithm requires two function evaluations: $f(x)$ and $f'(x)$.

5.8 Systems of Nonlinear Equations

Newton's method for systems of nonlinear equations follows the same strategy that was used for a single equation. Thus, we linearize and solve, repeating the steps as often as necessary. Let us illustrate with a pair of equations involving two variables:

$$\begin{cases}f_1(x_1,x_2)=0\\f_2(x_1,x_2)=0\end{cases}\tag{5.8.1}$$

Supposing that (x_1,x_2) is an approximate solution of (5.8.1), let us compute corrections h_1 and h_2 so that (x_1+h_1,x_2+h_2) will be better approximate solution. Using only linear terms in the Taylor expansion in two variables , we have

$$\begin{cases} 0 = f_1(x_1+h_1, x_2+h_2) \approx f_1(x_1,x_2) + h_1\dfrac{\partial f_1}{\partial x_1} + h_2\dfrac{\partial f_1}{\partial x_2} \\ 0 = f_2(x_1+h_1, x_2+h_2) \approx f_2(x_1,x_2) + h_1\dfrac{\partial f_2}{\partial x_1} + h_2\dfrac{\partial f_2}{\partial x_2} \end{cases} \tag{5.8.2}$$

The partial derivatives appearing in (5.8.2) are to be evaluated at (x_1, x_2). Equation (5.8.2) constitutes a pair of linear equations for determining h_1 and h_2. The coefficient matrix is the Jacobi matrix of f_1 and f_2:

$$\boldsymbol{J} = \begin{pmatrix} \dfrac{\partial f_1}{\partial x_1} & \dfrac{\partial f_1}{\partial x_2} \\ \dfrac{\partial f_2}{\partial x_1} & \dfrac{\partial f_2}{\partial x_2} \end{pmatrix}$$

To solve (5.8.2), we require $\boldsymbol{J}$ to be nonsingular. If this condition is met, the solution is

$$\begin{pmatrix} h_1 \\ h_2 \end{pmatrix} = -\boldsymbol{J}^{-1}\begin{pmatrix} f_1(x_1,x_2) \\ f_2(x_1,x_2) \end{pmatrix}$$

Hence, Newton's method for two nonlinear equations in two variables is

$$\begin{pmatrix} x_1^{(k+1)} \\ x_2^{(k+1)} \end{pmatrix} = \begin{pmatrix} x_1^{(k)} \\ x_2^{(k)} \end{pmatrix} + \begin{pmatrix} h_1^{(k)} \\ h_2^{(k)} \end{pmatrix}$$

To discuss larger system involving many more variables, no new ideas are needed. However, a matrix-vector formalism is very convenient in this situation. The system of equations

$$\begin{cases} f_1(x_1,\cdots,x_n)=0 \\ \cdots\cdots \\ f_n(x_1,\cdots,x_n)=0 \end{cases} \tag{5.8.3}$$

can be expressed simply as

$$F(x) = 0 \tag{5.8.4}$$

by letting $x = (x_1,\cdots,x_n)^{\mathrm{T}} \in \mathbf{R}^n$ and $F = (f_1,\cdots,f_n)^{\mathrm{T}}$. The **analogue** (类似) of equation (5.8.2) is

$$0 = F(x+H) \approx F(x) + F'(x)H \tag{5.8.5}$$

Infer that

$$x = x^{(k)} - F'(x^{(k)})^{-1}F(x^{(k)})$$

where $\boldsymbol{H} = (h_1, h_2, \cdots, h_n)^{\mathrm{T}}$ and $F'(x)$ is $n\times n$ Jacobi matrix $\boldsymbol{J}(x)$ with elements $\partial f_i / \partial x_j$, namely

$$F'(x)=\begin{pmatrix}\frac{\partial f_1(x)}{\partial x_1} & \frac{\partial f_1(x)}{\partial x_2} & \cdots & \frac{\partial f_1(x)}{\partial x_n}\\ \frac{\partial f_2(x)}{\partial x_1} & \frac{\partial f_2(x)}{\partial x_2} & \cdots & \frac{\partial f_2(x)}{\partial x_n}\\ \vdots & \vdots & & \vdots\\ \frac{\partial f_n(x)}{\partial x_1} & \frac{\partial f_n(x)}{\partial x_2} & \cdots & \frac{\partial f_n(x)}{\partial x_n}\end{pmatrix}$$

The correction vector H is obtained by solving the linear system of equations in (5.8.5). Theoretically, this means

$$H=-F'(x)^{-1}F(x) \tag{5.8.6}$$

but in practice H would usually be determined by Gauss elimination from (5.8.5), thus bypassing the more costly computation of the inverse in (5.8.6). Hence, Newton's method for n nonlinear equations in n variable is given by

$$x^{(k+1)}=x^{(k)}+H^{(k)}=x^{(k)}-F'(x^{(k)})^{-1}F(x^{(k)}) \tag{5.8.7}$$

Example 5.8.1 Starting with $x^{(0)}=(1.5,\ 1.0)^{\mathrm{T}}$, find the roots of the nonlinear system by using Newton's method.

$$\begin{cases}f_1(x_1,x_2)=x_1+2x_2-3=0\\ f_2(x_1,x_2)=2x_1^2+x_2^2-5=0\end{cases}$$

Solution Let

$$F(x)=\begin{pmatrix}x_1+2x_2-3\\ 2x_1^2+x_2^2-5\end{pmatrix}$$

Taking partial derivatives, we get Jacobi matrix

$$F'(x)=\begin{pmatrix}1 & 2\\ 4x_1 & 2x_2\end{pmatrix}$$

and

$$F'(x)^{-1}=\frac{1}{2x_2-8x_1}\begin{pmatrix}2x_2 & -2\\ -4x_1 & 1\end{pmatrix}$$

$$x^{(k+1)}=x^{(k)}-\frac{1}{2x_2^{(k)}-8x_1^{(k)}}\begin{pmatrix}2x_2^{(k)} & -2\\ -4x_1^{(k)} & 1\end{pmatrix}\begin{pmatrix}x_1^{(k)}+2x_2^{(k)}-3\\ 2(x_1^{(k)})^2+(x_2^{(k)})^2-5\end{pmatrix}$$

that is

$$\begin{cases}x_1^{(k+1)}=x_1^{(k)}-\dfrac{(x_2^{(k)})^2-2(x_1^{(k)})^2+x_1^{(k)}x_2^{(k)}-3x_2^{(k)}+5}{x_2^{(k)}-4x_1^{(k)}}\\ x_2^{(k+1)}=x_2^{(k)}-\dfrac{(x_2^{(k)})^2-2(x_1^{(k)})^2-8x_1^{(k)}x_2^{(k)}+12x_2^{(k)}-5}{2(x_2^{(k)}-4x_1^{(k)})}\end{cases}$$

Using the starting value $x^{(0)}=(1.5,\ 1.0)^{\mathrm{T}}$, by successive iteration, we obtain

$$x^{(1)}=(1.5, 0.75)^{\mathrm{T}}$$
$$x^{(2)}=(1.488\,095, 0.755\,952)^{\mathrm{T}}$$
$$x^{(3)}=(1.488\,034, 0.755\,983)^{\mathrm{T}}$$

Example 5.8.2 Starting with $(1,1,1)^{\mathrm{T}}$, carry out six iterations of Newton's method for finding a root of the nonlinear system

$$\begin{cases} xy = z^2 + 1 \\ xyz + y^2 = x^2 + 2 \\ \mathrm{e}^x + z = \mathrm{e}^y + 3 \end{cases}$$

Solution Let

$$F(x) = \begin{pmatrix} f_1(x_1, x_2, x_3) \\ f_2(x_1, x_2, x_3) \\ f_3(x_1, x_2, x_3) \end{pmatrix} = \begin{pmatrix} x_1x_2 - x_3^2 - 1 \\ x_1x_2x_3 - x_1^2 + x_2^2 - 2 \\ \mathrm{e}^{x_1} - \mathrm{e}^{x_2} + x_3 - 3 \end{pmatrix}$$

Taking partial derivatives, we get the Jacobi matrix

$$F'(x) = \begin{pmatrix} x_2 & x_1 & -2x_3 \\ x_2x_3 - 2x_1 & x_1x_3 + 2x_2 & x_1x_2 \\ \mathrm{e}^{x_1} & -\mathrm{e}^{x_2} & 1 \end{pmatrix}$$

$$x^{(k+1)} = x^{(k)} - F'(x^{(k)})^{-1} F(x^{(k)})$$

Using the starting value $x^{(0)} = (1,1,1)^{\mathrm{T}}$, we carry out the nonlinear Newton's method given in Equation (5.8.7) and obtain the following:

$$x^{(1)}=(2.189\,326\,0, 1.598\,475\,1, 1.393\,900\,6)^{\mathrm{T}}$$
$$x^{(2)}=(1.850\,589\,6, 1.444\,251\,4, 1.278\,224\,0)^{\mathrm{T}}$$
$$x^{(3)}=(1.780\,164\,7, 1.424\,435\,9, 1.239\,292\,4)^{\mathrm{T}}$$
$$x^{(4)}=(1.777\,671\,9, 1.423\,960\,9, 1.237\,473\,8)^{\mathrm{T}}$$
$$x^{(5)}=(1.777\,671\,9, 1.423\,960\,5, 1.237\,471\,1)^{\mathrm{T}}$$
$$x^{(6)}=(1.777\,671\,9, 1.423\,960\,5, 1.237\,471\,1)^{\mathrm{T}}$$

5.9 Computer Experiments

5.9.1 Functions Needed in the Experiments by Mathematica

1．D——求微分函数

格式 1：`D[f,x]`

功能：求$f(x)$的微分.

格式 2：`D[f,{x,n}]`

功能：求$f(x)$的 n 阶微分.

格式 3：`Dt[f,x]`

功能：求 $f(x)$的全微分 df/dt.

格式 4：`Dt[f]`

功能：求 $f(x)$的全微分 df.

2．If——条件分支语句

格式 1：`If[条件,语句组]`

功能：若条件（通常为关系表达式）成立, 则执行语句组.

格式 2：`If[条件,语句组 1,语句组 2]`

功能：若条件成立, 则执行语句组 1, 否则执行语句组 2.

3．Do——循环语句

格式 1：`Do[循环体,{循环变量,初值,终值,步长}]`

功能: 对循环范围内的每一个循环值, 执行一次循环体(表达式或语句组), 循环体各语句之间以分号分隔.初值或步长为 1 的项可省略.

格式 2：`Do[循环体,{n}]`

功能：执行 n 次循环体 black.

4．While——循环语句

格式：`While[判断条件,循环体]`

功能：当判断条件成立时开始, 反复执行循环体, 直到判断条件不成立时结束循环.

5．Break——循环控制语句

格式：`Break[]`

功能：结束本层循环.

5.9.2 Experiments by Mathematica

1. 二分法

（1）算法.

① 输入函数 $f(x)$及初始值 a, b, 精度 eps, 微小值 r;

② 求区间$[a,b]$的中点 x_k，计算该点的函数值 $f(x_k)$；

③ 判断 $f(x_k)<r$？

若是, 则输出实根 x_k，结束;

否则, 执行④;

④ 判断 $f(a)f(x_k)<0$?

若为真, 则 $b\leftarrow x_k$;

若为假, 则 $a\leftarrow x_k$;

⑤ 判断 $b-a$ 是否小于或等于计算精度 eps.

若为真, 则输出根的近似值$(a+b)/2$, 结束;

否则, 执行①.

（2）程序清单.

```
(*Bisection Method*)
Clear[xk,aa,bb]
f[x_]=Input["f(x)="];
a=Input["a="];aa=a;
b=Input["b="];bb=b;
r=10^(-10);
eps=Input["eps="];
n=0;
While[b-a>eps,
    xk=(a+b)/2;
    n=n+1;
    v=f[xk];
    If[Abs[v]<r,
       Print["n=",n,"    ","x=",xk,"    ","f[x]=",v];
       Break[]
    ];
    p=f[a]*v//N;
    If[p<0,b=xk,a=xk];
    Print["n=",n,"    ","x=",xk//N,"    ","eps=",(b-a)//N]
]
Print["方程",f[x],"=0 在区间[",aa,",",bb,"]内的根为 x=",xk//N,"    ","误
差 =",(b-a)//N]
```

（3）变量说明.

f[x_]：保存函数 $f(x)$(由键盘输入)；

a：保存含根区间左端点；

b：保存含根区间右端点；

eps：保存计算精度(由键盘输入)；

xk：保存含根区间中点 x_k；

v：保存 $f(x_k)$ 值；

r：刻画 $f(x_k)$的微小值，如果$|f(x_k)|<r$，则认为 $f(x_k)=0$；

n：保存迭代次数.

（4）计算实例.

Example 5.9.1 Find the root of $x^3-11.1x^2+38.8x-41.77=0$ in the interval [3,4] by using bisection method with error $<10^{-4}$.

操作步骤如下：

① 将光标定位在要执行的 Cell 中，按小键盘的【Enter】键；

② 在弹出的对话框中按提示分别输入

```
x^3-11.1x^2+38.8x-41.77, 3, 4, 10^(-4)
```

③ 每次输入后单击 OK 命令按钮，得图 5-15 所示的输出结果.

```
n=1     x=3.5       eps=0.5
n=2     x=3.75      eps=0.25
n=3     x=3.875     eps=0.125
n=4     x=3.9375    eps=0.0625
n=5     x=3.90625   eps=0.03125
n=6     x=3.92188   eps=0.015625
n=7     x=3.91406   eps=0.0078125
n=8     x=3.91797   eps=0.00390625
n=9     x=3.91602   eps=0.00195313
n=10    x=3.91699   eps=0.000976563
n=11    x=3.91748   eps=0.000488281
n=12    x=3.91772   eps=0.000244141
n=13    x=3.91785   eps=0.00012207
n=14    x=3.91779   eps=0.0000610352
方程-41.77+38.8x-11.1x²+x³=0在区间[3，4]
内的根为x=3.91779          误差=0.0000610352
```

图 5-15　二分法计算结果

2．简单迭代法

（1）算法.

① 输入迭代函数, 初值, 计算精度, 迭代次数上限;

② 对 k=0,1,2,⋯, $n_{\max}$,

计算 $x_{k+1} = g(x_k)$,

如果 $|x_{k+1} - x_k| < \text{eps}$, 则输出根 x, 停止,

$x_k \leftarrow x_{k+1}$;

③ 输出迭代失败, 停止.

（2）程序清单.

```
(*Simple Iteration*)
Clear[x0,x,xk,nmax,n];
g[x_]=Input["g(x)= "];
x0=Input["x0="];
eps=Input["eps="];
nmax= Input["nmax="];
Do[xk=N[g[x0],10];
   er=Abs[xk-x0]//N;
   Print["x=",xk,"n=",n,"er=",er];
    If[er<eps,
        Print["方程",g[x],"=x 的根为 x=",xk,"    ","误差 er=",er];
        Break[]
   ];
   x0=xk,
```

```
    {n,1,nmax}
];
If[er>eps,Print["迭代失败"]]
```

（3）变量说明.

g[x_]：保存迭代函数(由键盘输入)；

x0：保存初值及迭代过程中的 x_k；

xk：保存迭代过程中的 x_{k+1}；

nmax：保存迭代次数上限(键盘输入)；

er：保存积分误差值.

（4）计算实例.

Example 5.9.2 Find the root of $e^x+10x-2=0$ by using simple iterative method near x=0 with error <10^{-8}.

① 将光标定位在要执行的 Cell 中，按小键盘的【Enter】键；

② 在弹出的对话框中按提示分别输入

```
(2-Exp[x])/10,0,10^(-8)
```

③ 每次输入后单击 OK 命令按钮，得图 5-16 所示的输出结果.

x=0.1000000000	n=1	er=0.1
x=0.08948290819	n=2	er=0.0105171
x=0.09063913586	n=3	er=0.00115623
x=0.09051261667	n=4	er=0.000126519
x=0.09052646805	n=5	er=0.0000138514
x=0.09052495168	n=6	er=1.51637×10^{-6}
x=0.09052511769	n=7	er=1.66005×10^{-7}
x=0.09052509951	n=8	er=1.81733×10^{-8}
x=0.09052510150	n=9	er=1.98952×10^{-9}

方程 $\frac{1}{10}(2-e^x)=x$的根为x=0.09052510150　　误差er=1.98952×10^{-9}

图 5-16　简单迭代法计算结果

3. Newton 迭代法

（1）算法.

① 输入函数 $f(x)$，初值 x_0，迭代次数上限 n_{max}，计算精度 eps；

② 对 k=0,1,2,⋯,n_{max}；

如果 $|f'(x_0)|<r$，则输出“迭代失败”，停止；

$$x \leftarrow x_0-f(x_0)/f'(x_0)$$

如果$|x-x_0|<\text{eps}$，则输出根 x，停止；

$$x_0 \leftarrow x$$

③ 输出迭代失败, 停止.

（2）程序清单.

```
(*Newton Iteration*)
Clear[x,f,g];
f[x_]=Input["f[x]="];
g[x_]=D[f[x],x];
x0=Input["x0="];
r=10^(-11);
nmax=Input["nmax="];(*输入迭代次数上限*)
eps=Input["eps="];
Do[d0=g[x0];
If[Abs[d0//N]<r, Print["迭代失败"] ;Break[]];
x=x0-f[x0]/d0;
er=Abs[x-x0]//N;
Print["x=",x,"   ""n=",n,"   ""er=",er];
If[er<eps,Break[],x0=x], {n,1,nmax}];
If[er>eps,Print["迭代失败"]]
```

（3）变量说明.

f[x_]：保存函数 $f(x)$ (由键盘输入);

g[x_]：保存导函数 $f'(x)$；

x0：保存初始值及迭代过程的 x_k;

d0：保存 $f'(x_0)$；

r：保存微小值, 如果 $f'(x_0)<0$, 则认为 $f'(x_0)=0$.

（4）计算实例.

Example 5.9.3 Find the root of $x^3-x^2-1=0$ by using Newton iterative method near x=1.5 with error <10^{-12}.

操作步骤如下：

① 将光标定位在要执行的 Cell 中, 按小键盘的【Enter】键；

② 在弹出的对话框中按提示分别输入

```
x^3-x^2-1,1.5,100,10^(-12)
```

③ 每次输入后单击 OK 命令按钮, 得图 5-17 所示的输出结果.

Out[74]=$-1-x^2+x^3$

x	n	er
x=1.46667	n=5	er=0.0333333
x=1.46557	n=2	er=0.00109428
x=1.46557	n=3	er=1.15869×10^{-6}
x=1.46557	n=4	er=1.29807×10^{-12}
x=1.46557	n=5	er=2.22045×10^{-16}

图 5-17 Newton 迭代法计算结果

5.9.3 Experiments by Matlab

1．二分法

（1）函数语句.

```
[x,error]=bisection(fun,a,b,delta,n)
```

（2）参数说明.

fun：符号变量，输入参数，需求零点的函数；

a：实变量，输入参数，有根区间的左端点；

b：实变量，输入参数，有根区间的右端点；

delta：实变量，输入参数，需要达到的精度；

n：实变量，输入参数，允许的最高二分次数；

x：实变量，输出参数，所求根的近似值；

error：实变量，输出参数，计算误差.

（3）bisection.m 程序.

```
function [x,error]=bisection(fun,a,b,delta,n)
if nargin<4
   delta1=10^(-6);
end
if nargin<5
   n=100;
end
f=inline(fun);
if f(a)==0
   x=a;
   return
end
if f(b)==0
   x==b;
   return;
end
if f(a)*f(b)>0
   fprintf('函数在区间两端点的函数值不异号，故不满足二分法条件');
   return;
end
 fprintf('\n  k    a    b   mid   f(mid)   error\n')
for k=1:n
   mid=(b+a)/2;
   error=abs(a-b)/2;
   fprintf('%3g %14.10f %14.10f %14.10f %14.10f %14.10f
\n',k,a,b,mid,f(mid),error);
   if error<delta
```

```
        x=mid;
        fprintf('函数零点的近似值为 %4.3f \n',mid);
        return
    end
    if f(mid)==0
        x=mid;
        return
    elseif f(a)*f(mid)<0
        b=mid;
    else
        a=mid;
    end
end
fprintf('二分了%3d 次，还是没有达规定的误差要求',n);
```

（4）计算实例.

Example 5.9.4 Find the root of $x^3-11.1x^2+38.8x-41.77=0$ in the interval [3,4] by using bisection method with error$<10^{-4}$.

在命令窗口输入：

```
>> clear
>> fun='x^3-11.1*x^2+38.8*x-41.77';a=3;b=4;delta=1e-4;[x,error]=
bisection(fun,a,b,delta)
```

k	a	b	mid	f(mid)	error
1	3.0000000000	4.0000000000	3.5000000000	0.9300000000	0.5000000000
2	3.5000000000	4.0000000000	3.7500000000	0.3706250000	0.2500000000
3	3.7500000000	4.0000000000	3.8750000000	0.0921093750	0.1250000000
4	3.8750000000	4.0000000000	3.9375000000	-0.0417285156	0.0625000000
5	3.8750000000	3.9375000000	3.9062500000	0.0245861816	0.0312500000
6	3.9062500000	3.9375000000	3.9218750000	-0.0087336731	0.0156250000
7	3.9062500000	3.9218750000	3.9140625000	0.0078870583	0.0078125000
8	3.9140625000	3.9218750000	3.9179687500	-0.0004332852	0.0039062500
9	3.9140625000	3.9179687500	3.9160156250	0.0037244144	0.0019531250
10	3.9160156250	3.9179687500	3.9169921875	0.0016449438	0.0009765625
11	3.9169921875	3.9179687500	3.9174804688	0.0006056737	0.0004882813
12	3.9174804688	3.9179687500	3.9177246094	0.0000861553	0.0002441406
13	3.9177246094	3.9179687500	3.9178466797	-0.0001735747	0.0001220703
14	3.9177246094	3.9178466797	3.9177856445	-0.0000437121	0.0000610352

函数零点的近似值为 3.918

```
x =
    3.9178
```

```
error =
   6.1035e-005
```

2．牛顿迭代法

（1）函数语句.

```
y=newtondd(a,n,x0,nmax,eps1)
```

（2）参数说明.

a：n+1 个元素的实数组，输入参数，按降幂保存方程系数；

n：整变量，输入参数，方程阶数；

x0：实变量，输入参数，初始迭代值；

nmax：整变量，输入参数，允许的最大迭代次数；

eps1：实变量，输入参数，控制迭代精度；

y：实变量，输出参数，保存方程的根.

（3）编辑程序.

newtondd.m 程序：

```
function y=newtondd(a,n,x0,nmax,eps1)
x(1)=x0;
b=1;i=1;
while(abs(b)>eps1*x(i))
   i=i+1;
   x(i)=x(i-1)-ntf(a,n,x(i-1))/ntdf(a,n,x(i-1));
   b=x(i)-x(i-1);
   if(i>nmax) error('nmax is full');
      return;
   end
end
y=x(i);
i
```

ntf.m 程序 (生成待求根方程函数)：

```
function y=ntf(a,n,x)
y=0.0;
for i=1:(n+1)
   y=y+a(i)*x^(n+1-i);
end
```

ntdf.m 程序 (生成方程的一阶导数函数)：

```
function y=ntdf(a,n,x)
y=0.0;
for i=1:n
   y=y+a(i)*(n+1-i)*x^(n-i);
end
```

（4）计算实例.

Example 5.9.5 Find the root of $f(x)=x^3+2x^2+10x-20$ by using Newton iterative method near x=1 with error <10^{-8}.

操作步骤如下：

① 编辑并保存 newtondd.m、ntf.m、ntdf.m 程序；

② 在命令窗口输入

```
>> a=[1,2,10,-20];
>> n=3;
>> x0=1;
>> nmax=1000;
>> eps1=1e-8;
>> y=newtondd(a,n,x0,nmax,eps1)
i =
    6
y =
    1.36880810782137
```

Exercises 5

Questions

1. To find a root of the equation $x^3-x^2-1=0$ in a neighborhood of $x_0=1.5$, rewrite the equation in the following equivalent forms, and structure the iteration formulas.

(1) $x=1+1/x^2$, the iteration formula $x_{k+1}=1+1/x_k^2$.

(2) $x^3=1+x^2$, the iteration formula $x_{k+1}=\sqrt[3]{1+x_k^2}$.

(3) $x^2=\dfrac{1}{x-1}$, the iteration formula $x_{k+1}=1/\sqrt{x_k-1}$.

Determine the convergence for each iteration formula.

2. Compare the number of computations for finding the root of $e^x+10x-2=0$ with accuracy 10^{-3}.

(1) Use the bisection method starting with the interval [0,1].

(2) Use the iteration method $x_{k+1}=(2-e^{x_k})/10$, the initial value $x_0=0$.

3. Given the function $f(x)$, let $f'(x)$ exist for all x and $0<m\leqslant f'(x)\leqslant M$, prove that the iteration method $x_{k+1}=x_k-\lambda f(x_k)$ converge to the root x^* of $f(x)=0$ for arbitrary λ within $0<\lambda<2/M$.

4. Perform four iterations of Newton's method for the polynomial

$$P(x)=x^2-3x+1$$

starting with $x_0=2.5$. Use a hand calculator.

5. Discuss Newton's formula for computing the square root of a

$$x_{k+1} = \frac{1}{2}\left(x_k + \frac{a}{x_k}\right), \quad x_0 > 0, k = 0,1,2,\cdots$$

Prove that $x_k \geqslant \sqrt{a}$ for all $k = 0,1,2,\cdots$, and the sequence $\{x_k\}$ is **monotone decreasing** (单调递减).

6. If Newton's method is used on $x^3 - a = 0$, deduce the iteration formula for solving the cubic root $\sqrt[3]{a}$, and discuss its convergence.

7. If Newton's method is used on $f(x) = 1 - \frac{a}{x^2} = 0$, deduce the iteration formula for solving the square root $\sqrt{a}$, and compute $\sqrt{115}$ by using this formula.

8. If Newton's method is used on $f(x) = x^n - a = 0$ and $f(x) = 1 - \frac{a}{x^n} = 0$, deduce the iteration formula respectively for solving $\sqrt[n]{a}$, and compute

$$\lim_{k\to\infty}(\sqrt[n]{a} - x_{k+1})/(\sqrt[n]{a} - x_k)^2$$

9. Show that

$$x_{k+1} = \frac{x_k(x_k^2 + 3a)}{3x_k^2 + a}$$

is a method of order three for computing the square root of a positive number a.

Assume that the initial value x_0 sufficiently approximates to the root x^*, compute

$$\lim_{k\to\infty}(\sqrt{a} - x_{k+1})/(\sqrt{a} - x_k)^3$$

10. Starting with $x^{(0)} = (1.6,\ 1.2)^{\mathrm{T}}$, carry out an iteration of Newton's method for nonlinear system on

$$\begin{cases} x^2 + y^2 = 4 \\ x^2 - y^2 = 1 \end{cases}$$

Explain the results.

11. Starting with $x^{(0)} = (0,\ 0,\ 1)^{\mathrm{T}}$, carry out an iteration of Newton's method for nonlinear system on

$$\begin{cases} xy - z^2 = 1 \\ xyz - x^2 + y^2 = 2 \\ \mathrm{e}^x - \mathrm{e}^y + z = 3 \end{cases}$$

Explain the results.

12. Show that

$$\lim_{k\to\infty}\underbrace{\sqrt{2+\sqrt{2+\sqrt{2+\cdots+\sqrt{2+\sqrt{2}}}}}}_{\text{k square roots}} = 2.$$

13. Show that Newton's method for the function $f(x) = x^n - a$, $x > 0$, where $n>1$

Computer Questions

1. Find a real root of $x^3+2x^2+10x-20=0$ with accuracy 10^{-4}.

2. Compare the number of computations for finding the root of $e^x+10x-2=0$ with accuracy 10^{-5}.

(1) Use the bisection method starting with the interval [0,1].

(2) Use the iteration method $x_{k+1}=(2-e^{x_k})/10$, the initial value $x_0=0$.

3. The equation $2x^4+24x^3+61x^2-16x+1=0$ has two roots near 0.1. Determine them by means of Newton's method.

4. The polynomial $P(x)=x^3+94x^2-389x+294$ has zeros 1,3 and −98. The point $x_0=2$ should therefore be a good starting point for computing either of the small zeros by the Newton iteration. Carry out the calculation and explain what happens.

6 Direct Methods for Solving Linear Systems

提　要

工程技术和自然科学中的许多问题，如电路网络问题、结构设计问题、三次样条问题及最小二乘法问题等，都需要求解线性代数方程组

$$\begin{cases} a_{11}x_1 + a_{12}x_2 + \cdots + a_{1n}x_n = b_1 \\ a_{21}x_1 + a_{22}x_2 + \cdots + a_{2n}x_n = b_2 \\ \cdots\cdots \\ a_{n1}x_1 + a_{n2}x_2 + \cdots + a_{nn}x_n = b_n \end{cases}$$

或写成矩阵形式

$$\boldsymbol{Ax}=\boldsymbol{b}$$

其中

$$\boldsymbol{A} = \begin{pmatrix} a_{11} & a_{12} & \cdots & a_{1n} \\ a_{21} & a_{22} & \cdots & a_{2n} \\ \vdots & \vdots & & \vdots \\ a_{n1} & a_{n2} & \cdots & a_{nn} \end{pmatrix}$$

$$\boldsymbol{x} = (x_1, x_2, \cdots, x_n)^{\mathrm{T}},\ \boldsymbol{b} = (b_1, b_2, \cdots, b_n)^{\mathrm{T}}$$

如果矩阵 $\boldsymbol{A}$ 非奇异，则可用 Gramer 法则求解方程组，但它只适合于阶数 n 很小的情况，而不适合于高阶方程组。因此，需要研究适用于计算机的解方程组的数值算法。

线性方程组的数值解法一般有两类：直接法与迭代法。所谓直接法是能在有限的计算步数内求得精确解（不计舍入误差）的方法，它仅适用于中小型方程组；而迭代法一般适用于解大型稀疏矩阵方程组。本章主要介绍解线性方程组的直接法，包括 Gaussian 消去法、三角分解法、平方根法和追赶法等。

词　汇

direct method	直接法	dense matrix	稠密矩阵
iterative method	迭代法	elimination	消元法

pivot	主元素	square root method	平方根法
backward substitution	回代	diagonal element	对角元
triangular decomposition	三角分解	method of speedup	追赶法
lower triangular matrix	下三角阵	recursive formula	递推公式
upper triangular matrix	上三角阵	homogeneity	齐次性
full pivoting	完全主元素	triangle inequality	三角不等式
column pivoting	列主元素	eigenvalue	特征值
eigenvector	特征向量	reduced pivot element	约化主元素
normal equation	法方程	multiplier	乘数
large and sparse matrix	大型稀疏矩阵	determinant	行列式

6.1 Introduction

Linear systems of equations are associated with many problems in engineering and science, as well as with applications of mathematics to the social sciences and the quantitative study of business and economic problems. For example, Kirchhoff's laws of electrical **circuits** (回路) state that both the net flow of current through each junction and the net voltage drop around each closed **loop** (闭路) of a circuit are zero. Suppose that a potential of V volts is applied between the points A and G in the circuit and that i_1, i_2, i_3, i_4 and i_5 represent current flow as shown in the **diagram** (图表) (see Figure 6-1). Using G as a reference point, Kirchhoff's laws imply that the currents satisfy the following system of linear equations:

$$
\begin{aligned}
&5i_1 + 5i_2 = V \\
&i_3 - i_4 - i_5 = 0 \\
&2i_4 - 3i_5 = 0 \\
&i_1 - i_2 - i_3 = 0 \\
&5i_2 - 7i_3 - 2i_4 = 0
\end{aligned}
$$

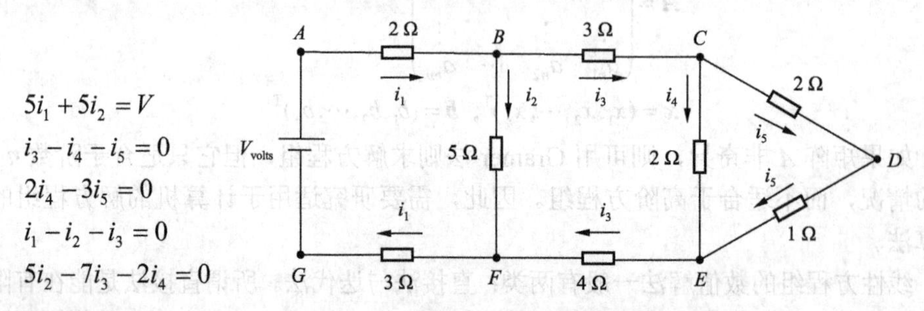

Figure 6-1 Electrical circuits

In principle, we have to distinguish between two groups of methods for the solution of linear systems:

(1) In the so-called direct methods (Chapter 6), the exact solution, in principle, is determined through a finite number of arithmetic operations (in real arithmetic leaving aside the influence of round-off errors).

(2) In contrast to this, iterative methods (Chapter 7) generate a sequence of

approximations to the solution by repeating the application of the same computational procedure at each step of the iteration. Usually, they are applied for large systems with special structures that ensure convergence of the successive approximations.

A key consideration for the selection of a solution method for a linear system is its structure. In some problems, the matrix of the linear system may be a **dense matrix**(稠密矩阵), i.e., it has few zero entries. And in other problems, the matrix may be very **large and sparse** (大型稀疏), i.e., only a small fraction of the entries are different from zero. Roughly speaking, direct methods are best for dense matrices, whereas iterative methods are best for very large and sparse matrices.

When you finish this chapter you will be able to efficiently solve linear algebraic systems of equations. You will also be able to compute the determinant and the inverse of a matrix. You will learn how to represent linear systems of equations using compact vector notation, and then solve them using direct method based on elementary row operations. You will be able to compare the computational effort required for the direct methods including Gaussian elimination and ***LU*** decomposition.

Objectives

- Know how to formulate linear algebraic systems of equations using vector and matrix notation .
- Be able to solve a linear algebraic system using Gaussian elimination.
- Understand how to define and compute the matrix determinant (行列式).
- Be able to factor a matrix into lower-triangular and upper-triangular parts, and solve a linear algebraic system using *LU* decomposition.
- Be able to efficiently solve tridiagonal systems of equations.
- Know how to compute the norm and condition number of a matrix.

6.2 Gaussian Elimination（高斯消去法）

6.2.1 Basic Gaussian Elimination

We proceed with describing the Gaussian elimination method for a system of linear equations

$$\boldsymbol{Ax} = \boldsymbol{b}$$

here $\boldsymbol{A}$ is a given $n \times n$ matrix $\boldsymbol{A}=(a_{ij})$ with real (or complex) entries, $\boldsymbol{b}$ is a given vector $\boldsymbol{b}=(b_1,\cdots,b_n)^{\mathrm{T}} \in \mathbf{R}^n$ (or $\mathbf{C}^n$), and we are looking for a solution vector $\boldsymbol{x}=(x_1,\cdots,x_n)^{\mathrm{T}} \in \mathbf{R}^n$ (or $\mathbf{C}^n$). More explicitly, our system of equations can be written in the form

$$\sum_{k=1}^{n} a_{jk} x_k = b_j, \quad j = 1,\cdots,n$$

that is

$$\begin{cases} a_{11}x_1 + a_{12}x_2 + \cdots + a_{1n}x_n = b_1 \\ a_{21}x_1 + a_{22}x_2 + \cdots + a_{2n}x_n = b_2 \\ \cdots\cdots \\ a_{n1}x_1 + a_{n2}x_2 + \cdots + a_{nn}x_n = b_n \end{cases} \tag{6.2.1}$$

Assuming that the reader is familiar with basic linear algebra, we recall the following various ways of saying that the matrix $\boldsymbol{A}$ is nonsingular:

(1) The inverse matrix $\boldsymbol{A}^{-1}$ exists.

(2) For each $\boldsymbol{b}$ the linear system $\boldsymbol{Ax}=\boldsymbol{b}$ has a unique solution.

(3) The **homogenous system** (齐次方程组) $\boldsymbol{Ax}=\boldsymbol{0}$ has only the **trivial solution** (平凡解).

(4) The determinant of $\boldsymbol{A}$ satisfies $\det \boldsymbol{A} \neq 0$.

(5) The rows (columns) of $\boldsymbol{A}$ are linearly independent.

The very basic idea of the Gaussian elimination method is to use the first equation to eliminate the first unknown from the last $n-1$ equations, then use the new second equation to eliminate the second unknown from the last $n-2$ equations, etc. This way, by $n-1$ such eliminations the given linear system is transformed into an equivalent linear system that is of triangular form.

The System (6.2.1) is denoted

$$\boldsymbol{A}^{(1)}\boldsymbol{x} = \boldsymbol{b}^{(1)}$$

where $\boldsymbol{A}^{(1)} = (a_{ij}^{(1)}) = (a_{ij})$, $\boldsymbol{b}^{(1)} = \boldsymbol{b}$.

We begin by considering a nonsingular matrix $\boldsymbol{A}$. Let $a_{11}^{(1)} \neq 0$, and compute **multiplier** (乘数) $m_{i1} = a_{i1}^{(1)} / a_{11}^{(1)}$. To eliminate the unknown x_1, for $i = 2,3,\cdots,n$, we multiply the first equation by m_{i1} and subtract the result from the ith equation. This procedure leads to a system of the form

$$\begin{pmatrix} a_{11}^{(1)} & a_{12}^{(1)} & \cdots & a_{1n}^{(1)} \\ 0 & a_{22}^{(2)} & \cdots & a_{2n}^{(2)} \\ \vdots & \vdots & & \vdots \\ 0 & a_{n2}^{(2)} & \cdots & a_{nn}^{(2)} \end{pmatrix} \begin{pmatrix} x_1 \\ x_2 \\ \vdots \\ x_n \end{pmatrix} = \begin{pmatrix} b_1^{(1)} \\ b_2^{(2)} \\ \vdots \\ b_n^{(2)} \end{pmatrix}$$

denoted

$$\boldsymbol{A}^{(2)}\boldsymbol{x} = \boldsymbol{b}^{(2)}$$

where

$$\begin{cases} a_{ij}^{(2)} = a_{ij}^{(1)} - m_{i1}a_{1j}^{(1)} & (i = 2,3,\cdots,n; j = 2,3,\cdots,n) \\ b_i^{(2)} = b_i^{(1)} - m_{i1}b_1^{(1)} & (i = 2,3,\cdots,n) \end{cases}$$

Proceeding in this way, the given $n \times n$ system for the unknowns $x_1,\cdots,x_n$ is equivalently transformed into an $(n-1)\times(n-1)$ system for the unknowns $x_2,\cdots,x_n$. Adding a multiple of one row of a matrix to another row does not change the value of its determinant. Therefore, in the above elimination the determinant of the system remains (with the exception of a possible change of its sign if the order of rows or columns is changed). Hence, the resulting $(n-1)\times(n-1)$ system for $x_2,\cdots,x_n$ again has a **nonvanishing** (非零) determinant, and we can apply precisely the same procedure to eliminate the second unknown x_2 from the remaining $(n-1)\times(n-1)$ system.

We have finished the elimination of step1 down to step **(***k*−1**)**, and obtained the equivalent system of linear equations

$$\begin{pmatrix} a_{11}^{(1)} & a_{12}^{(1)} & \cdots & a_{1k}^{(1)} & \cdots & a_{1n}^{(1)} \\ & a_{22}^{(2)} & \cdots & a_{2k}^{(2)} & \cdots & a_{2n}^{(2)} \\ & & \ddots & \vdots & & \vdots \\ & & & a_{kk}^{(k)} & \cdots & a_{kn}^{(k)} \\ & & & \vdots & & \vdots \\ & & & a_{nk}^{(k)} & \cdots & a_{nn}^{(k)} \end{pmatrix} \begin{pmatrix} x_1 \\ x_2 \\ \vdots \\ x_k \\ \vdots \\ x_n \end{pmatrix} = \begin{pmatrix} b_1^{(1)} \\ b_2^{(2)} \\ \vdots \\ b_k^{(k)} \\ \vdots \\ b_n^{(k)} \end{pmatrix}$$

denote

$$\boldsymbol{A}^{(k)}\boldsymbol{x} = \boldsymbol{b}^{(k)}$$

Let $a_{kk}^{(k)} \neq 0$, and compute multiplier

$$m_{ik} = a_{ik}^{(k)} / a_{kk}^{(k)} \qquad (i = k+1,\cdots,n)$$

To eliminate the unknown x_k, for $i = k+1,\cdots,n$ we multiply the *k*th equation by m_{ik} and subtract the result from the *i*th $(i = k+1,\cdots,n)$ equation.

For this we have to require that $a_{11} \neq 0$, since we assume the matrix to be nonsingular, this can be achieved by recording the rows or the columns of the given system.

By *n*−1 recursive elimination steps of the form

$$\begin{cases} a_{ij}^{(k+1)} = a_{ij}^{(k)} - m_{ik}a_{kj}^{(k)} & (i = k+1,\cdots,n;\ j = k+1,\cdots,n) \\ b_i^{(k+1)} = b_i^{(k)} - m_{ik}b_k^{(k)} & (i = k+1,\cdots,n) \end{cases}$$

We can obtain an equivalent triangular system of equations

$$\begin{pmatrix} a_{11}^{(1)} & a_{12}^{(1)} & \cdots & a_{1n}^{(1)} \\ & a_{22}^{(2)} & \cdots & a_{2n}^{(2)} \\ & & \ddots & \vdots \\ & & & a_{nn}^{(n)} \end{pmatrix} \begin{pmatrix} x_1 \\ x_2 \\ \vdots \\ x_n \end{pmatrix} = \begin{pmatrix} b_1^{(1)} \\ b_2^{(2)} \\ \vdots \\ b_n^{(n)} \end{pmatrix} \tag{6.2.2}$$

The triangular system can be solved recursively by first obtaining x_n from the last equation, then obtaining x_{n-1} from the second to last equation, etc. This procedure is known as **backward substitution** (回代). Explicitly, it is described by

$$\begin{cases} x_n = b^{(n)} / a_{nn}^{(n)} \\ x_k = (b_k^{(k)} - \sum\limits_{j=k+1}^{n} a_{kj}^{(k)}x_j) / a_{kk}^{(k)} & (k = n-1, n-2,\cdots,1) \end{cases} \tag{6.2.3}$$

Example 6.2.1 Solve the linear system by Gaussian elimination

$$\begin{cases} 2x_1+2x_2+2x_3=1 & (1) \\ 3x_1+2x_2+4x_3=1/2 & (2) \\ x_1+3x_2+9x_3=5/2 & (3) \end{cases}$$

Solution　By performing $(2)-\frac{3}{2}\times(1),\ (3)-\frac{1}{2}\times(1)$, we obtain

$$\begin{cases} 2x_1+2x_2+2x_3=1 \\ -x_2+x_3=-1 & (4) \\ 2x_2+8x_3=2 & (5) \end{cases}$$

Compute $(5)+2\times(4)$ to obtain

$$\begin{cases} 2x_1+2x_2+2x_3=1 \\ -x_2+x_3=-1 \\ 10x_3=0 & (6) \end{cases}$$

By backward substitution, we have

$$x_1=-\frac{1}{2},\quad x_2=1,\quad x_3=0$$

Written in matrices：

$$[\boldsymbol{A},\boldsymbol{b}]=\left(\begin{array}{ccc:c} 2 & 2 & 2 & 1 \\ 3 & 2 & 4 & 1/2 \\ 1 & 3 & 9 & 5/2 \end{array}\right) \xrightarrow[(3)-\frac{1}{2}\times(1)]{(2)-\frac{3}{2}\times(1)} \left(\begin{array}{ccc:c} 2 & 2 & 2 & 1 \\ 0 & -1 & 1 & -1 \\ 0 & 2 & 8 & 2 \end{array}\right) \xrightarrow{(3)-(-2)\times(2)} \left(\begin{array}{ccc:c} 2 & 2 & 2 & 1 \\ 0 & -1 & 1 & -1 \\ 0 & 0 & 10 & 0 \end{array}\right)$$

$a_{kk}^{(k)}$ is called **elimination element,** or **pivot element** (主元素).

Theorem 6.2.1　The pivot elements $a_{ii}^{(i)}\neq 0(i=1,2,\cdots,k)$ if and only if the **leading principal minor** (前主子式) $D_i\neq 0(i=1,2,\cdots,k)$. i.e.,

$$D_1=a_{11}\neq 0$$

$$D_i=\begin{vmatrix} a_{11} & \cdots & a_{1i} \\ \vdots & & \vdots \\ a_{i1} & \cdots & a_{ii} \end{vmatrix}\neq 0\quad (i=1,2,\cdots,k)$$

Proof　We use the method of **induction** (归纳). Obviously, the Theorem 6.2.1 holds for k=1. Assume that the Theorem 6.2.1 holds for $k-1$, and $D_i\neq 0(i=1,2,\cdots,k-1)$, then by **induction assumption** (归纳假设) $a_{ii}^{(i)}\neq 0\quad(i=1,2,\cdots,k-1)$, we can reduce $\boldsymbol{A}^{(1)}$ to $\boldsymbol{A}^{(k)}$ with Gaussian elimination, i.e.,

$$
A^{(1)} \to A^{(k)} = \begin{pmatrix} a_{11}^{(1)} & a_{12}^{(1)} & \cdots & a_{1k}^{(1)} & \cdots & a_{1n}^{(1)} \\ & a_{22}^{(2)} & \cdots & a_{2k}^{(2)} & \cdots & a_{2n}^{(2)} \\ & & \ddots & \vdots & & \vdots \\ & & & a_{kk}^{(k)} & \cdots & a_{kn}^{(k)} \\ & & & \vdots & & \vdots \\ & & & a_{nk}^{(k)} & \cdots & a_{nn}^{(k)} \end{pmatrix}
$$

and

$$
D_2 = \begin{vmatrix} a_{11}^{(1)} & a_{12}^{(1)} \\ 0 & a_{22}^{(2)} \end{vmatrix} = a_{11}^{(1)} a_{22}^{(2)}
$$

$$
D_k = \begin{vmatrix} a_{11}^{(1)} & \cdots & a_{1k}^{(1)} \\ & \ddots & \vdots \\ & & a_{kk}^{(k)} \end{vmatrix} = a_{11}^{(1)} a_{22}^{(2)} \cdots a_{kk}^{(k)} \tag{6.2.4}
$$

from $D_i \neq 0 (i = 1,2,\cdots,k)$, we can deduce $a_{kk}^{(k)} \neq 0$.

Obviously, let $a_{ii}^{(i)} \neq 0 \ (i = 1,2,\cdots,k)$, using (6.2.4) we can deduce $D_i \neq 0 (i = 1,2,\cdots,k)$.

□

Corollary 6.2.1 If the leading principal minors of order 1 to $n-1$ of A are nonzero, i.e., $D_k \neq 0 \ (k = 1,2,\cdots,n-1)$, then

$$
\begin{cases} a_{11}^{(1)} = D_1 \\ a_{kk}^{(k)} = D_k / D_{k-1} \qquad (k = 2,\cdots,n) \end{cases}
$$

6.2.2 Triangular Decomposition（三角分解）

In the sequel we will indicate how Gaussian elimination provides an ***LU*** decomposition (or factorization) of a given matrix.

Definition 6.2.1 A factorization of a matrix A into a **product** (乘积)

$$
A = LU
$$

of a lower (left) triangular matrix L and an upper (right) triangular matrix U is called an ***LU*** **decomposition of *A*.**

A matrix $A = (a_{ij})$ is called **lower triangular** or **left triangular** if $a_{jk} = 0$ for $j < k$, it is called **upper triangular** or **right triangular** if $a_{jk} = 0$ for $j > k$. The product of two lower (upper) triangular matrices again is lower (upper) triangular, lower (upper) triangular matrices with nonvanishing diagonal elements are nonsingular, and the inverse matrix of a lower (upper) triangular matrix again is lower (upper) triangular.

Theorem 6.2.2 For a nonsingular matrix A, Gaussian elimination (without reordering rows and columns) yields an ***LU*** decomposition.

Proof In the first elimination step we multiply the first equation by $m_{i1}=a_{i1}/a_{11}$ $(i=2,3,\cdots,n)$ and subtract the result from the ith equation, i.e., the matrix $\boldsymbol{A}^{(1)}=\boldsymbol{A}$ is multiplied from the left by the lower triangular matrix

$$\boldsymbol{L}_1=\begin{pmatrix}1 & & & & \\ -m_{21} & 1 & & & \\ -m_{31} & & 1 & & \\ \vdots & & & \ddots & \\ -m_{n1} & & & & 1\end{pmatrix}$$

The resulting matrix $\boldsymbol{A}^{(2)}=\boldsymbol{L}_1\boldsymbol{A}^{(1)}$ is of the form

$$\boldsymbol{A}^{(2)}=\begin{pmatrix}a_{11} & * \\ \boldsymbol{O} & \overline{\boldsymbol{A}}_{n-1}\end{pmatrix}$$

where $\overline{\boldsymbol{A}}_{n-1}$ is an $(n-1)\times(n-1)$ matrix. In the second step the same procedure is repeated for the $(n-1)\times(n-1)$ matrix $\overline{\boldsymbol{A}}_{n-1}$. The corresponding $(n-1)\times(n-1)$ elimination matrix is completed as an $n\times n$ triangular matrix $\boldsymbol{L}_2$ by setting the diagonal element in the first row equal to one. In this way, $n-1$ elimination steps lead to

$$\boldsymbol{L}_{n-1}\cdots\boldsymbol{L}_1\boldsymbol{A}=\boldsymbol{U}$$

$$\boldsymbol{A}=(\boldsymbol{L}_{n-1}\cdots\boldsymbol{L}_1)^{-1}\boldsymbol{U}=\boldsymbol{L}_1^{-1}\boldsymbol{L}_2^{-1}\cdots\boldsymbol{L}_{n-1}^{-1}\boldsymbol{U}=\boldsymbol{L}\boldsymbol{U}$$

with nonsingular lower triangular matrices $\boldsymbol{L}_1,\cdots,\boldsymbol{L}_{n-1}$ and an upper triangular matrix $\boldsymbol{U}$. Where

$$\boldsymbol{L}_k=\begin{pmatrix}1 & & & & & \\ & \ddots & & & & \\ & & 1 & & & \\ & & -m_{k+1,k} & 1 & & \\ & & \vdots & & \ddots & \\ & & -m_{nk} & & & 1\end{pmatrix}$$

$$\boldsymbol{L}=\boldsymbol{L}_1^{-1}\boldsymbol{L}_2^{-1}\cdots\boldsymbol{L}_{n-1}^{-1}=\begin{pmatrix}1 & & & & \\ m_{21} & 1 & & & \\ m_{31} & m_{32} & 1 & & \\ \vdots & \vdots & \vdots & \ddots & \\ m_{n1} & m_{n2} & m_{n3} & \cdots & 1\end{pmatrix}$$

□

We wish to point out that not every nonsingular matrix allows an $\boldsymbol{LU}$ decomposition. For example

$$\begin{pmatrix}0 & 1\\ 1 & 0\end{pmatrix}$$

has no $\boldsymbol{LU}$ decomposition.

Example 6.2.2

$$\begin{cases} 2x_1+2x_2+2x_3=1 & (1) \\ 3x_1+2x_2+4x_3=1/2 & (2) \\ x_1+3x_2+9x_3=5/2 & (3) \end{cases}$$

Solution Since

$$[\boldsymbol{A},\boldsymbol{b}]=\left(\begin{array}{ccc:c} 2 & 2 & 2 & 1 \\ 3 & 2 & 4 & 1/2 \\ 1 & 3 & 9 & 5/2 \end{array}\right) \xrightarrow[(3)-\frac{1}{2}\times(1)]{(2)-\frac{3}{2}\times(1)} \left(\begin{array}{ccc:c} 2 & 2 & 2 & 1 \\ 0 & -1 & 1 & -1 \\ 0 & 2 & 8 & 2 \end{array}\right)$$

$$\xrightarrow{(3)-(-2)\times(2)} \left(\begin{array}{ccc:c} 2 & 2 & 2 & 1 \\ 0 & -1 & 1 & -1 \\ 0 & 0 & 10 & 0 \end{array}\right)$$

Therefore

$$\begin{pmatrix} 2 & 2 & 2 \\ 3 & 2 & 4 \\ 1 & 3 & 9 \end{pmatrix} = \begin{pmatrix} 1 & & \\ 3/2 & 1 & \\ 1/2 & -2 & 1 \end{pmatrix} \begin{pmatrix} 2 & 2 & 2 \\ & -1 & 1 \\ & & 10 \end{pmatrix}$$

Theorem 6.2.3 (Theorem on *LU* decomposition) If all n leading principal minors of the $n\times n$ matrix $\boldsymbol{A}$ are nonzero, then $\boldsymbol{A}$ has a unique $\boldsymbol{LU}$ decomposition.

Proof It needs only to prove the uniqueness. Let $\boldsymbol{A}$ be nonsingular and

$$\boldsymbol{A}=\boldsymbol{LU}=\boldsymbol{L}_1\boldsymbol{U}_1$$

where $\boldsymbol{L}$ and $\boldsymbol{L}_1$ are **unit (identity) lower triangular matrices** (单位下三角阵), $\boldsymbol{U}$ and $\boldsymbol{U}_1$ are **upper triangular matrices** (上三角阵).

Since $\boldsymbol{U}_1^{-1}$ exists, therefore

$$\boldsymbol{L}^{-1}\boldsymbol{L}_1=\boldsymbol{U}\boldsymbol{U}_1^{-1}$$

and $\boldsymbol{L}^{-1}\boldsymbol{L}_1$ is a lower triangular matrix, $\boldsymbol{U}\boldsymbol{U}_1^{-1}$ is an upper triangular matrix.

Hence

$$\boldsymbol{L}^{-1}\boldsymbol{L}_1=\boldsymbol{U}\boldsymbol{U}_1^{-1}=\boldsymbol{I}$$

where $\boldsymbol{I}$ represents $n\times n$ identity matrix.

Therefore

$$\boldsymbol{U}=\boldsymbol{U}_1,\ \boldsymbol{L}=\boldsymbol{L}_1 \qquad \square$$

The algorithm based on the preceding analysis is known as **Doolittle's decomposition** (杜利特分解) when $\boldsymbol{L}$ is unit lower triangular ($l_{ii}=1$ for $1\leqslant i\leqslant n$) and as **Crout's decomposition** (克劳特分解) when $\boldsymbol{U}$ is unit upper triangular ($u_{ii}=1$ for $1\leqslant i\leqslant n$).

Example 6.2.3 Find the Doolittle and Crout factorizations of the matrix

$$A=\begin{pmatrix}60&30&20\\30&20&15\\20&15&12\end{pmatrix}$$

Solution The Doolittle factorization from Theorem 6.2.2 is

$$A=\begin{pmatrix}1&&\\1/2&1&\\1/3&1&1\end{pmatrix}\begin{pmatrix}60&30&20\\&5&5\\&&1/3\end{pmatrix}=LU$$

Rather than computing the next factorization directly, we can obtain it from the Doolittle factorization above. By putting the diagonal elements of U into a diagonal matrix D, we can write

$$A=\begin{pmatrix}1&&\\1/2&1&\\1/3&1&1\end{pmatrix}\begin{pmatrix}60&&\\&5&\\&&1/3\end{pmatrix}\begin{pmatrix}1&1/2&1/3\\&1&1\\&&1\end{pmatrix}=LD\overline{U}$$

By putting $\overline{L}=LD$, we obtain the Crout factorization

$$A=\begin{pmatrix}60&&\\30&5&\\20&5&1/3\end{pmatrix}\begin{pmatrix}1&1/2&1/3\\&1&1\\&&1\end{pmatrix}=\overline{LU}$$

6.3 Gaussian Elimination with Column Pivoting（高斯主元素消去法）

We first see the following example.

Example 6.3.1 The linear system

$$\begin{cases}0.0001x_1+x_2=1 & (1)\\ x_1+x_2=2 & (2)\end{cases}$$

has the exact solution $x^*=(1.000\,100\,01,\ \ 0.999\,899\,99)^{\mathrm{T}}$. Suppose Gaussian elimination is performed on this system using three-digit arithmetic with rounding.

The first pivot element, $a_{11}^{(1)}=0.000\,1$, is small, and its associated multiplier,

$$m_{21}=\frac{1}{0.000\,1}=10\,000$$

is the large number 100 00. Performing $(2)-m_{21}\times(1)$ and the approximate rounding gives

$$[A,b]=\left(\begin{array}{cc:c}0.000\,1&1&1\\1&1&2\end{array}\right)\xrightarrow{m_{21}=100\,00}\left(\begin{array}{cc:c}0.000\,1&1&1\\0&-0.1\times10^5&-0.1\times10^5\end{array}\right)$$

By backward substitution, we obtain $x_2=1,\ \ x_1=0$.

Obviously, x_2 is a close approximation to the actual value, $x_2=0.999\,899\,99$.

However, because of the small pivot $a_{11}^{(1)}=0.0001$ yields $x_1 \approx \frac{1-1.00}{0.0001}=0$.

This ruins the approximation to the actual value $x_1 = 1.00010001$.

Reconsider the system. We use Gaussian elimination with reordering the rows of the given system.

$$(\boldsymbol{A},\boldsymbol{b}) \xrightarrow{r_1 \Leftrightarrow r_2} \left(\begin{array}{cc:c} 1 & 1 & 2 \\ 0.000100 & 1 & 1 \end{array}\right) \xrightarrow{m_{21}=0.000100} \left(\begin{array}{cc:c} 1 & 1 & 2 \\ 0 & 1.00 & 1.00 \end{array}\right)$$

By backward substitution, we obtain

$$x_2 = 1.00, \quad x_1 = 1.00$$

This is a better solution.

In order to control the influence of round-off errors, we want to keep the quotient (multiplier) $m_{ik}=a_{ik}^{(k)}/a_{kk}^{(k)}$ small, i.e., we want to have a large pivot element $a_{kk}^{(k)}$. Therefore, instead of only requiring $a_{kk}^{(k)} \neq 0$, in practice, either **full pivoting** (全主元素) or **column pivoting** (行或列主元素) is employed. For full pivoting , both the rows and the columns are reordered such that $a_{kk}^{(k)}$ has maximal absolute value in the $(n-k+1)\times(n-k+1)$ matrix remaining for the kth forward elimination step. In order to minimize the additional computational **cost** (代价) caused by pivoting, for column pivoting the rows (or columns) are reordered such that $a_{kk}^{(k)}$ has maximal absolute value in the elimination column, i.e., in the kth column. For instance (e.g.), in the first elimination, if $\left|a_{i_1,1}\right| = \max\limits_{1\leqslant i\leqslant n}\left|a_{i1}\right| \neq 0$, then interchange the rows of 1 and i_1. Of course, in the actual **implementation** (实现) of the Gaussian elimination algorithm the reordering of rows and columns need not be done explicitly. Instead , the interchange may be done only implicitly by leaving the pivot element at its original location and keeping track of the interchange of rows and columns through the associated **permutation matrix** (置换矩阵,排列矩阵).

Example 6.3.2 Solve the system by using column pivoting

$$\begin{pmatrix} 3 & 1 & 6 \\ 2 & 1 & 3 \\ 1 & 1 & 1 \end{pmatrix}\begin{pmatrix} x_1 \\ x_2 \\ x_3 \end{pmatrix} = \begin{pmatrix} 2 \\ 7 \\ 4 \end{pmatrix}$$

Solution

$$\begin{pmatrix} 3 & 1 & 6 & 2 \\ 2 & 1 & 3 & 7 \\ 1 & 1 & 1 & 4 \end{pmatrix} \xrightarrow[m_{31}=\frac{1}{3}]{m_{21}=\frac{2}{3}} \begin{pmatrix} 3 & 1 & 6 & 2 \\ & \frac{1}{3} & -1 & \frac{17}{3} \\ & \frac{2}{3} & -1 & \frac{10}{3} \end{pmatrix} \xrightarrow[m_{32}=\frac{1}{2}]{(r_2 \leftrightarrow r_3)} \begin{pmatrix} 3 & 1 & 6 & 2 \\ & \frac{2}{3} & -1 & \frac{10}{3} \\ & & -\frac{1}{2} & 4 \end{pmatrix}$$

We obtain the equivalent system

$$\begin{pmatrix} 3 & 1 & 6 \\ & \frac{2}{3} & -1 \\ & & -\frac{1}{2} \end{pmatrix}\begin{pmatrix} x_1 \\ x_2 \\ x_3 \end{pmatrix}=\begin{pmatrix} 2 \\ \frac{10}{3} \\ 4 \end{pmatrix}$$

This implies

$$\boldsymbol{x}=(19,-7,-8)^{\mathrm{T}}$$

We have $\boldsymbol{PA}=\boldsymbol{LU}$, where

$$\boldsymbol{P}=\begin{pmatrix} 1 & 0 & 0 \\ 0 & 0 & 1 \\ 0 & 1 & 0 \end{pmatrix},\ \boldsymbol{L}=\begin{pmatrix} 1 & & \\ \frac{1}{3} & 1 & \\ \frac{2}{3} & \frac{1}{2} & 1 \end{pmatrix},\ \boldsymbol{U}=\begin{pmatrix} 3 & 1 & 6 \\ & \frac{2}{3} & -1 \\ & & -\frac{1}{2} \end{pmatrix}$$

Theorem 6.3.1 If matrix $\boldsymbol{A}$ is nonsingular, then there exists a permutation matrix $\boldsymbol{P}$, such that

$$\boldsymbol{PA}=\boldsymbol{LU}$$

Where $\boldsymbol{L}$ is an identity lower triangular matrix and $\boldsymbol{U}$ is an upper triangular matrix.

6.4 Methods of the Triangular Decomposition

6.4.1 The Direct Methods of The Triangular Decomposition

Suppose that $\boldsymbol{A}$ can be factored into the product of a lower triangular matrix $\boldsymbol{L}$ and an upper triangular matrix $\boldsymbol{U}$: $\boldsymbol{A}=\boldsymbol{LU}$. Then, to solve the system of equations $\boldsymbol{Ax}=\boldsymbol{b}$, it is enough to solve this problem in two stages:

$$\boldsymbol{Ly}=\boldsymbol{b} \quad \text{solve for } \boldsymbol{y}$$

$$\boldsymbol{Ux}=\boldsymbol{y} \quad \text{solve for } \boldsymbol{x}$$

Our previous analysis indicates that solving these two triangular systems is simple.

In the case of Doolittle's decomposition, we have

$$(1)\ \begin{cases} y_1=b_1 \\ y_i=b_i-\sum\limits_{k=1}^{i-1} l_{ik}y_k \quad (i=2,3,\cdots,n) \end{cases}$$

$$(2)\ \begin{cases} x_n=y_n/u_{nn} \\ x_i=(y_i-\sum\limits_{k=i+1}^{n} u_{ik}x_k)/u_{ii} \quad (i=n-1,n-2,\cdots,1) \end{cases}$$

Example 6.4.1 Solve the system by using triangular decomposition

$$\begin{pmatrix}1 & 2 & -1\\1 & -1 & 5\\4 & 1 & -2\end{pmatrix}\begin{pmatrix}x_1\\x_2\\x_3\end{pmatrix}=\begin{pmatrix}3\\0\\2\end{pmatrix}$$

Solution Let

$$\begin{pmatrix}l_{11} & & \\l_{21} & l_{22} & \\l_{31} & l_{32} & l_{33}\end{pmatrix}\begin{pmatrix}1 & u_{12} & u_{13}\\ & 1 & u_{23}\\ & & 1\end{pmatrix}=\begin{pmatrix}1 & 2 & -1\\1 & -1 & 5\\4 & 1 & -2\end{pmatrix}.$$

By comparing the elements in the two sides, one obtain

$$\begin{pmatrix}l_{11} & & \\l_{21} & l_{22} & \\l_{31} & l_{32} & l_{33}\end{pmatrix}=\begin{pmatrix}1 & 0 & 0\\1 & -3 & 0\\4 & -7 & -12\end{pmatrix}$$

$$\begin{pmatrix}1 & u_{12} & u_{13}\\ & 1 & u_{23}\\ & & 1\end{pmatrix}=\begin{pmatrix}1 & 2 & -1\\0 & 1 & -2\\0 & 0 & 1\end{pmatrix}$$

We have the equivalent systems

$$\begin{pmatrix}1 & 0 & 0\\1 & -3 & 0\\4 & -7 & -12\end{pmatrix}\begin{pmatrix}y_1\\y_2\\y_3\end{pmatrix}=\begin{pmatrix}3\\0\\2\end{pmatrix}$$

$$\begin{pmatrix}1 & 2 & -1\\0 & 1 & -2\\0 & 0 & 1\end{pmatrix}\begin{pmatrix}x_1\\x_2\\x_3\end{pmatrix}=\begin{pmatrix}y_1\\y_2\\y_3\end{pmatrix}$$

These imply $y_1=3,\ y_2=1,\ y_3=\frac{1}{4}$ and $x_1=\frac{1}{4},\ x_2=\frac{3}{2},\ x_3=\frac{1}{4}$.

6.4.2 The Square Root Method（平方根法）

Recall that an $n\times n$ matrix A is called **symmetric** (对称的) if it has real coefficients and $A=A^{\mathrm{T}}$. A symmetric matrix A is called **positive definite** (正定的) if $x^{\mathrm{T}}Ax>0$ for all $x\in\mathbf{R}^n$ with $x\neq 0$. Positive definite matrices have positive diagonal elements . Let A be a symmetric matrix and its all leading principal minors be non-zero, then the ***LU*** decomposition is always possible.

$$A=\begin{pmatrix}1 & & & \\l_{21} & 1 & & \\\vdots & \vdots & \ddots & \\l_{n1} & l_{n2} & \cdots & 1\end{pmatrix}\begin{pmatrix}u_{11} & u_{12} & \cdots & u_{1n}\\ & u_{22} & \cdots & u_{2n}\\ & & \ddots & \vdots\\ & & & u_{nn}\end{pmatrix}=LU\equiv LDU_0$$

where D is a diagonal matrix, U_0 is a unit upper triangular matrix, which are of the following forms respectively.

$$D = \begin{pmatrix} u_{11} & & & \\ & u_{22} & & \\ & & \ddots & \\ & & & u_{nn} \end{pmatrix}, \quad \text{and} \quad U_0 = \begin{pmatrix} 1 & \frac{u_{12}}{u_{11}} & \cdots & \cdots & \frac{u_{1n}}{u_{11}} \\ & 1 & \frac{u_{23}}{u_{22}} & \cdots & \frac{u_{2n}}{u_{22}} \\ & & 1 & & \vdots \\ & & & \ddots & \\ & & & & 1 \end{pmatrix}$$

Since

$$A = A^{\mathrm{T}} = U_0^{\mathrm{T}}(DL^{\mathrm{T}})$$

From the uniqueness of the decomposition, we have

$$U_0^{\mathrm{T}} = L$$

therefore

$$A = LDL^{\mathrm{T}}$$

Theorem 6.4.1 Let A be an $n \times n$ symmetric matrix , if its all n leading principal minors are nonzero, then A can be uniquely decomposed

$$A = LDL^{\mathrm{T}}$$

where L is a unit lower triangular matrix and D is a diagonal matrix.

Now, let A be a positive definite matrix, then it has positive diagonal elements.

$$D = \begin{pmatrix} d_1 & & \\ & \ddots & \\ & & d_n \end{pmatrix} = \begin{pmatrix} \sqrt{d_1} & & \\ & \ddots & \\ & & \sqrt{d_n} \end{pmatrix} \begin{pmatrix} \sqrt{d_1} & & \\ & \ddots & \\ & & \sqrt{d_n} \end{pmatrix} = D^{\frac{1}{2}} D^{\frac{1}{2}}$$

hence

$$A = LDL^{\mathrm{T}} = LD^{\frac{1}{2}} D^{\frac{1}{2}} L^{\mathrm{T}} = \left(LD^{\frac{1}{2}} \right) \left(LD^{\frac{1}{2}} \right)^{\mathrm{T}} = L_1 L_1^{\mathrm{T}}$$

where $L_1 = LD^{\frac{1}{2}}$ is a lower triangular matrix.

Theorem 6.4.2 (Cholesky decomposition, 乔来斯基分解) Let A be an $n \times n$ positive definite matrix, then there exists a real nonsingular lower triangular matrix L such that $A = LL^{\mathrm{T}}$. If we restrict that the diagonal elements are positive, then the decomposition is unique.

Let

$$A = \begin{pmatrix} l_{11} & & & \\ l_{21} & l_{22} & & \\ \vdots & \vdots & \ddots & \\ l_{n1} & l_{n2} & \cdots & l_{nn} \end{pmatrix} \begin{pmatrix} l_{11} & l_{21} & \cdots & l_{n1} \\ & l_{22} & \cdots & l_{n2} \\ & & \ddots & \vdots \\ & & & l_{nn} \end{pmatrix} = LL^{\mathrm{T}}$$

where $l_{ii} > 0 (i = 1, 2, \cdots, n)$, we obtain the computational formula of the square root method for solving a system with positive definite matrix.

(1) Solve Cholesky decomposition for $\boldsymbol{A}$, we have

$$l_{11}=\sqrt{a_{11}},\ l_{i1}=a_{i1}/l_{11}\quad (i=2,\cdots,n)$$

$$\begin{cases} l_{jj}=\left(a_{jj}-\sum_{k=1}^{j-1} l_{jk}^{2}\right)^{1/2} \\ l_{ij}=(a_{ij}-\sum_{k=1}^{j-1} l_{ik}l_{jk})\Big/l_{jj} \quad (i=j+1,\cdots,n,\ j\neq n) \end{cases}$$

To solve the system of equations $\boldsymbol{Ax}=\boldsymbol{b}$, it is enough to solve this problem in two stages:

$$\boldsymbol{Ly}=\boldsymbol{b} \qquad \text{solve for } \boldsymbol{y}$$

$$\boldsymbol{L}^{\mathrm{T}}\boldsymbol{x}=\boldsymbol{y} \qquad \text{solve for } \boldsymbol{x}$$

(2) Solve $\boldsymbol{Ly}=\boldsymbol{b}$, we have

$$\begin{cases} y_1=b_1/l_{11} \\ y_i=(b_i-\sum_{k=1}^{i-1} l_{ik}y_k)\Big/l_{ii} \quad (i=2,\cdots,n) \end{cases}$$

(3) Solve $\boldsymbol{L}^{\mathrm{T}}\boldsymbol{x}=\boldsymbol{y}$, we have

$$\begin{cases} x_n=y_n/l_{nn} \\ x_i=(y_i-\sum_{k=i+1}^{n} l_{ki}x_k)\Big/l_{ii} \quad (i=n-1,\cdots,2,1) \end{cases}$$

Example 6.4.2 Find the Cholesky factorization of the matrix in Example 6.2.3.

Solution From Example 6.2.3, $\boldsymbol{A}$ has the factorization in the following

$$\boldsymbol{A}=\begin{pmatrix} 1 & & \\ 1/2 & 1 & \\ 1/3 & 1 & 1 \end{pmatrix}\begin{pmatrix} 60 & & \\ & 5 & \\ & & 1/3 \end{pmatrix}\begin{pmatrix} 1 & 1/2 & 1/3 \\ & 1 & 1 \\ & & 1 \end{pmatrix}=\boldsymbol{LD\overline{U}}$$

The Cholesky factorization is obtained by splitting (分裂) D into the form $\boldsymbol{D}^{1/2}\boldsymbol{D}^{1/2}$ in the $\boldsymbol{LD\overline{U}}$-factorization and associating one factor with L and the other with $\overline{\boldsymbol{U}}$. Thus,

$$\boldsymbol{A}=\begin{pmatrix} 1 & & \\ 1/2 & 1 & \\ 1/3 & 1 & 1 \end{pmatrix}\begin{pmatrix} \sqrt{60} & & \\ & \sqrt{5} & \\ & & 1/\sqrt{3} \end{pmatrix}\begin{pmatrix} \sqrt{60} & & \\ & \sqrt{5} & \\ & & 1/\sqrt{3} \end{pmatrix}\begin{pmatrix} 1 & 1/2 & 1/3 \\ & 1 & 1 \\ & & 1 \end{pmatrix}$$

$$=\begin{pmatrix} \sqrt{60} & & \\ \frac{1}{2}\sqrt{60} & \sqrt{5} & \\ \frac{1}{3}\sqrt{60} & \sqrt{5} & \frac{\sqrt{3}}{3} \end{pmatrix}\begin{pmatrix} \sqrt{60} & \frac{1}{2}\sqrt{60} & \frac{1}{3}\sqrt{60} \\ & \sqrt{5} & \sqrt{5} \\ & & \frac{\sqrt{3}}{3} \end{pmatrix}=\boldsymbol{\overline{L}\overline{L}}^{\mathrm{T}}$$

Example 6.4.3 Solve the system of linear equations by Cholesky decomposition with four-decimal-digit .

$$\begin{cases} 5x_1 - 3x_2 + x_3 = 3 \\ -3x_1 + 2x_2 - x_3 = -2 \\ x_1 - x_2 + 4x_3 = 4 \end{cases}$$

Solution (1) $l_{11} = \sqrt{a_{11}} \approx 2.236\,1$

$$l_{21} = a_{21} / l_{11} \approx -1.341\,6$$

$$l_{31} = a_{31} / l_{11} \approx 0.447\,2$$

$$l_{22} = (a_{22} - l_{21}^2)^{\frac{1}{2}} \approx 0.447\,3$$

$$l_{32} = (a_{32} - l_{31}l_{21}) / l_{22} \approx -0.894\,3$$

$$l_{33} = (a_{33} - l_{31}^2 - l_{32}^2)^{\frac{1}{2}} \approx 1.732\,1$$

(2) Solve $\boldsymbol{Ly} = \boldsymbol{b}$, we have

$$y_1 \approx 1.341\,6, \quad y_2 \approx -0.447\,4, \quad y_3 \approx 1.732\,0$$

(3) Solve $\boldsymbol{L}^{\mathrm{T}}\boldsymbol{x} = \boldsymbol{y}$, we have

$$x_1 \approx 0.999\,3, \quad x_2 \approx 0.998\,9, \quad x_3 \approx 0.999\,9$$

The exact solution

$$x_1 = 1, \quad x_2 = 1, \quad x_3 = 1$$

6.4.3 The Speedup Method（追赶法）

In applications, systems of equations often arise in which the coefficient matrix has a special structure. We consider one example of this, the tridiagonal (三对角) system.

Given a tridiagonal system of equations of diagonal dominance

$$\boldsymbol{Ax} = \boldsymbol{f}$$

or

$$\begin{pmatrix} b_1 & c_1 & & & & & \\ a_2 & b_2 & c_2 & & & & \\ & \ddots & \ddots & \ddots & & & \\ & & a_i & b_i & c_i & & \\ & & & \ddots & \ddots & \ddots & \\ & & & & a_{n-1} & b_{n-1} & c_{n-1} \\ & & & & & a_n & b_n \end{pmatrix} \begin{pmatrix} x_1 \\ x_2 \\ \vdots \\ x_n \end{pmatrix} = \begin{pmatrix} f_1 \\ f_2 \\ \vdots \\ f_n \end{pmatrix}$$

Where , when $|i - j| > 1$, $a_{ij} = 0$, and:

(a) $|b_1| > |c_1| > 0$;

(b) $|b_i| \geqslant |a_i| + |c_i| > 0,\ a_i, c_i \neq 0\ (i = 2,3,\cdots,n-1)$;

(c) $|b_n| > |a_n| > 0$.

Theorem 6.4.3 Given a tridiagonal system of equations $\boldsymbol{Ax}=\boldsymbol{f}$ and $\boldsymbol{A}$ satisfies the conditions (a), (b) and (c) , then $\boldsymbol{A}$ is nonsingular.

Proof By the **method of induction** (归纳法). Obviously, if $n = 2$, then

$$\det(\boldsymbol{A}) = \begin{vmatrix} b_1 & c_1 \\ a_2 & b_2 \end{vmatrix} = b_1 b_2 - c_1 a_2 \neq 0$$

Assume that the theorem holds for order $n-1$, in the following, we prove that the theorem holds for order n. By the condition (a) $b_1 \neq 0$, and from the first elimination step we have

$$\boldsymbol{A} \to \begin{pmatrix} b_1 & c_1 & & & \\ & b_2 - \dfrac{c_1}{b_1} a_2 & c_2 & & \\ & a_3 & b_3 & c_3 & \\ & & \ddots & \ddots & \\ & & & a_n & b_n \end{pmatrix} \equiv \boldsymbol{A}^{(2)}$$

Obviously

$$\det(\boldsymbol{A}) = b_1 \det(\boldsymbol{B})$$

where

$$\boldsymbol{B} = \begin{pmatrix} \alpha_2 & c_2 & & \\ a_3 & b_3 & c_3 & \\ & \ddots & \ddots & \\ & & a_n & b_n \end{pmatrix}, \quad \alpha_2 = b_2 - \frac{c_1}{b_1} a_2$$

and

$$|\alpha_2| = \left| b_2 - \frac{c_1}{b_1} a_2 \right| \geqslant |b_2| - \left| \frac{c_1}{b_1} \right| |a_2| > |b_2| - |a_2| \geqslant |c_2| \neq 0$$

By **inductive assumption** (归纳假设), we have $\det(\boldsymbol{B}) \neq 0$, hence

$$\det(\boldsymbol{A}) \neq 0$$

□

Theorem 6.4.4 Given a tridiagonal system of equations $\boldsymbol{Ax} = \boldsymbol{f}$, and $\boldsymbol{A}$ satisfies the conditions (a), (b) and (c), then the all leading principal minors of $\boldsymbol{A}$ are nonzero.

The factorization algorithms can be simplified considerably in the case of tridiagonal matrices because a large number of zeros appear in these matrices in regular patterns. It is particularly interesting to observe the form the Crout or Doolittle method assumes in this case.

To illustrate the situation, suppose a tridiagonal matrix $\boldsymbol{A}$ can be factored into the triangular matrices $\boldsymbol{L}$ and $\boldsymbol{U}$. Since $\boldsymbol{A}$ has only $(3n-2)$ nonzero entries, there are only $(3n-2)$

conditions to be applied to determine the entries of $\boldsymbol{L}$ and $\boldsymbol{U}$, provided, of course, that the zero entries of $\boldsymbol{A}$ are also obtained. Suppose that the matrices can be found in the form

$$\boldsymbol{L}=\begin{pmatrix} p_1 & & & & & \\ r_2 & p_2 & & & & \\ & \ddots & \ddots & & & \\ & & r_i & p_i & & \\ & & & \ddots & \ddots & \\ & & & & r_n & p_n \end{pmatrix} \text{ and } \boldsymbol{U}=\begin{pmatrix} 1 & q_1 & & & \\ & \ddots & \ddots & & \\ & & 1 & q_i & \\ & & & \ddots & \ddots & \\ & & & & 1 & q_{n-1} \\ & & & & & 1 \end{pmatrix}$$

That is

$$\boldsymbol{A}=\begin{pmatrix} b_1 & c_1 & & & & & \\ a_2 & b_2 & c_2 & & & & \\ & \ddots & \ddots & \ddots & & & \\ & & a_i & b_i & c_i & & \\ & & & \ddots & \ddots & \ddots & \\ & & & & a_{n-1} & b_{n-1} & c_{n-1} \\ & & & & & a_n & b_n \end{pmatrix}$$

$$=\begin{pmatrix} p_1 & & & & & \\ r_2 & p_2 & & & & \\ & \ddots & \ddots & & & \\ & & r_i & p_i & & \\ & & & \ddots & \ddots & \\ & & & & r_n & p_n \end{pmatrix}\begin{pmatrix} 1 & q_1 & & & & \\ & \ddots & \ddots & & & \\ & & 1 & q_i & & \\ & & & \ddots & \ddots & \\ & & & & 1 & q_{n-1} \\ & & & & & 1 \end{pmatrix}$$

There are $(2n-1)$ undetermined entries of $\boldsymbol{L}$ and $(n-1)$ undetermined entries of $\boldsymbol{U}$, which totals the number of conditions, $(3n-2)$. The 0 entries of $\boldsymbol{A}$ are obtained automatically.

The multiplication involved with $\boldsymbol{A}=\boldsymbol{LU}$ gives, in addition to the 0 entries:

(1) $b_1 = p_1$, $c_1 = p_1 q_1$, $q_1 = c_1/p_1$;

(2) $a_i = r_i$, $b_i = p_i + r_i q_{i-1} = p_i + a_i q_{i-1}$ $\quad (i = 2,\cdots,n)$;

(3) $c_i = p_i q_i$ $\quad (i = 2,\cdots,n-1)$.

A solution to tridiagonal system is found by the following method of speedup:

(1) The formula of decomposition

$$\begin{cases} q_1 = c_1 / b_1 \\ p_i = b_i - a_i q_{i-1} \quad (i = 2,\cdots,n) \\ q_i = c_i / p_i = c_i /(b_i - a_i q_{i-1}) \quad (i = 2,\cdots,n-1) \end{cases}$$

To solve the system of equations $\boldsymbol{Ax}=\boldsymbol{f}$, it is enough to solve this problem in two stages:

$$\boldsymbol{Ly}=\boldsymbol{f} \qquad \text{solve for } \boldsymbol{y}$$
$$\boldsymbol{Ux}=\boldsymbol{y} \qquad \text{solve for } \boldsymbol{x}$$

(2) The recursive formula of solving $\boldsymbol{Ly=f}$

$$\begin{cases} y_1 = f_1 / b_1 \\ y_i = (f_i - a_i y_{i-1}) / (b_i - a_i q_{i-1}) \quad (i = 2,3,\cdots,n) \end{cases}$$

(3) The recursive formula of solving $\boldsymbol{Ux=y}$

$$\begin{cases} x_n = y_n \\ x_i = y_i - q_i x_{i+1} \quad (i = n-1,\cdots,2,1) \end{cases}$$

Theorem 6.4.5 Given a tridiagonal system of equations $\boldsymbol{Ax}=\boldsymbol{f}$, where $\boldsymbol{A}$ satisfies the conditions (a), (b) and (c), then $\boldsymbol{A}$ is nonsingular and $\{p_i\},\{q_i\}$ in the speedup formula satisfy:

(1) $0<|q_i|<1 \quad (i=1,2,\cdots,n-1)$;

(2) $0<|c_i|\leqslant|b_i|-|a_i|<|p_i|<|b_i|+|a_i| \quad (i=2,\cdots,n-1)$,

$0<|b_n|-|a_n|<|p_n|<|b_n|+|a_n|$.

Proof By the method of induction. When n = 1, from $p_1=b_1\neq 0$ and $|b_1|>|c_1|>0$, we obtain $0<|q_1|<1$.

Assume (1) holds for $i-1$, i.e., $0<|q_{i-1}|<1$, then

$$|p_i|=|b_i-a_i q_{i-1}|\geqslant|b_i|-|a_i q_{i-1}|>|b_i|-|a_i|\geqslant|c_i|\neq 0$$

Note that $q_i=c_i/p_i$, we obtain $0<|q_i|<1$ and

$$|p_i|=|b_i-a_i q_{i-1}|\leqslant|b_i|+|a_i q_{i-1}|<|b_i|+|a_i|$$

We have finished (1) and (2). □

Example 6.4.4 Solve the system by the speedup method.

$$\begin{pmatrix} 2 & -1 & & \\ -1 & 2 & -1 & \\ & -1 & 2 & -1 \\ & & -1 & 2 \end{pmatrix}\begin{pmatrix} x_1 \\ x_2 \\ x_3 \\ x_4 \end{pmatrix}=\begin{pmatrix} 1 \\ 0 \\ 0 \\ 1 \end{pmatrix}$$

Solution (1) Compute $\{q_i\}$

$$\begin{cases} q_1 = c_1/b_1 \\ q_i = c_i / (b_i - a_i q_{i-1}) \quad (i=2,\cdots,n-1) \end{cases}$$

$$q_1=-\frac{1}{2},\quad q_2=-\frac{2}{3},\quad q_3=-\frac{3}{4}$$

(2) Compute $\{y_i\}$

$$\begin{cases} x_n = y_n \\ x_i = y_i - q_i x_{i+1} \quad (i = n-1,\cdots,2,1) \end{cases}$$

$$x_1 = 1,\ x_2 = 1,\ x_3 = 1,\ x_4 = 1.$$

Where we have the decomposition:

$$\begin{pmatrix} 2 & -1 & & \\ -1 & 2 & -1 & \\ & -1 & 2 & -1 \\ & & -1 & 2 \end{pmatrix} = \begin{pmatrix} 2 & & & \\ -1 & 3/2 & & \\ & -1 & 4/3 & \\ & & -1 & 5/4 \end{pmatrix} \begin{pmatrix} 1 & -1/2 & & \\ & 1 & -2/3 & \\ & & 1 & -3/4 \\ & & & 1 \end{pmatrix}$$

6.5 Analysis of Round-off Errors

6.5.1 Condition Number

Consider the linear system

$$\boldsymbol{Ax}=\boldsymbol{b}$$

where matrix $\boldsymbol{A}$ is nonsingular, $\boldsymbol{x}$ is the exact solution of the system. Since the data are obtained by measurement or computation, therefore there exist errors in $\boldsymbol{A}$ or $\boldsymbol{b}$. We will discuss the influence of a small error in the data of $\boldsymbol{A}$ (or $\boldsymbol{b}$) to the solution.

We begin with an example

$$\begin{cases} 5x_1 + 7x_2 = 0.7 \\ 7x_1 + 10x_2 = 1 \end{cases}$$

The exact solution $\boldsymbol{x} = (0.0,\ 0.1)^{\mathrm{T}}$

Let

$$\boldsymbol{b} \to \boldsymbol{b}+\delta\boldsymbol{b}=(0.69, 1.01)^{\mathrm{T}}, \quad \text{i.e., } \delta\boldsymbol{b}=(-0.01, 0.01)^{\mathrm{T}}$$

Denote

$$\begin{cases} 5\tilde{x}_1 + 7\tilde{x}_2 = 0.69 \\ 7\tilde{x}_1 + 10\tilde{x}_2 = 1.01 \end{cases}$$

Then

$$\tilde{\boldsymbol{x}} = (-0.17,\ 0.22)^{\mathrm{T}}$$

Obviously, we obtain **drastic** (猛烈，激烈) changes in the solution.

Given a linear system

$$\boldsymbol{Ax}=\boldsymbol{b} \tag{6.5.1}$$

where matrix $\boldsymbol{A}$ is nonsingular, $\boldsymbol{x}$ is the exact solution. Suppose that $\boldsymbol{A}$ is exact, $\boldsymbol{b}$ has error $\delta\boldsymbol{b}$, the solution is $\boldsymbol{x}+\delta\boldsymbol{x}$, then

$$\boldsymbol{A}(\boldsymbol{x}+\delta\boldsymbol{x}) = \boldsymbol{b}+\delta\boldsymbol{b}, \quad \delta\boldsymbol{x} = \boldsymbol{A}^{-1}\delta\boldsymbol{b}$$

$$\|\delta x\| \leqslant \|A^{-1}\|\|\delta b\| \tag{6.5.2}$$

from (6.5.1), we have

$$\|b\| \leqslant \|A\|\|x\|$$

That is

$$\frac{1}{\|x\|} \leqslant \frac{\|A\|}{\|b\|} \quad (\text{ let}\|b\| \neq 0 \ ) \tag{6.5.3}$$

From (6.5.2) and (6.5.3), we obtain:

$$\frac{\|\delta x\|}{\|x\|} \leqslant \|A^{-1}\|\|A\|\frac{\|\delta b\|}{\|b\|} \tag{6.5.4}$$

Inequality (6.5.4) tells us that the relative error in x is no greater than $\|A^{-1}\|\|A\|$ times the relative error in b. From Inequality (6.5.4), we see that if $\|A^{-1}\|\|A\|$ is small, then small **perturbations** (扰动,摄动) in b lead to small perturbations in x.

Definition 6.5.1 Let matrix A be nonsingular, we call

$$\text{cond}(A)_v = \|A^{-1}\|_v\|A\|_v \quad (v = 1,2 \text{ or } \infty)$$

condition number (条件数) *of A*.

A matrix with large condition number is said to be **ill conditioned** (病态). For an ill conditioned matrix A, there will be cases in which the solution of a system $Ax = b$ will be very sensitive to small changes in the vector b. In other words, to **attain** (达到) a certain precision in the determination of x, we shall require significantly higher precision in b. If the condition number of A is of **moderate** (适度的) size, the matrix is said to be **well conditioned** (良态).

Now, suppose that b is exact, A has small error δA, the solution is $x + \delta x$, then

$$\begin{aligned} (A + \delta A)(x + \delta x) &= b \\ (A + \delta A)\delta x &= -(\delta A)x \end{aligned} \tag{6.5.5}$$

If δA is not restricted, then $A + \delta A$ may be singular, and

$$(A + \delta A) = A(I + A^{-1}\delta A) \tag{6.5.6}$$

From Theorem 3.3.3, if $\|A^{-1}\delta A\| < 1$, then $(I + A^{-1}\delta A)^{-1}$ exists, and from (6.5.5), (6.5.6), we have

$$\delta x = -(I + A^{-1}\delta A)^{-1}A^{-1}(\delta A)x$$

hence

$$\|\delta x\| \leqslant \frac{\|A^{-1}\|\|\delta A\|\|x\|}{1 - \|A^{-1}(\delta A)\|}$$

Let $\|A^{-1}\|\|\delta A\| < 1$, we obtain

$$\frac{\|\delta x\|}{\|x\|} \leqslant \frac{\|A^{-1}\|\|A\|\frac{\|\delta A\|}{\|A\|}}{1-\|A^{-1}\|\|A\|\frac{\|\delta A\|}{\|A\|}} \tag{6.5.7}$$

Example 6.5.1 Given the system

$$\begin{pmatrix} 3 & 1.001 \\ 6 & 1.997 \end{pmatrix}\begin{pmatrix} x_1 \\ x_2 \end{pmatrix} = \begin{pmatrix} 1.999 \\ 4.003 \end{pmatrix}$$

compute $\text{cond}(A)_\infty$.

Solution

$$A = \begin{pmatrix} 3 & 1.001 \\ 6 & 1.997 \end{pmatrix}, \quad A^{-1} = \frac{1}{-0.015}\begin{pmatrix} 1.997 & -1.001 \\ -6 & 3 \end{pmatrix}$$

and
$$\|A\|_\infty = 7.997 \ , \quad \|A^{-1}\|_\infty = 600$$

therefore

$$\text{cond}(A)_\infty = 4\,798.2$$

Hence the given system is ill-conditioned linear system.

Usually, we use the following condition numbers:

(1) $\text{cond}(A)_\infty = \|A^{-1}\|_\infty \|A\|_\infty$.

(2) spectral condition number

$$\text{cond}(A)_2 = \|A^{-1}\|_2 \|A\|_2 = \sqrt{\frac{\lambda_{\max}(A^{\mathrm{T}}A)}{\lambda_{\min}(A^{\mathrm{T}}A)}}$$

If A is symmetric, then

$$\text{cond}(A)_2 = \frac{|\lambda_1|}{|\lambda_n|}$$

where λ_1 and λ_n are eigenvalues with the greatest and smallest absolute values respectively.

Properties of the condition number:

(1) For any nonsingular matrix A, we have $\text{cond}(A)_v \geqslant 1$.

(2) If matrix A is nonsingular and $c \neq 0$ (const), then

$$\text{cond}(cA)_v = \text{cond}(A)_v$$

(3) If A is an **orthogonal matrix** (正交阵) ($A^{\mathrm{T}}A = I$), then $\text{cond}(A)_2 = 1$; If A is nonsingular matrix, R is an orthogonal Matrix, then

$$\text{cond}(RA)_2 = \text{cond}(AR)_2 = \text{cond}(A)_2$$

Example 6.5.2 Given Hilbert matrix

$$H_n=\begin{pmatrix}1 & \frac{1}{2} & \cdots & \frac{1}{n}\\ \frac{1}{2} & \frac{1}{3} & \cdots & \frac{1}{n+1}\\ \vdots & \vdots & & \vdots\\ \frac{1}{n} & \frac{1}{n+1} & \cdots & \frac{1}{2n-1}\end{pmatrix}$$

compute the condition number of H_3.

Solution

$$H_3=\begin{pmatrix}1 & \frac{1}{2} & \frac{1}{3}\\ \frac{1}{2} & \frac{1}{3} & \frac{1}{4}\\ \frac{1}{3} & \frac{1}{4} & \frac{1}{5}\end{pmatrix},\quad H_3^{-1}=\begin{pmatrix}9 & -36 & 30\\ -36 & 192 & -180\\ 30 & -180 & 180\end{pmatrix}$$

$$\|H_3\|_\infty=11/6,\quad \|H_3^{-1}\|_\infty=408$$

therefore

$$\text{cond}(H_3)=748$$

Analogously

$$\text{cond}\,(H_6)_\infty=2.9\times10^7,\quad \text{cond}\,(H_7)_\infty=9.85\times10^8$$

We consider the linear system

$$H_3x=(11/6,\ \ 13/12,\ \ 47/60)^{\mathrm{T}}=b$$

Let H_3 and b have small errors (take three significant digits), then

$$\begin{pmatrix}1.00 & 0.500 & 0.333\\ 0.500 & 0.333 & 0.250\\ 0.333 & 0.250 & 0.200\end{pmatrix}\begin{pmatrix}x_1+\delta x_1\\ x_2+\delta x_2\\ x_3+\delta x_3\end{pmatrix}=\begin{pmatrix}1.83\\ 1.08\\ 0.783\end{pmatrix}\tag{6.5.8}$$

denoted $(H_3+\delta H_3)(x+\delta x)=b+\delta b$, the exact solutions of the system $H_3x=b$ and (6.5.8) are $x=(1,\ 1,\ 1)^{\mathrm{T}}$ and $x+\delta x=(\ 1.089\,512\,538\ ,\ 0.487\,967\,062\ ,\ 1.491\,002\,798\ )^{\mathrm{T}}$ respectively, hence

$$\delta x=(\ 0.089\,5\ ,\ -0.512\,0\ ,\ 0.491\,0\ )^{\mathrm{T}}$$

$$\frac{\|\delta H_3\|_\infty}{\|H_3\|_\infty}\approx0.18\times10^{-3}<0.02\%$$

$$\frac{\|\delta b\|_\infty}{\|b\|_\infty}\approx0.182\%,\qquad \frac{\|\delta x\|_\infty}{\|x\|_\infty}\approx51.2\%$$

That is, the relative errors of H_3 and b are less than 0.2%, the relative error of the solution is greater than 50%.

6.5.2 Iterative Refinement（迭代改善）

If $\boldsymbol{x}^{(0)}$ is an approximate solution of the equation

$$\boldsymbol{Ax}=\boldsymbol{b}$$

then the **precise solution** (精确解) x is given by

$$\boldsymbol{x}=\boldsymbol{x}^{(0)}+(\boldsymbol{x}-\boldsymbol{x}^{(0)})=\boldsymbol{x}^{(0)}+\boldsymbol{A}^{-1}(\boldsymbol{b}-\boldsymbol{Ax}^{(0)})=\boldsymbol{x}^{(0)}+\boldsymbol{e}^{(0)}$$

where $\boldsymbol{e}^{(0)}=\boldsymbol{A}^{-1}(\boldsymbol{b}-\boldsymbol{Ax}^{(0)})$ and is called the **error vector.** The **residual vector** (剩余向量) corresponding to the approximate solution $\boldsymbol{x}^{(0)}$ is $\boldsymbol{r}^{(0)}=\boldsymbol{b}-\boldsymbol{Ax}^{(0)}$. It is computable. Of course, we do not want to compute $\boldsymbol{A}^{-1}$, but the vector $\boldsymbol{e}^{(0)}=\boldsymbol{A}^{-1}\boldsymbol{r}^{(0)}$ can be obtained by solving the equation

$$\boldsymbol{Ae}^{(0)}=\boldsymbol{r}^{(0)}$$

These remarks lead to a numerical procedure called **iterative improvement** or **iterative refinement**, which we now describe in more detail.

Suppose that the equation $\boldsymbol{Ax}=\boldsymbol{b}$ has been solved by Gaussian elimination. Since the result is not expected to be the exact solution (because of round-off error), we denote it by $\boldsymbol{x}^{(0)}$ and then compute $\boldsymbol{r}^{(0)},\boldsymbol{e}^{(0)}$, and $\boldsymbol{x}^{(1)}$ by the three equations

$$\begin{cases}\boldsymbol{r}^{(0)}=\boldsymbol{b}-\boldsymbol{Ax}^{(0)}\\ \boldsymbol{Ae}^{(0)}=\boldsymbol{r}^{(0)}\\ \boldsymbol{x}^{(1)}=\boldsymbol{x}^{(0)}+\boldsymbol{e}^{(0)}\end{cases} \tag{6.5.9}$$

To obtain better solutions $\boldsymbol{x}^{(2)},\boldsymbol{x}^{(3)},\cdots$, this process can be repeated. The success of the method depends on computing the residuals $\boldsymbol{r}^{(i)}$ in double precision to avoid the loss of significance expected in the subtraction. Thus, the expression $b_i-\sum_{j=1}^{n}a_{ij}x_j^{(0)}$ is evaluated in double precision.

Example 6.5.3 Solve the system by using iterative refinement

$$\begin{pmatrix}3.3330 & 15920 & -10.333\\ 2.2220 & 16.710 & 9.6120\\ 1.5611 & 5.1791 & 1.6852\end{pmatrix}\begin{pmatrix}x_1\\ x_2\\ x_3\end{pmatrix}=\begin{pmatrix}15913\\ 28.544\\ 8.4254\end{pmatrix}$$

Solution The given system has the exact solution $\boldsymbol{x}=(1,1,1)^{\mathrm{T}}$.

First, using Gaussian elimination with five-digit rounding arithmetic leads to the approximate solution

$$\boldsymbol{x}^{(0)}=(1.2001,\ 0.99991,\ 0.92538)^{\mathrm{T}}$$

The residual vector corresponding to $\boldsymbol{x}^{(0)}$ is computed to be

$$\boldsymbol{r}^{(0)}=\boldsymbol{b}-\boldsymbol{Ax}^{(0)}=(-0.00518,0.27412914,-0.186160367)^{\mathrm{T}}$$

and the solution to $\boldsymbol{Ae}^{(0)}=\boldsymbol{r}^{(0)}$ to be

$$\boldsymbol{e}^{(0)}=(-0.20008,8.9987\times10^{-5},0.074607)^{\mathrm{T}}$$

we obtain

$$\boldsymbol{x}^{(1)}=\boldsymbol{x}^{(0)}+\boldsymbol{e}^{(0)}=(1.0000,1.0000,0.99999)^{\mathrm{T}}$$

and the actual error in this approximation is

$$\left\|x-x^{(1)}\right\|_{\infty}=1\times10^{-5}$$

Repeat this process, we compute $r^{(1)}=b-Ax^{(1)}$ and solve the system $Ae^{(1)}=r^{(1)}$, which gives

$$e^{(1)}=(1.500\,2\times10^{-9},\ \ 2.095\,1\times10^{-10},\ \ 1.000\,0\times10^{-5})^{\mathrm{T}}$$

we conclude that

$$x^{(2)}=x^{(1)}+e^{(1)}=(1.000\,0,\ \ 1.000\,0,\ \ 1.000\,0)^{\mathrm{T}}$$

is sufficiently accurate, which is certainly correct.

6.6 Computer Experiments

6.6.1 Functions Needed in the Experiments by Mathematica

（1）Table——建表函数.

格式 1：`Table [expr,{n}]`

功能：生成具有 n 个相同项 expr 的一维表.

格式 2：`Table[f,{i,imin,imax},{j,jmin,jmax}]`

功能：生成 f 的二维数值表，i 从 $i$min 变到 $i$max，j 从 $j$min 变到 $j$max，步长均为 1.

说明：

① 对应 i 从 $i$min 到 $i$max 的每一个取值，j 优先从 $j$min 到 $j$max 跑一趟；

② 一般地，表需先定义后使用，如 Table[0,{n},{n}] 为定义了一个 n 阶二维表；

③ 通过变量赋值的方法（即变量={表 1，表 2，…，表 n}）也可生成表，但若表 i（i=1,2,…,n）中的元素个数不等，则进行矩阵运算时会出错。

（2）IdentityMatrix[n]　　生成 n 阶单位阵（表输出方式）.

（3）MatrixForm[list]]　　按矩阵形式输出表 list.

6.6.2 Experiments by Mathematica

1．Gauss 消元法

（1）算法.

① 输入：变量个数 n，系数矩阵 A，常数矩阵 b；

② 消元：作 n−1 次循环：

a. 判断对角线元素 $|a_{kk}|<r=10^{-10}$ ？

如果为真，输出“消元失败”，结束；

否则，执行 b；

b. 作 n−k 次循环：

$a_{ik}\longleftarrow a_{ik}/a_{kk}$；

$a_{ij}\leftarrow a_{ij}+a_{ik}a_{kj}$ ($j=k+1,k+2,\cdots,n$)；

$b_i\leftarrow b_i+a_{ik}b_k$；

③ $x_n \leftarrow b_n / a_{nn}$；

④ 回代：作 n−1 次逆向循环（$k=n-1,n-2,\cdots,1$），

$$x_k \leftarrow (b_k - \sum_{j=k+1}^{n} a_{kj} x_j) / a_{kk}；$$

⑤ 输出：方程组的解 x_k；

（2）程序清单.

```
(*Gaussian Elimination*)
Clear[a, b, x];
n=Input["n ="];
a=Input["A="];
b=Input["b="];
r=10^(-10);
t=1;
Do[If [Abs[a[[k, k]]]<r,
   Print["Gauss 消元失败"];
   t=0;
   Break[]
   ];
   Do[m=-a[[i,k]]/a[[k,k]];
   Do[a[[i,j]]=a[[i,j]]+m*a[[k,j]],{j,k+1,n}];
   b[[i]]=b[[i]]+m*b[[k]],
   {i,k+1,n}],{k,1,n}
];
x=Table[0,{n}];
If[t==1,x[[n]]=b[[n]]/a[[n, n]];
Do[x[[k]]=(b[[k]]-Sum[a[[k, j]]*x[[j]],{j,k+1,n}])/a[[k,k]], {k,n-1,1,
-1} ];
   Print["Ax=b 的解为", x]
]
```

（3）变量说明.

n：保存方程组变量个数；

x：保存线性方程组的解；

a：保存线性方程组的系数矩阵及计算的中间结果；

b：保存线性方程组的常数矩阵；

r：微小值，可修改.

（4）计算实例.

Example 6.6.1　Use the Gaussian elimination to solve the following linear system

$$\begin{pmatrix} -3 & 2 & -1 \\ 6 & -6 & 7 \\ 3 & -4 & 4 \end{pmatrix} \begin{pmatrix} x_1 \\ x_2 \\ x_3 \end{pmatrix} = \begin{pmatrix} -1 \\ -7 \\ -6 \end{pmatrix}$$

操作步骤如下：

① 将光标定位在要执行的 Cell 中，按小键盘的【Enter】键；

② 在弹出的对话框中按提示分别输入

```
3,{{-3,2,-1},{6,-6,7},{3,-4,4}},{-1,-7,-6}
```

③ 每次输入后单击 OK 命令按钮，输出结果为：

```
Ax=b 的解为 {2,2,-1}
```

2. Doolittle 分解法

（1）算法.

① 输入：变量个数 n，系数矩阵 $\boldsymbol{A}$，常数矩阵 $\boldsymbol{b}$；

② 判断 $a_{11}=0$？

如果为真，输出“***LU*** 分解失败”，结束；

否则，$a_{j1} \leftarrow a_{j1}/a_{11}$（$j=2,\cdots,n$）；

③ 作 $n-1$ 次循环：

$$a_{ii} \leftarrow a_{ii} - \sum_{k=1}^{i-1} a_{ik} a_{ki};$$

④ 判断 $a_{ii}=0$ ？

如果为真，输出“不能进行 ***LU*** 分解”，结束；

否则，执行⑤；

⑤ 作 $n-i$ 次循环($j=i+1, i=2,\cdots,n$)：

$$a_{ij} \leftarrow a_{ij} - \sum_{k=1}^{i-1} a_{ik} a_{kj};$$

$$a_{ji} \leftarrow (a_{ji} - \sum_{k=1}^{i-1} a_{jk} a_{ki}) / a_{ii};$$

⑥ 作 n 次逆向循环（$k=n,n-1,\cdots,1$）

$$x_k \leftarrow (b_k - \sum_{j=k+1}^{n} a_{kj} x_j) / a_{kk};$$

⑦ 输出.

（2）程序清单.

```
(*Doolittle Factorization*)
Clear[a,b,x,r,n];
n=Input["n="];
a=Input["A="];
b=Input["b="];
t=1;
r=0.000000001;
```

```
  If[Abs[a[[1,1]]]<r,
     t=0; Break[],
     Do[a[[j,1]]=a[[j,1]]/a[[1,1]],
     {j,2,n}]
   ];
  If[t==1,
   Do[a[[i,i]]=a[[i,i]]-Sum[a[[i,k]]*a[[k,i]],{k,1,i-1}];
      If [ Abs[a[[i,i]]]< r,
         t=0;Print["不能分解"]; Break[],
         Do[a[[i,j]]=a[[i,j]]-Sum[a[[i,k]]*a[[k,j]],{k,1,i-1}];
            a[[j,i]]=a[[j,i]]-Sum[a[[j,k]]*a[[k,i]],{k,1,i-1}];
            a[[j,i]]=a[[j,i]]/a[[i,i]],
            {j,i+1,n}
            ]
        ],
     {i,2,n}
      ]
   ];
  If[ t==1,
    Do[b[[i]]=b[[i]]-Sum[a[[i,k]]*b[[k]],{k,1,i-1}],{i,1,n}];
    Print["Ly=b 的解 y=",b];
    Do[b[[i]]=(b[[i]]-Sum[a[[i, k]]*b[[k]],{k, i+1,n}])/a[[i, i]],{i,
n,1,-1}];
    u=IdentityMatrix[n];
   Do[Do[u[[i,j]]=a[[i,j]],{i,1,j}],{j,1,n}];
    l=a-u+IdentityMatrix[n];
    Print["LU 分解紧凑格式=",MatrixForm[a]];
    Print["矩阵 L=",MatrixForm[l]];
    Print["矩阵 U=",MatrixForm[u]];
    Print["解向量 x=",b]
    ]
```

（3）变量说明.

n：保存方程组变量个数；

x：保存线性方程组的解；

a：保存线性方程组的系数矩阵及计算的中间结果；

b：保存线性方程组的常数矩阵；

r：微小值，可修改；

l：保存单位下三角矩阵 $\boldsymbol{L}$；

u：保存上三角矩阵 $\boldsymbol{U}$.

（4）计算实例.

Example 6.6.2 Use the Doolittle factorization to solve the following linear system

$$\begin{pmatrix} 2 & 1 & 0 & 1 \\ 2 & 3 & -1 & 1 \\ -1 & 2 & 3 & -1 \\ 3 & -1 & -1 & 2 \end{pmatrix}\begin{pmatrix} x_1 \\ x_2 \\ x_3 \\ x_4 \end{pmatrix}=\begin{pmatrix} 2 \\ 1 \\ 4 \\ -3 \end{pmatrix}$$

操作步骤如下：

① 光标定位在要执行的 Cell 中，按小键盘的【Enter】键；

② 在弹出的对话框中按提示分别输入

```
4,{{2,1,0,1},{2,3,-1,1},{-1,2,3,-1},{3,-1,-1,2}},{2,1,4,-3}
```

③ 每次输入后单击 OK 命令按钮，输出结果如图 6-2 所示.

Ly=b的解y= $\left\{2, -1, \frac{25}{4}, -\frac{67}{17}\right\}$

LU分解紧凑格式= $\begin{pmatrix} 2 & 1 & 0 & 1 \\ 1 & 2 & -1 & 0 \\ -\frac{1}{2} & \frac{5}{4} & \frac{17}{4} & -\frac{1}{2} \\ \frac{3}{2} & -\frac{5}{4} & -\frac{9}{17} & \frac{4}{17} \end{pmatrix}$

矩阵L = $\begin{pmatrix} 1 & 0 & 0 & 0 \\ 1 & 1 & 0 & 0 \\ -\frac{1}{2} & \frac{5}{4} & 1 & 0 \\ \frac{3}{2} & -\frac{5}{4} & -\frac{9}{17} & 1 \end{pmatrix}$

矩阵U= $\begin{pmatrix} 2 & 1 & 0 & 1 \\ 0 & 2 & -1 & 0 \\ 0 & 0 & \frac{17}{4} & -\frac{1}{2} \\ 0 & 0 & 0 & \frac{4}{17} \end{pmatrix}$

解向量x= $\left\{\frac{39}{4}, -\frac{3}{4}, -\frac{1}{2}, -\frac{67}{4}\right\}$

图 6-2 Doolittle 分解法计算结果

3. 追赶法

（1）算法.

① 输入：变量个数 n，系数矩阵对应的三个向量 $\boldsymbol{a}$，$\boldsymbol{b}$，$\boldsymbol{c}$，常数矩阵 $\boldsymbol{d}$；

② $p_1 \leftarrow b_1$；

$q_1 \leftarrow c_1/b_1$；

③ 作 n−1 次循环（$k=2, 3,\cdots, n$）：

判断 $b_{k-1}=0$ ？

如果为真，输出“追赶法失败”，结束；

否则，计算：

$p_k \leftarrow b_k - a_k q_{k-1}$；

$q_k \leftarrow c_k / (b_k - a_k q_{k-1})$；

④ $y_1 \leftarrow d_1/b_1$；

⑤ 作 $n-1$ 次循环（$k=2,3,\cdots,n$）

$y_k \leftarrow (d_k - a_k y_{k-1}) / (b_k - a_k q_{k-1})$；

⑥ $x_n \leftarrow y_n$；

⑦ 作 $n-1$ 次逆向循环（$k=n-1,n-2,\cdots,1$）

$x_k \leftarrow y_k - q_k x_{k+1}$；

⑧ 输出 $\boldsymbol{L}$、$\boldsymbol{U}$ 及方程的解。

（2）程序清单.

```
(*Speedup Method *)
Clear[a, b, c, d, x, y];
n=Input["n="];
a=Input["a="];
b=Input["b="];
c=Input["c="];
d=Input["d="];
r=0.00001;
flag=1;
p=Table[0,{n}]; q=Table[0,{n}];
p[[1]]=b[[1]]; q[[1]]=c[[1]]/b[[1]];
Do[ If [Abs[p[[k-1]]]<r, Print["追赶法失败"]; flag=0; Break[]];
    p[[k]]=b[[k]]-a[[k]]*q[[k-1]];
    q[[k]]=c[[k]]/(b[[k]]-a[[k]]*q[[k-1]]),
    {k,2,n}
   ];
If[ flag==1,
   x=Table[0,{n}]; y=Table[0,{n}];
   y[[1]]=d[[1]]/b[[1]];
   Do[y[[k]]=(d[[k]]-a[[k]]*y[[k-1]])/(b[[k]]-a[[k]]*q[[k-1]]),
   {k,2,n}];
   x[[n]]=y[[n]];
   Do[x[[k]]=y[[k]]-q[[k]]*x[[k+1]],{k,n-1,1,-1}];
       sp=Table[0,{n},{n}]; sq=Table[0,{n},{n}];
       Do[sp[[k,k]]=p[[k]]; sq[[k,k]]=1,{k,1,n}];
    Do[sp[[k+1,k]]=a[[k+1]]; sq[[k,k+1]]=q[[k]],{k,1,n-1}];
    Print["L =", MatrixForm[sp], "  ", "U =", MatrixForm[sq] ];
    Print["Ax=d 的解为",  x]
  ]
```

（3）变量说明.

n：保存方程组变量个数；

a：保存三对角系数矩阵 ***A*** 的向量 ***a***；

b：保存三对角系数矩阵 ***A*** 的向量 ***b***；

c：保存三对角系数矩阵 ***A*** 的向量 ***c***；

d：保存线性方程组的常数矩阵；

r：微小值，可修改；

flag：解判断标志；

p：保存 ***LU*** 分解中 ***L*** 的主对角线向量；

q：保存 ***LU*** 分解中 ***U*** 的主对角线上方向量；

sp：保存下三角矩阵 ***L***；

sq：保存单位上三角矩阵 ***U***；

y：保存下三角线性方程组的解；

x：保存线性方程组的解.

（4）计算实例.

Example 6.6.3 Use the speedup method to solve the following linear system

$$\begin{pmatrix} 2 & -1 & & & \\ -1 & 2 & -1 & & \\ & -1 & 2 & -1 & \\ & & -1 & 2 & -1 \\ & & & 2 & -1 \end{pmatrix}\begin{pmatrix} x \\ x_2 \\ x_3 \\ x_4 \\ x_5 \end{pmatrix}=\begin{pmatrix} 1 \\ 1 \\ 1 \\ 1 \\ 1 \end{pmatrix}$$

操作步骤如下：

① 光标定位在要执行的 Cell 中，按小键盘的【Enter】键；

② 在弹出的对话框中按提示分别输入

```
5,{0,-1,-1,-1,2},{2,2,2,2,-1},{-1,-1,-1,-1,0},{1,1,1,1,1}
```

③ 每次输入后单击 OK 命令按钮，输出结果如图 6-3 所示.

$$L=\begin{pmatrix} 2 & 0 & 0 & 0 & 0 \\ -1 & \frac{3}{2} & 0 & 0 & 0 \\ 0 & -1 & \frac{4}{3} & 0 & 0 \\ 0 & 0 & -1 & \frac{5}{4} & 0 \\ 0 & 0 & 0 & 2 & \frac{3}{5} \end{pmatrix} \quad U=\begin{pmatrix} 1 & -\frac{1}{2} & 0 & 0 & 0 \\ 0 & 1 & -\frac{2}{3} & 0 & 0 \\ 0 & 0 & 1 & -\frac{3}{4} & 0 \\ 0 & 0 & 0 & 1 & -\frac{4}{5} \\ 0 & 0 & 0 & 0 & 1 \end{pmatrix}$$

Ax=d 的解为{1, 1, 0, -2, -5}

图 6-3 追赶法计算结果

6.6.3 Functions Needed in the Experiments by Matlab

见表 6-1.

表 6-1 函数

函 数	意 义
rank(A)	求矩阵A的秩
lu(A)	对矩阵A进行LU分解
inv(A)	求矩阵A的逆

6.6.4 Experiments by Matlab

1. Gauss 消元法

（1）函数语句.

```
x=gauss(A,b)
```

（2）参数说明.

A：矩阵变量，输入参数，方程组的系数矩阵（方阵）；

b：向量变量，输入参数，方程组的常数项向量；

x：向量变量，输出参数，方程组的解.

（3）gauss.m 程序.

```
function x=gauss(A,b)
[n,m]=size(A);
if n~=m
   fprintf('系数矩阵的行列不相等');
   return
end
if n~=length(b)
  fprintf('系数矩阵与常数项矩阵的行数不相等');
   return
end
for k=1:n-1
   if abs(A(k,k))<1e-20
      fprintf('Gauss 消元失效');
      return
   end
   for i=k+1:n
      for j=k+1:n
         A(i,j)=A(i,j)-A(k,j)*A(i,k)/A(k,k);
      end
      b(i)=b(i)-b(k)*A(i,k)/A(k,k);
   end
end
x(n)=b(n)/A(n,n);
for k=n-1:-1:1
   x(k)=b(k);
   for i=k+1:n
      x(k)=x(k)-A(k,i)*x(i);
   end
  x(k)=x(k)/A(k,k);
end
```

（4）计算实例.

Example 6.6.4 Use the Gaussian elimination to solve the following linear system

$$\begin{pmatrix} -3 & 2 & -1 \\ 6 & -6 & 7 \\ 3 & -4 & 4 \end{pmatrix}\begin{pmatrix} x_1 \\ x_2 \\ x_3 \end{pmatrix}=\begin{pmatrix} -1 \\ -7 \\ -6 \end{pmatrix}$$

在命令窗口输入：

```
>> A=[-3,2,-1; 6,-6,7; 3,-4,4]; b=[-1;-7;-6]; x=gauss(A,b)
x =
     2     2    -1
```

2. Doolittle 分解法

（1）函数语句.

```
[x,B,C]=TriangularDec(A,b)
```

（2）TriangularDec.m 程序.

```
function [x,B,C]=TriangularDec(A,b)
[B,C]=DoolFactorization(A);
y=lowerTriangular(B,b);
x=upperTriangular(C,y);
```

子程序 1：

```
function [B,C]=DoolFactorization(A)
%对 A 矩阵进行 Doolittle 分解
[n,m]=size(A);
if n~=m
    fprintf('矩阵的行列不相等');
    return
end
C(1,:)=A(1,:);B=eye(n);
B(2:n,1)=A(2:n,1)/C(1,1);
for k=2:n
    C(k,k:n)=A(k,k:n)-B(k,1:k-1)*C(1:k-1,k:n);
    B(k+1:n,k)=(A(k+1:n,k)-B(k+1:n,1:k-1)*C(1:k-1,k))/C(k,k);
end
```

子程序 2：

```
function x=lowerTriangular(A,b);
%解系数矩阵为下三角矩阵 A 的线性方程组 AX=b
[m,n]=size(A);
if m~=n
    fprintf('A 必须是方阵');
    return
end
if n~=length(b)
    error('系数矩阵和常数项矩阵的维数不相等');
end
```

```
if A(1,1)==0
   error('系数矩阵不是可逆的');
end
x(1)=b(1)/A(1,1);
for k=2:n
   if A(k,k)==0
      error('系数矩阵不是可逆的');
   end
   x(k)=(b(k)-A(k,1:k-1)*x(1:k-1)')/A(k,k);
end
```

子程序 3:

```
function x=upperTriangular(A,b);
%解系数矩阵为上三角矩阵 A 的线性方程组 AX=b
[m,n]=size(A);
if m~=n
   fprintf('A 必须是方阵');
   return
end
if n~=length(b)
   error('系数矩阵和常数项矩阵的维数不相等');
end
if A(n,n)==0
   error('系数矩阵不是可逆的');
end
x(n)=b(n)/A(n,n);
for k=n-1:-1:1
   if A(k,k)==0
      error('系数矩阵不是可逆的');
   end
   x(k)=(b(k)-A(k,k+1:n)*x(k+1:n)')/A(k,k);
end
```

（3）计算实例.

Example 6.6.5 Use the Doolittle factorization to solve the following linear system

$$\begin{pmatrix} 2 & 1 & 0 & 1 \\ 2 & 3 & -1 & 1 \\ -1 & 2 & 3 & -1 \\ 3 & -1 & -1 & 2 \end{pmatrix}\begin{pmatrix} x_1 \\ x_2 \\ x_3 \\ x_4 \end{pmatrix}=\begin{pmatrix} 2 \\ 1 \\ 4 \\ -3 \end{pmatrix}$$

在命令窗口输入:

```
>> clear
>> A=[2,1,0,1;2,3,-1,1;-1,2,3,-1;3,-1,-1,2];b=[2;1;4;-3];[x,B,C]=
TriangularDec(A,b)
x =
   9.7500  -0.7500  -0.5000  -16.7500
```

```
B =
    1.0000         0         0         0
    1.0000    1.0000         0         0
   -0.5000    1.2500    1.0000         0
    1.5000   -1.2500   -0.5294    1.0000
C =
    2.0000    1.0000         0    1.0000
         0    2.0000   -1.0000         0
         0         0    4.2500   -0.5000
         0         0         0    0.2353
```

3. 追赶法

（1）函数语句.

```
x=zgTridiagonal(a,b,c,d)
```

（2）参数说明.

a：向量变量，输入参数，系数矩阵的下对角线元素数组；

b：向量变量，输入参数，系数矩阵的主对角线元素数组；

c：向量变量，输入参数，系数矩阵的上对角线元素数组；

d：向量变量，输入参数，常数项矩阵；

x：向量变量，输出参数，方程组的解.

（3）zgTridiagonal.m 程序.

```
function x=zgTridiagonal(a,b,c,d)
n=length(b);
if length(b)~=length(a)+1|length(b)~=length(c)+1|length(b)~=length(d)
    fprintf('失败信息：输入的三对角矩阵在维数方面存在问题或是系数矩阵与常数项矩阵在维数上不匹配');
    return
end
for k=2:n
    if b(k-1)==0
        fprintf('失败信息：化简后系数矩阵的第 %3d 个主对角线元素为 0，不能应用追赶法求解。',k-1);
        return
    end
    b(k)=b(k)-c(k-1)*a(k-1)/b(k-1);
    d(k)=d(k)-d(k-1)*a(k-1)/b(k-1);
end
if b(n)==0
    fprintf('失败信息：化简后系数矩阵的第 %3d 个主对角线元素为 0，不能应用追赶法求解。',n);
    return
end
x(n)=d(n)/b(n);
for k=n-1:-1:1
```

```
    x(k)=(d(k)-c(k)*x(k+1))/b(k);
end
```

（4）计算实例.

Example 6.6.6 Use the speedup method to solve the following linear system

$$\begin{pmatrix} 2 & -1 & & & \\ -1 & 2 & -1 & & \\ & -1 & 2 & -1 & \\ & & -1 & 2 & -1 \\ & & & 2 & -1 \end{pmatrix}\begin{pmatrix} x_1 \\ x_2 \\ x_3 \\ x_4 \\ x_5 \end{pmatrix}=\begin{pmatrix} 1 \\ 1 \\ 1 \\ 1 \\ 1 \end{pmatrix}$$

在命令窗口输入：

```
>> clear
>> a=[-1,-1,-1,2];b=[2,2,2,2,-1];c=[-1,-1,-1,-1];d=[1,1,1,1,1];x=
zgTridiagonal(a,b,c,d)
x =
    1.0000    1.0000    0.0000   -2.0000   -5.0000
```

Example 6.6.7 Solve the linear system $\boldsymbol{Ax}=\boldsymbol{b}$, where

$$\boldsymbol{A}=\begin{pmatrix} 4 & 5 & 8 & 7 \\ 8 & 2 & 1 & 3 \\ 9 & 1 & 0 & 2 \\ 6 & 5 & 7 & 3 \end{pmatrix},\quad \boldsymbol{b}=\begin{pmatrix} 7 \\ 6 \\ 2 \\ 1 \end{pmatrix}$$

```
>> clear
>> A=[4 5 8 7;8 2 1 3;9 1 0 2;6 5 7 3];
>> b=[7 6 2 1]';
>> B=[A b]
B =
    4     5     8     7     7
    8     2     1     3     6
    9     1     0     2     2
    6     5     7     3     1
>> RA=rank(A)
RA =
    4
>> RB=rank(B)
RB =
    4
>> [L,U]=lu(A)
L =
    0.4444    1.0000         0         0
    0.8889    0.2439    1.0000         0
    1.0000         0         0         0
    0.6667    0.9512    0.6410    1.0000
U =
```

```
    9.0000    1.0000         0    2.0000
         0    4.5556    8.0000    6.1111
         0         0   -0.9512   -0.2683
         0         0         0   -3.9744
>> x=U\(L\b)
x =
   -0.7419
    4.7032
   -3.4323
    1.9871
```

Exercises 6

Questions

1. Find the ***LU*** decomposition of the following matrices.

(1) $\begin{pmatrix} 2 & -1 & 1 \\ 3 & 3 & 9 \\ 3 & 3 & 5 \end{pmatrix}$; (2) $\begin{pmatrix} 2 & 0 & 0 & 0 \\ 1 & 1.5 & 0 & 0 \\ 0 & -3 & 0.5 & 0 \\ 2 & -2 & 1 & 1 \end{pmatrix}$.

2. Solve the following linear systems using the Gaussian elimination.

(1) $\begin{cases} 2x_1+3x_2+4x_3=6 \\ 3x_1+5x_2+2x_3=5 \\ 4x_1+3x_2+30x_3=32 \end{cases}$; (2) $\begin{cases} x_1-x_2+2x_3-x_4=-8 \\ 2x_1-2x_2+3x_3-3x_4=-20 \\ x_1+x_2+x_3=-2 \\ x_1-x_2+4x_3+3x_4=4 \end{cases}$.

3. Solve the following linear system using the column pivoting.

$$\begin{pmatrix} -3 & 2 & 6 \\ 10 & -7 & 0 \\ 5 & -1 & 5 \end{pmatrix}\begin{pmatrix} x_1 \\ x_2 \\ x_3 \end{pmatrix}=\begin{pmatrix} 4 \\ 7 \\ 6 \end{pmatrix}$$

4. Use Gaussian elimination with column pivoting to find the determinant of the following matrices.

(1) $\boldsymbol{A}=\begin{pmatrix} 3 & -1 & 4 \\ -1 & 2 & -2 \\ 2 & -3 & -2 \end{pmatrix}$; (2) $\boldsymbol{B}=\begin{pmatrix} 0 & -1 & 0 & 1 \\ 0 & 1 & 0 & 1 \\ 1 & 1 & 2 & 0 \\ 2 & 0 & 1 & 0 \end{pmatrix}$.

5. Solve the following linear systems using the square root method (Cholesky decomposition).

(1) $\begin{pmatrix} 6 & 7 & 5 \\ 7 & 13 & 8 \\ 5 & 8 & 6 \end{pmatrix}\begin{pmatrix} x_1 \\ x_2 \\ x_3 \end{pmatrix} = \begin{pmatrix} 9 \\ 10 \\ 9 \end{pmatrix}$; (2) $\begin{cases} 4x_1 - x_2 + x_3 = 6 \\ -x_1 + 4.25x_2 + 2.75x_3 = -0.5 \\ x_1 + 2.75x_2 + 3.5x_3 = 1.25 \end{cases}$.

6. Use speedup method for tridiagonal systems to solve the following linear systems:

(1) $\begin{cases} 3x_1 + x_2 = -1 \\ 2x_1 + 4x_2 + x_3 = 7 \\ 2x_2 + 5x_3 = 9 \end{cases}$; (2) $\begin{cases} 2x_1 - x_2 = 3 \\ -x_1 + 2x_2 - x_3 = -3 \\ -x_2 + 2x_3 = 1 \end{cases}$.

7. Consider the system

$$\begin{cases} x_2 + 2x_3 = 1 \\ 2x_1 - x_2 = 2 \\ 2x_2 + x_3 = 3 \end{cases}$$

Determine the factorization $\boldsymbol{PA} = \boldsymbol{LU}$, where $\boldsymbol{P}$ is a permutation matrix. Use this factorization to obtain $\det\boldsymbol{A}$.

8. Find α so that the following matrices are positive definite.

(1) $\boldsymbol{A} = \begin{pmatrix} \alpha & 1 & -1 \\ 1 & 2 & 1 \\ -1 & 1 & 4 \end{pmatrix}$; (2) $\boldsymbol{B} = \begin{pmatrix} 2 & \alpha & -1 \\ \alpha & 2 & 1 \\ -1 & 1 & 4 \end{pmatrix}$.

9. Show that if $\boldsymbol{A}$ is orthogonal matrix, then $\operatorname{cond}(\boldsymbol{A})_2 = 1$.

10. Let $\boldsymbol{A}, \boldsymbol{B} \in \mathbf{R}^{n\times n}$, and $\|\cdot\|$ is an operator norm on $\mathbf{R}^{n\times n}$, show that

$$\operatorname{cond}(\boldsymbol{AB}) \leqslant \operatorname{cond}(\boldsymbol{A}) \operatorname{cond}(\boldsymbol{B})$$

11. Solve the following system, and apply three steps of iterative refinement.

$$\begin{pmatrix} 60 & 30 & 20 \\ 30 & 20 & 15 \\ 20 & 15 & 12 \end{pmatrix}\begin{pmatrix} x_1 \\ x_2 \\ x_3 \end{pmatrix} = \begin{pmatrix} 110 \\ 65 \\ 47 \end{pmatrix}$$

Computer Questions

1. Find Doolittle factorization of the following matrix.

$$\boldsymbol{A} = \begin{pmatrix} 25 & 0 & 0 & 0 & 1 \\ 0 & 27 & 4 & 3 & 2 \\ 0 & 54 & 58 & 0 & 0 \\ 0 & 108 & 116 & 0 & 0 \\ 100 & 0 & 0 & 0 & 24 \end{pmatrix}$$

2. Write and test a program to solve the system by using the Cholesky method.

$$\begin{cases} 0.05x_1+0.07x_2+0.06x_3+0.05x_4=0.23 \\ 0.07x_1+0.10x_2+0.08x_3+0.07x_4=0.32 \\ 0.06x_1+0.08x_2+0.10x_3+0.09x_4=0.33 \\ 0.05x_1+0.07x_2+0.09x_3+0.10x_4=0.31 \end{cases}$$

3. Use the speedup method to solve the following system.

$$\begin{cases} 0.5x_1+0.25x_2+ \qquad =0.35 \\ 0.35x_1+0.8x_2+0.4x_3 \qquad =0.77 \\ 0.25x_2+x_3+0.5x_4=-0.5 \\ x_3-2x_4=-2.25 \end{cases}$$

4. Write and test programs to solve Problems 1-6.

5. Solve the following system, and apply three steps of iterative refinement. Print $\boldsymbol{r},\boldsymbol{e}$ and $\boldsymbol{x}$ after each iteration.

$$\begin{pmatrix} 420 & 210 & 140 & 105 \\ 210 & 140 & 105 & 84 \\ 140 & 105 & 84 & 70 \\ 105 & 84 & 70 & 60 \end{pmatrix}\begin{pmatrix} x_1 \\ x_2 \\ x_3 \\ x_4 \end{pmatrix}=\begin{pmatrix} 875 \\ 539 \\ 399 \\ 319 \end{pmatrix}$$

7 Iterative Techniques for Solving Linear Systems

提　要

在工程计算中常常要求解大型稀疏矩阵方程组，即系数矩阵的阶数很高但零元素很多的方程组．对于大型稀疏矩阵方程组一般采用迭代解法，即用某种极限过程去逐步逼近线性方程组的精确解．

将线性方程组 $\boldsymbol{Ax}=\boldsymbol{b}$ 变形为等价的方程组

$$\boldsymbol{x}=\boldsymbol{Bx}+\boldsymbol{f}$$

由此建立迭代公式

$$\boldsymbol{x}^{(k+1)}=\boldsymbol{Bx}^{(k)}+\boldsymbol{f},\ k=0,1,2,\cdots$$

给定初始向量 $\boldsymbol{x}^{(0)}$，按此公式计算得近似解向量序列 $\{\boldsymbol{x}^{(k)}\}_{k=0}^{\infty}$．若对任意初始向量 $\boldsymbol{x}^{(0)}$，当迭代次数无限增加时，序列 $\{\boldsymbol{x}^{(k)}\}_{k=0}^{\infty}$ 都有相同的极限 $\boldsymbol{x}^*$，即

$$\lim_{k\to\infty}\boldsymbol{x}^{(k)}=\boldsymbol{x}^*$$

则称迭代公式是收敛的，否则是发散的．称矩阵 $\boldsymbol{B}$ 为迭代矩阵．对于不同的迭代矩阵得到不同的迭代格式．本章主要介绍雅可比(Jacobi)迭代法、高斯-塞得尔(Gauss-Seidel)迭代法和超松弛(SOR)迭代法．

词　汇

direct method	直接法
iterative method	迭代法
component	分量
splitting matrix	分裂矩阵
relaxation factor	松弛系数
underrelaxation method	低松弛法
reducible	可约
irreducible	不可约
diagonally dominant	对角占优
asymptotic rate of convergence	渐近收敛速率
successive over relaxation method	逐次超松弛迭代法

7.1 Introduction

In Chapter 6, we discussed direct methods for solving the matrix problem $\boldsymbol{Ax}=\boldsymbol{b}$. They proceed through a finite number of steps and produce a solution x that would be completely accurate were it not for round-off errors. An iterative method, by contrast, produces a sequence of vectors that ideally converges to the solution. The computation is halted when an approximate solution is obtained having some specified (给定,指定) accuracy or after a certain number of iterations. Indirect methods are almost always iterative in nature, i.e., a simple process is applied repeatedly to generate the sequence referred to previously.

For large linear systems containing thousands of equations, iterative methods often have **decisive** (决定性) advantages over direct methods in terms of speed and demands on computer memory. Sometimes, a **modest** (适当的) number of iterations will **suffice** (满足,足够) to produce an acceptable solution. For **sparse** (稀疏) systems (in which a large proportion of the elements in $\boldsymbol{A}$ are 0), iterative methods are often very efficient. In sparse problems, the nonzero elements of $\boldsymbol{A}$ sometimes stored in a sparse-storage format. In other case, it is not necessary to store $\boldsymbol{A}$ at all.

An iterative technique to solve the $n\times n$ linear system $\boldsymbol{Ax}=\boldsymbol{b}$ starts with an initial approximation $\boldsymbol{x}^{(0)}$ to the solution $\boldsymbol{x}$ and generates a sequence of vectors $\{\boldsymbol{x}^{(k)}\}_0^{\infty}$ that converges to $\boldsymbol{x}$. Iterative techniques involve a process that converts the system $\boldsymbol{Ax}=\boldsymbol{b}$ into an equivalent system of the form $\boldsymbol{x}=\boldsymbol{Bx}+\boldsymbol{f}$ for some fixed matrix $\boldsymbol{B}$ and vector $\boldsymbol{f}$.

Example 7.1.1 Consider the linear system

$$\begin{pmatrix} 5 & -1 \\ -2 & 7 \end{pmatrix}\begin{pmatrix} x_1 \\ x_2 \end{pmatrix}=\begin{pmatrix} 4 \\ -3 \end{pmatrix}$$

How can it be solved by an iterative process?

Solution We first convert the given system to the form $\boldsymbol{x}=\boldsymbol{Bx}+\boldsymbol{f}$, solve the ith equation for the ith unknown , for i=1,2, to obtain

$$\begin{cases} x_1=\dfrac{1}{5}x_2+\dfrac{4}{5} \\ x_2=\dfrac{2}{7}x_1-\dfrac{3}{7} \end{cases}$$

Then the given system can be written in the equivalent form $\boldsymbol{x}=\boldsymbol{Bx}+\boldsymbol{f}$, with

$$\boldsymbol{B}=\begin{pmatrix} 0 & \dfrac{1}{5} \\ \dfrac{2}{7} & 0 \end{pmatrix} \quad \text{and} \quad \boldsymbol{f}=\begin{pmatrix} \dfrac{4}{5} \\ -\dfrac{3}{7} \end{pmatrix}$$

For an initial approximation, we let $\boldsymbol{x}^{(0)}=(0,0)^{\mathrm{T}}$. Then the iteration scheme is obtained.

$$\begin{cases} x_1^{(k+1)}=\dfrac{1}{5}x_2^{(k)}+\dfrac{4}{5} \\ x_2^{(k+1)}=\dfrac{2}{7}x_1^{(k)}-\dfrac{3}{7} \end{cases}$$

This is known as the Jacobi method or iteration. Initially, we select for $x_1^{(0)}$ and $x_2^{(0)}$ the best available guess for the solution, or simply set them to 0. The equations above then generate what we hope are improved values, $x_1^{(1)}$ and $x_2^{(1)}$. The process is repeated a prescribed number of times or until a certain precision appears to have been achieved in the vector $(x_1^{(k)}, x_2^{(k)})^{\mathrm{T}}$. Here are some results of the iterates of the Jacobi method for this example, where err=$\|\boldsymbol{x}^{(k)}-\boldsymbol{x}^{(k-1)}\|_\infty$.

See Table 7-1.

Table 7-1 Some results by using Jacobi method

solutions	iterations	errors
$\boldsymbol{x}$ = {0.8, −0.428 571}	k = 1	err = 0.8
$\boldsymbol{x}$ = {0.714 286, −0.2}	k = 2	err = 0.22 8571
$\boldsymbol{x}$ = {0.76, −0.224 49}	k = 3	err = 0.045 714 3
$\boldsymbol{x}$ = {0.755 102, −0.211 429}	k = 4	err = 0.013 061 2
$\boldsymbol{x}$ = {0.757 714, −0.212 828}	k = 5	err = 0.002 6122 4
$\boldsymbol{x}$ = {0.757 434, −0.212 082}	k = 6	err = 0.000 746 356
$\boldsymbol{x}$ = {0.757 584, −0.212 162}	k = 7	err = 0.000 149 271
$\boldsymbol{x}$ = {0.757 568, −0.212 119}	k = 8	err = 0.000 042 648 9

This iterative process could be **modified** (修改) so the newest value for $x_1^{(k)}$ is used immediately in the second equation. The resulting method is called the **Gauss-Seidel method**. Its equations are

$$\begin{cases} x_1^{(k+1)}=\dfrac{1}{5}x_2^{(k)}+\dfrac{4}{5} \\ x_2^{(k+1)}=\dfrac{2}{7}x_1^{(k+1)}-\dfrac{3}{7} \end{cases}$$

some of the output from the Gauss-Seidel method follows Table 7-2.

Table 7-2 Some results by using Gauss-Seidel method

solutions	iterations	errors
$\boldsymbol{x}$ = {0.8, −0.2}	k = 1	err = 0.8
$\boldsymbol{x}$ = {0.76, −0.211 429}	k = 2	err = 0.04
$\boldsymbol{x}$ = {0.757 714, −0.212 082}	k = 3	err = 0.002 285 71
$\boldsymbol{x}$ = {0.757 584, −0.212 119}	k = 4	err = 0.000 130 612
$\boldsymbol{x}$ = {0.757 576, −0.212 121}	k = 5	err = 0.000 007 463 56

From the results, we can see, both the Jacobi and the Gauss-Seidel iterates seem to be converging to the same limit, and the latter is converging faster. Also, notice that, in contract to a direct method, the precision we obtain in the solution depends on when the iterative process is halted. When you finish this chapter, you will be able to efficiently solve very large sparse linear algebraic systems of equations using iterative methods including the Jacibi, Gauss Seidel, and successive relaxation methods. These overall goals will be achieved by mastering the following chapter objectives.

Objectives

- Know how to improve the accuracy of a solution using iterative correction.
- Be able to solve a linear algebraic system using the Jacobi, Gauss-Seidel, and successive relaxation methods.
- Understand the conditions under which the iterative methods converge.
- Understand how linear algebraic system techniques can be used to solve practical engineering problems.

7.2 Basic Iterative Methods

In this section we consider iterative methods in a more general mathematical setting. Given a linear system

$$\boldsymbol{Ax}=\boldsymbol{b} \tag{7.2.1}$$

where $\boldsymbol{A}=(a_{ij})\in\mathbf{R}^{n\times n}$ is a nonsingular matrix .

The method is written in the form $\boldsymbol{x}^{(k+1)}=\boldsymbol{Bx}^{(k)}+\boldsymbol{f}$ by splitting $\boldsymbol{A}$ into the form:

$$\boldsymbol{A}=\boldsymbol{M}-\boldsymbol{N}$$

where $\boldsymbol{M}$ is a nonsingular matrix.

The original problem $\boldsymbol{Ax}=\boldsymbol{b}$ is then rewritten into the equivalent form

$$\boldsymbol{x}=\boldsymbol{M}^{-1}\boldsymbol{Nx}+\boldsymbol{M}^{-1}\boldsymbol{b}$$

This suggests an iterative process, defined by writing

$$\begin{cases}\boldsymbol{x}^{(0)}\ \text{(initial vector)}\\ \boldsymbol{x}^{(k+1)}=\boldsymbol{Bx}^{(k)}+\boldsymbol{f}\quad (k=0,1,\cdots)\end{cases} \tag{7.2.2}$$

where $\boldsymbol{B}=\boldsymbol{M}^{-1}\boldsymbol{N}=\boldsymbol{M}^{-1}(\boldsymbol{M}-\boldsymbol{A})=\boldsymbol{I}-\boldsymbol{M}^{-1}\boldsymbol{A},\ \boldsymbol{f}=\boldsymbol{M}^{-1}\boldsymbol{b}.$

We call $\boldsymbol{B}=\boldsymbol{I}-\boldsymbol{M}^{-1}\boldsymbol{A}$ **iterative matrix**. The initial value $\boldsymbol{x}^{(0)}$ is arbitrary if a good guess of the solution is available, it should be used for $\boldsymbol{x}^{(0)}$. We shall say that the iterative method in Equation (7.2.2) is convergent if the sequence of vectors $\{\boldsymbol{x}^{(k)}\}_0^{\infty}$ converges to the same limit for any initial vector $\boldsymbol{x}^{(0)}$.

Let $\boldsymbol{D}$ be the diagonal matrix whose diagonal entries are those of $\boldsymbol{A}$, $-\boldsymbol{L}$ be the strictly lower-triangular part of $\boldsymbol{A}$, and $-\boldsymbol{U}$ be the strictly upper-triangular part of $\boldsymbol{A}$. With this notation,

$$A=\begin{pmatrix} a_{11} & a_{12} & \cdots & a_{1n} \\ a_{21} & a_{22} & \cdots & a_{2n} \\ \vdots & \vdots & & \vdots \\ a_{n1} & a_{n2} & \cdots & a_{nn} \end{pmatrix}$$

is split into

$$A=\begin{pmatrix} a_{11} & & & \\ & a_{22} & & \\ & & \ddots & \\ & & & a_{nn} \end{pmatrix}-\begin{pmatrix} 0 & & & \\ -a_{21} & 0 & & \\ \vdots & \ddots & \ddots & \\ -a_{n1} & \cdots & -a_{n,n-1} & 0 \end{pmatrix}-\begin{pmatrix} 0 & -a_{12} & \cdots & -a_{1n} \\ & \ddots & \ddots & \vdots \\ & & 0 & -a_{n-1,n} \\ & & & 0 \end{pmatrix} \tag{7.2.3}$$

$$\equiv D-L-U$$

7.2.1 Jacobi Method

Assume that all the diagonal entries of A are different from zero, hence the inverse of D exists. Take $M=D$, $A=D-N$. This results in the matrix form of Jacobi iterative technique:

$$\begin{cases} x^{(0)} \text{ (initial vector)} \\ x^{(k+1)} = Bx^{(k)} + f \quad (k=0,1,\cdots) \end{cases} \tag{7.2.4}$$

where $B = I - D^{-1}A = D^{-1}(L+U) \equiv J, \quad f = D^{-1}b$, J is called **Jacobi matrix**.

From

$$x^{(k+1)} = D^{-1}(L+U)x^{(k)} + D^{-1}b, k=1,2,\cdots$$

that is

$$Dx^{(k+1)} = (L+U)x^{(k)} + b, k=1,2,\cdots$$

Denote $x^{(k)} = (x_1^{(k)}, x_2^{(k)}, \cdots, x_n^{(k)})^{\mathrm{T}}$, then

$$a_{ii}x_i^{(k+1)} = b_i - \sum_{\substack{j=1 \\ j\neq i}}^{n} a_{ij}x_j^{(k)}$$

Written in components, one step of the Jacobi iteration scheme reads

$$x_i^{(k+1)} = \frac{1}{a_{ii}}(b_i - \sum_{\substack{j=1 \\ j\neq i}}^{n} a_{ij}x_j^{(k)}), \quad i=1,2,\cdots,n$$

The formula of the Jacobi iteration for solving $Ax=b$ can be written in **components** (分量):

$$\begin{cases} x^{(0)} = (x_1^{(0)}, x_2^{(0)}, \cdots, x_n^{(0)})^{\mathrm{T}} \\ x_i^{(k+1)} = \dfrac{1}{a_{ii}}(b_i - \sum_{\substack{j=1 \\ j\neq i}}^{n} a_{ij}x_j^{(k)}) \quad (i=1,2\cdots,n; k=0,1,\cdots) \end{cases} \tag{7.2.5}$$

Example 7.2.1 Solve the following linear system by Jacobi iteration.

$$\begin{cases} 10x_1 - x_2 + 2x_3 = 3 \\ -x_1 + 11x_2 - x_3 + 3x_4 = 15 \\ 2x_1 - x_2 + 10x_3 - x_4 = -8 \\ 3x_2 - x_3 + 8x_4 = 6 \end{cases}$$

Solution Solve the ith equation for the ith unknown , for $i = 1,2,3,4$, to obtain

$$\begin{cases} x_1 = \dfrac{1}{10}(3 + x_2 - 2x_3) \\ x_2 = \dfrac{1}{11}(15 + x_1 + x_3 - 3x_4) \\ x_3 = \dfrac{1}{10}(-8 - 2x_1 + x_2 + x_4) \\ x_4 = \dfrac{1}{8}(6 - 3x_2 + x_3) \end{cases}$$

Construct the Jacobi iteration as follows :

$$\begin{cases} x_1^{(k+1)} = \dfrac{1}{10}(3 + x_2^{(k)} - 2x_3^{(k)}) \\ x_2^{(k+1)} = \dfrac{1}{11}(15 + x_1^{(k)} + x_3^{(k)} - 3x_4^{(k)}) \\ x_3^{(k+1)} = \dfrac{1}{10}(-8 - 2x_1^{(k)} + x_2^{(k)} + x_4^{(k)}) \\ x_4^{(k+1)} = \dfrac{1}{8}(6 - 3x_2^{(k)} + x_3^{(k)}) \end{cases} \quad (k = 0,1,2,\cdots)$$

Take $\boldsymbol{x}^{(0)} = (0,0,0,0)^{\mathrm{T}}$, and generate the iterates in Table 7-3.

Table 7-3 Some results by using Jacobi iteration

solutions	errors
$\boldsymbol{x}^{(1)}$ = {0.3, 1.363 64, −0.8, 0.75}	err = 1.363 64
$\boldsymbol{x}^{(2)}$ = {0.596 364, 1.113 64, −0.648 636, 0.138 636}	err = 0.611 364
$\boldsymbol{x}^{(3)}$ = {0.541 091, 1.321 07, −0.794 045, 0.251 307}	err = 0.207 438
$\boldsymbol{x}^{(4)}$ = {0.590 917, 1.272 1, −0.750 98, 0.155 341}	err = 0.095 965 4
$\boldsymbol{x}^{(5)}$ = {0.577 406, 1.306 72, −0.775 439, 0.179 089}	err = 0.034 617
$\boldsymbol{x}^{(6)}$ = {0.585 76, 1.296 79, −0.766 9, 0.163 05}	err = 0.016 038 7
$\boldsymbol{x}^{(7)}$ = {0.583 059, 1.302 7, −0.771 168, 0.167 841}	err = 0.005 909 84
$\boldsymbol{x}^{(8)}$ = {0.584 504, 1.300 76, −0.769 558, 0.165 091}	err = 0.002 749 62
$\boldsymbol{x}^{(9)}$ = {0.583 988, 1.301 79, −0.770 316, 0.166 02}	err = 0.001 027 59
$\boldsymbol{x}^{(10)}$ = {0.584 242, 1.301 42, −0.770 017, 0.165 54}	err = 0.000 480 076

Since

$$\left\|\boldsymbol{x}^{(10)} - \boldsymbol{x}^{(9)}\right\|_\infty = 0.000\,480\,076 < 10^{-3}$$

$\boldsymbol{x}^{(10)}$ is accepted as a reasonable approximation to the solution with accuracy 10^{-3}.

7.2.2 Gauss-Seidel Method

Take

$$\boldsymbol{M}=\boldsymbol{D}-\boldsymbol{L} \text{ (lower triangular matrix)}, \quad \text{and } \boldsymbol{A}=\boldsymbol{M}-\boldsymbol{N}$$

This results in the matrix form of Gauss-Seidel technique:

$$\begin{cases} \boldsymbol{x}^{(0)} \text{ (initial vector)} \\ \boldsymbol{x}^{(k+1)} = \boldsymbol{B}\boldsymbol{x}^{(k)} + \boldsymbol{f} \quad (k=0,1,\cdots) \end{cases} \tag{7.2.6}$$

where $\boldsymbol{B}=\boldsymbol{I}-(\boldsymbol{D}-\boldsymbol{L})^{-1}\boldsymbol{A}=(\boldsymbol{D}-\boldsymbol{L})^{-1}\boldsymbol{U}\equiv\boldsymbol{G}, \quad \boldsymbol{f}=(\boldsymbol{D}-\boldsymbol{L})^{-1}\boldsymbol{b}$.

$\boldsymbol{G}=(\boldsymbol{D}-\boldsymbol{L})^{-1}\boldsymbol{U}$ is called Gauss-Seidel matrix.

From $(\boldsymbol{D}-\boldsymbol{L})\boldsymbol{x}^{(k+1)}=\boldsymbol{U}\boldsymbol{x}^{(k)}+\boldsymbol{b}$, to obtain

$$\boldsymbol{D}\boldsymbol{x}^{(k+1)}=\boldsymbol{L}\boldsymbol{x}^{(k+1)}+\boldsymbol{U}\boldsymbol{x}^{(k)}+\boldsymbol{b}$$

Written in components, one step of the Gauss-Seidel iteration scheme reads

$$a_{ii}x_i^{(k+1)}=-\sum_{j=1}^{i-1}a_{ij}x_j^{(k+1)}-\sum_{j=i+1}^{n}a_{ij}x_j^{(k)}+\boldsymbol{b}_i$$

hence

$$\begin{cases} \boldsymbol{x}^{(0)}=(x_1^{(0)},x_2^{(0)},\cdots,x_n^{(0)})^{\mathrm{T}} \\ x_i^{(k+1)}=\dfrac{1}{a_{ii}}(b_i-\sum_{j=1}^{i-1}a_{ij}x_j^{(k+1)}-\sum_{j=i+1}^{n}a_{ij}x_j^{(k)}) \end{cases} \quad (i=1,\cdots,n;k=0,1,\cdots) \tag{7.2.7}$$

Gauss-Seidel method commonly **abbreviated** (缩写) as G-S method.

Example 7.2.2 Solve the following linear system by G-S method.

$$\begin{cases} 10x_1-x_2+2x_3=3 \\ -x_1+11x_2-x_3+3x_4=15 \\ 2x_1-x_2+10x_3-x_4=-8 \\ 3x_2-x_3+8x_4=6 \end{cases}$$

Solution Construct the G-S iteration as follows :

$$\begin{cases} x_1^{(k+1)}=\dfrac{1}{10}(3+x_2^{(k)}-2x_3^{(k)}) \\ x_2^{(k+1)}=\dfrac{1}{11}(15+x_1^{(k+1)}+x_3^{(k)}-3x_4^{(k)}) \\ x_3^{(k+1)}=\dfrac{1}{10}(-8-2x_1^{(k+1)}+x_2^{(k+1)}+x_4^{(k)}) \\ x_4^{(k+1)}=\dfrac{1}{8}(6-3x_2^{(k+1)}+x_3^{(k+1)}) \end{cases} \quad (k=0,1,2,\cdots)$$

Take $\boldsymbol{x}^{(0)}=(0,0,0,0)^{\mathrm{T}}$, and generate the iterates in Table 7-4.

Table 7-4 Some results by using G-S method

solutions	errors
$x^{(1)}=\{0.3,1.390\,91,-0.720\,909,0.138\,295\}$	err = 1.390 91
$x^{(2)}=\{0.583\,273,1.313\,41,-0.771\,484,0.161\,037\}$	err = 0.283 273
$x^{(3)}=\{0.585\,638,1.302\,82,-0.770\,742,0.165\,099\}$	err = 0.010 585
$x^{(4)}=\{0.584\,431,1.301\,67,-0.770\,209,0.165\,597\}$	err = 0.001 207 03
$x^{(5)}=\{0.584\,209,1.301\,56,-0.770\,126,0.165\,648\}$	err = 0.000 221 532

Since

$$\left\|\boldsymbol{x}^{(5)}-\boldsymbol{x}^{(4)}\right\|=0.000\,221\,532<10^{-3}$$

$\boldsymbol{x}^{(5)}$ is accepted as a reasonable approximation to the solution with accuracy 10^{-3}.

7.2.3 SOR Method（超松弛方法）

Take

$$\boldsymbol{M}=\frac{1}{\omega}(\boldsymbol{D}-\omega\boldsymbol{L})\quad\text{(lower triangular matrix with parameter)}$$

where the weight factor $\omega>0$ is called **relaxation factor** (松弛因子). From (7.2.2), we can structure an iterative method which iterative matrix is

$$\begin{aligned}\boldsymbol{L}_{\omega}&\equiv\boldsymbol{I}-\omega(\boldsymbol{D}-\omega\boldsymbol{L})^{-1}\boldsymbol{A}\\&=(\boldsymbol{D}-\omega\boldsymbol{L})^{-1}((1-\omega)\boldsymbol{D}+\omega\boldsymbol{U})\end{aligned}$$

This results in the matrix form of successive over relaxation method, commonly abbreviated as SOR method .

$$\begin{cases}\boldsymbol{x}^{(0)}\ \text{(initial vector)}\\\boldsymbol{x}^{(k+1)}=\boldsymbol{L}_{\omega}\boldsymbol{x}^{(k)}+\boldsymbol{f}\quad(k=0,1,\cdots)\end{cases}\tag{7.2.8}$$

where

$$\boldsymbol{L}_{\omega}=(\boldsymbol{D}-\omega\boldsymbol{L})^{-1}((1-\omega)\boldsymbol{D}+\omega\boldsymbol{U}),\quad\boldsymbol{f}=\omega(\boldsymbol{D}-\omega\boldsymbol{L})^{-1}\boldsymbol{b}$$

From (7.2.8)

$$(\boldsymbol{D}-\omega\boldsymbol{L})\boldsymbol{x}^{(k+1)}=((1-\omega)\boldsymbol{D}+\omega\boldsymbol{U})\boldsymbol{x}^{(k)}+\omega\boldsymbol{b}$$

or

$$\boldsymbol{D}\boldsymbol{x}^{(k+1)}=\boldsymbol{D}\boldsymbol{x}^{(k)}+\omega(\boldsymbol{b}+\boldsymbol{L}\boldsymbol{x}^{(k+1)}+\boldsymbol{U}\boldsymbol{x}^{(k)}-\boldsymbol{D}\boldsymbol{x}^{(k)})$$

Written in components, one step of the SOR method scheme reads

$$\begin{cases}\boldsymbol{x}^{(0)}=(x_1^{(0)},x_2^{(0)},\cdots,x_n^{(0)})^{\mathrm{T}}\\x_i^{(k+1)}=x_i^{(k)}+\dfrac{\omega}{a_{ii}}(b_i-\sum\limits_{j=1}^{i-1}a_{ij}x_j^{(k+1)}-\sum\limits_{j=i}^{n}a_{ij}x_j^{(k)})\end{cases}\quad(i=1,2,\cdots,n;k=0,1,\cdots)\tag{7.2.9}$$

(1) When $\omega=1$, SOR method is Gauss-Seidel method.

(2) When $\omega>1$, it is called **overrelaxation method**.

(3) When $\omega<1$ it is called **underrelaxation method.**

Example 7.2.3 Solving the following linear system by SOR method.

$$\begin{pmatrix} -4 & 1 & 1 & 1 \\ 1 & -4 & 1 & 1 \\ 1 & 1 & -4 & 1 \\ 1 & 1 & 1 & -4 \end{pmatrix}\begin{pmatrix} x_1 \\ x_2 \\ x_3 \\ x_4 \end{pmatrix}=\begin{pmatrix} 1 \\ 1 \\ 1 \\ 1 \end{pmatrix}$$

Solution The exact solution is $\boldsymbol{x}^*=(-1,-1,-1,-1)^{\mathrm{T}}$.

Take $\boldsymbol{x}^{(0)}=(0.0,\ 0.0,\ 0.0,\ 0.0)^{\mathrm{T}}$, then

$$\begin{cases} x_1^{(k+1)}=x_1^{(k)}-\dfrac{\omega}{4}(1+4x_1^{(k)}-x_2^{(k)}-x_3^{(k)}-x_4^{(k)}) \\ x_2^{(k+1)}=x_2^{(k)}-\dfrac{\omega}{4}(1-x_1^{(k+1)}+4x_2^{(k)}-x_3^{(k)}-x_4^{(k)}) \\ x_3^{(k+1)}=x_3^{(k)}-\dfrac{\omega}{4}(1-x_1^{(k+1)}-x_2^{(k+1)}+4x_3^{(k)}-x_4^{(k)}) \\ x_4^{(k+1)}=x_4^{(k)}-\dfrac{\omega}{4}(1-x_1^{(k+1)}-x_2^{(k+1)}-x_3^{(k+1)}+4x_4^{(k)}) \end{cases}$$

(1) Take $\omega=1.3$

$$\boldsymbol{x}^{(11)}=(-0.999\,996\,46,\ -1.000\,003\,10,\ -0.999\,999\,53,\ -0.999\,999\,12)^{\mathrm{T}}$$

and

$$\left\|\boldsymbol{\varepsilon}^{(11)}\right\|_2=\left\|\boldsymbol{x}^*-\boldsymbol{x}^{(11)}\right\|_2\leqslant 0.46\times10^{-5}=\varepsilon\ ,\quad k=11$$

(2) Take $\omega=1.0$, the accuracy is equal to ε, then $k=22$.

(3) Take $\omega=1.7$, the accuracy is equal to ε, then $k=33$.

7.3 Iterative Method Convergence

7.3.1 Basic Theorems

Given a linear system

$$\boldsymbol{Ax=b} \tag{7.3.1}$$

where $\boldsymbol{A}=(a_{ij})\in\mathbf{R}^{n\times n}$ is a nonsingular matrix, $\boldsymbol{x}^*$ is the exact solution.

The equivalent system

$$\boldsymbol{x}=\boldsymbol{Bx}+\boldsymbol{f} \tag{7.3.2}$$

then

$$\boldsymbol{x}^*=\boldsymbol{Bx}^*+\boldsymbol{f} \tag{7.3.3}$$

The iterative scheme

$$\boldsymbol{x}^{(k+1)}=\boldsymbol{Bx}^{(k)}+\boldsymbol{f} \tag{7.3.4}$$

denote

$$\boldsymbol{\varepsilon}^{(k)}=\boldsymbol{x}^{(k)}-\boldsymbol{x}^*\quad (k=0,1,2,\cdots)$$

then from (7.3.3) and (7.3.4), we obtain the recursive formula

$$\boldsymbol{\varepsilon}^{(k+1)} = \boldsymbol{B}\boldsymbol{\varepsilon}^{(k)} \quad (k = 0,1,2,\cdots)$$

$$\boldsymbol{\varepsilon}^{(k)} = \boldsymbol{B}\boldsymbol{\varepsilon}^{(k-1)} = \cdots = \boldsymbol{B}^k\boldsymbol{\varepsilon}^{(0)}$$

$$\left\|\boldsymbol{\varepsilon}^{(k)}\right\| = \left\|\boldsymbol{B}^k\boldsymbol{\varepsilon}^{(0)}\right\| \leqslant \left\|\boldsymbol{B}\right\|^k \left\|\boldsymbol{\varepsilon}^{(0)}\right\| = q^k \left\|\boldsymbol{\varepsilon}^{(0)}\right\|$$

Hence, if the iterative matrix $\boldsymbol{B}$ satisfies the condition $\|\boldsymbol{B}\| = q < 1$, then

$$\boldsymbol{x}^{(k)} - \boldsymbol{x}^* = \boldsymbol{\varepsilon}^{(k)} \to 0 \quad (k \to \infty)$$

i.e., the iterative method converges.

Theorem 7.3.1 Given the system $\boldsymbol{x} = \boldsymbol{B}\boldsymbol{x} + \boldsymbol{f}$, and $\{\boldsymbol{x}^{(k)}\}$ is a sequence yielded by the iteration $\boldsymbol{x}^{(k+1)} = \boldsymbol{B}\boldsymbol{x}^k + \boldsymbol{f}$ ($\boldsymbol{x}^{(0)}$ is an arbitrary initial vector), if some norm of matrix $\boldsymbol{B}$ satisfies $\|\boldsymbol{B}\| = q < 1$, then $\{\boldsymbol{x}^{(k)}\}$ converges to the unique solution of the given system.

Example 7.3.1 Consider the convergence of Jacobi method for solving the following system.

$$\begin{cases} 10x_1 - x_2 + 2x_3 = 6 \\ -x_1 + 11x_2 - x_3 + 3x_4 = 25 \\ 2x_1 - x_2 + 10x_3 - x_4 = -11 \\ 3x_1 - x_3 + 8x_4 = 15 \end{cases}$$

Solution

$$\boldsymbol{A} = \begin{pmatrix} 10 & -1 & 2 & 0 \\ -1 & 11 & -1 & 3 \\ 2 & -1 & 10 & -1 \\ 3 & 0 & -1 & 8 \end{pmatrix} = \begin{pmatrix} 10 & & & \\ & 11 & & \\ & & 10 & \\ & & & 8 \end{pmatrix} - \begin{pmatrix} 0 & & & \\ 1 & 0 & & \\ -2 & 1 & 0 & \\ -3 & 0 & 1 & 0 \end{pmatrix} - \begin{pmatrix} 0 & 1 & -2 & 0 \\ & 0 & 1 & -3 \\ & & 0 & 1 \\ & & & 0 \end{pmatrix}$$

$$\equiv \boldsymbol{D} - \boldsymbol{L} - \boldsymbol{U}$$

The iterative matrix of Jacobi

$$\boldsymbol{J} = \boldsymbol{D}^{-1}(\boldsymbol{L} + \boldsymbol{U}) = \begin{pmatrix} 0 & \frac{1}{10} & -\frac{2}{10} & 0 \\ \frac{1}{11} & 0 & \frac{1}{11} & -\frac{3}{11} \\ -\frac{2}{10} & \frac{1}{10} & 0 & \frac{1}{10} \\ -\frac{3}{8} & 0 & \frac{1}{8} & 0 \end{pmatrix}$$

$$\|\boldsymbol{J}\|_\infty = \max\left\{\frac{3}{10}, \frac{5}{11}, \frac{4}{10}, \frac{4}{8}\right\} = \frac{1}{2} < 1$$

Therefore, the Jacobi method is convergent.

Definition 7.3.1 Given a sequence of matrices $\boldsymbol{A}_k = (a_{ij}^{(k)}) \in \mathbf{R}^{n \times n}$ and, if $\lim_{k\to\infty} a_{ij}^{(k)} = a_{ij}$ $(i, j = 1,2,\cdots,n)$, then we call that $\{\boldsymbol{A}_k\}$ converges to $\boldsymbol{A}$, and denote

$$\lim_{k\to\infty} \boldsymbol{A}_k = \boldsymbol{A}$$

Example 7.3.2 Given a sequence of matrices

$$\boldsymbol{A}=\begin{pmatrix}\lambda & 1\\ 0 & \lambda\end{pmatrix},\ \boldsymbol{A}^2=\begin{pmatrix}\lambda^2 & 2\lambda\\ 0 & \lambda^2\end{pmatrix},\cdots,\ \boldsymbol{A}^k=\begin{pmatrix}\lambda^k & k\lambda^{k-1}\\ 0 & \lambda^k\end{pmatrix},\cdots$$

if $|\lambda|<1$, then

$$\lim_{k\to\infty}\boldsymbol{A}_k=\lim_{k\to\infty}\boldsymbol{A}^k=\begin{pmatrix}0 & 0\\ 0 & 0\end{pmatrix}$$

Theorem 7.3.2 $\lim\limits_{k\to\infty}\boldsymbol{A}_k=\boldsymbol{A}\Leftrightarrow\lim\limits_{k\to\infty}\|\boldsymbol{A}_k-\boldsymbol{A}\|=0,$ where $\|\cdot\|$ is an arbitrary operator norm of matrix.

Theorem 7.3.3 $\lim\limits_{k\to\infty}\boldsymbol{A}_k=\boldsymbol{A}\Leftrightarrow\lim\limits_{k\to\infty}\boldsymbol{A}_k\boldsymbol{x}=\boldsymbol{A}\boldsymbol{x},$ for any vector $\boldsymbol{x}\in\mathbf{R}^n$.

Theorem 7.3.4 Let $\boldsymbol{B}=(b_{ij})\in\mathbf{R}^{n\times n}$, then $\lim\limits_{k\to\infty}\boldsymbol{B}^k=0$ (zero matrix) if and only if the

$$\rho(\boldsymbol{B})<1$$

holds.

Theorem 7.3.5 Let $\boldsymbol{B}$ be an $n\times n$ matrix. Then the successive approximations

$$\boldsymbol{x}^{(k+1)}=\boldsymbol{B}\boldsymbol{x}^{(k)}+\boldsymbol{f},\quad k=0,1,2,\cdots \tag{7.3.5}$$

converge for each initial vector $\boldsymbol{x}^{(0)}$ if and only if

$$\rho(\boldsymbol{B})<1$$

for the **spectral radius** (谱半径) of $\boldsymbol{B}$.

Proof Sufficiency Let $\rho(\boldsymbol{B})<1$, then $\boldsymbol{Ax}=\boldsymbol{f}$ $(\boldsymbol{A}=\boldsymbol{I}-\boldsymbol{B})$ has a unique solution, denoted $\boldsymbol{x}^*$, i.e.,

$$\boldsymbol{x}^*=\boldsymbol{B}\boldsymbol{x}^*+\boldsymbol{f}$$

the error vectors

$$\boldsymbol{\varepsilon}^{(k)}=\boldsymbol{x}^{(k)}-\boldsymbol{x}^*=\boldsymbol{B}^k\boldsymbol{\varepsilon}^{(0)},\quad \boldsymbol{\varepsilon}^{(0)}=\boldsymbol{x}^{(0)}-\boldsymbol{x}^*$$

Since $\rho(\boldsymbol{B})<1$, from Theorem 7.3.4, we have $\lim\limits_{k\to\infty}\boldsymbol{B}^k=0$. Hence for any $\boldsymbol{x}^{(0)}$, $\lim\limits_{k\to\infty}\boldsymbol{\varepsilon}^{(k)}=0.$ i.e., $\lim\limits_{k\to\infty}\boldsymbol{x}^{(k)}=\boldsymbol{x}^*$.

Necessity Assume for any $\boldsymbol{x}^{(0)}$, $\lim\limits_{k\to\infty}\boldsymbol{x}^{(k)}=\boldsymbol{x}^*$ holds, where $\boldsymbol{x}^{(k+1)}=\boldsymbol{B}\boldsymbol{x}^{(k)}+\boldsymbol{f}$.

Obviously, $\boldsymbol{x}^*$ is the solution of system (7.3.5), i.e., $\boldsymbol{x}^*=\boldsymbol{B}\boldsymbol{x}^*+\boldsymbol{f}$ holds for any $\boldsymbol{x}^{(0)}$, and

$$\boldsymbol{\varepsilon}^{(k)}=\boldsymbol{x}^{(k)}-\boldsymbol{x}^*=\boldsymbol{B}^k\boldsymbol{\varepsilon}^{(0)}\to 0\ (k\to\infty)$$

From Theorem 7.3.3

$$\lim_{k\to\infty}\boldsymbol{B}^k=0$$

And from Theorem 7.3.4, we have $\rho(\boldsymbol{B})<1$. □

Corollary Let $Ax=b$, where $A=D-L-U$ is nonsingular and D is nonsingular, then
(1)Jacobi method is convergent if and only if $\rho(J)<1$, where $J=D^{-1}(L+U)$.
(2)Gauss-Seidel method is convergent if and only if $\rho(G)<1$,where $G=(D-L)^{-1}U$.
(3)SOR method is convergent if and only if $\rho(L_\omega)<1$, where

$$L_\omega=(D-\omega L)^{-1}\quad((1-\omega)D+\omega U)$$

Example 7.3.3 Let

$$\begin{cases}x_1+2x_2-2x_3=1\\x_1+x_2+x_3=1\\2x_1+2x_2+x_3=1\end{cases}$$

Consider the convergence by Jacobi and G-S iterations.

Solution

$$A=\begin{pmatrix}1&&\\&1&\\&&1\end{pmatrix}-\begin{pmatrix}0&&\\-1&0&\\-2&-2&0\end{pmatrix}-\begin{pmatrix}0&-2&2\\&0&-1\\&&0\end{pmatrix}=D-L-U$$

Jacobi iterative matrix

$$J=D^{-1}(L+U)=\begin{pmatrix}0&-2&2\\-1&0&-1\\-2&-2&0\end{pmatrix}$$

The **characteristic equation** (特征方程) is of the form

$$\det(\lambda I-J)=\lambda^3=0$$

We obtain $\rho(J)=0<1$.

Therefore, Jacobi iteration is convergent.

G-S iterative matrix

$$G=(D-L)^{-1}U=\begin{pmatrix}0&-2&2\\0&2&-3\\0&0&2\end{pmatrix}$$

The characteristic equation is

$$\det(\lambda I-G)=\lambda(\lambda-2)(\lambda-2)=0$$

The eigenvalues are

$$\lambda_1=0,\quad \lambda_2=\lambda_3=2$$

and $\rho(G)=2>1$.

Therefore , G-S iteration is not convergent.

Example 7.3.4 Consider the convergence of the following system with Jacobi iteration.

$$\begin{cases}8x_1-3x_2+2x_3=20\\4x_1+11x_2-x_3=33\\6x_1+3x_2+12x_3=36\end{cases}$$

Solution The characteristic equation of the iterative matrix

$$\det(\lambda \boldsymbol{I}-\boldsymbol{J})=\lambda^3+0.034\,090\,909\lambda+0.039\,772\,727=0$$

The eigenvalues

$$\lambda_1=-0.308\,2,\ \ \lambda_{2,3}=0.154\,1\pm i0.324\,5$$

$$|\lambda_1|<1,\ \ |\lambda_2|=|\lambda_3|=0.359\,2<1$$

i.e., $\rho(\boldsymbol{J})<1$, therefore Jacobi iteration is convergent.

Example 7.3.5 Consider the convergence of the following system with iterative method.

$$\boldsymbol{x}^{(k+1)}=\boldsymbol{B}\boldsymbol{x}^k+\boldsymbol{f}$$

where $\boldsymbol{B}=\begin{pmatrix}0&2\\3&0\end{pmatrix}$, $\boldsymbol{f}=\begin{pmatrix}5\\3\end{pmatrix}$.

Solution The characteristic equation of the iterative matrix

$$\det(\lambda \boldsymbol{I}-\boldsymbol{B})=\lambda^2-6=0$$

The eigenvalues

$$\lambda_{1,2}=\pm\sqrt{6}$$

i.e., $\rho(\boldsymbol{J})>1$, therefore the iterative method is not convergent.

Theorem 7.3.6 Given a system

$$\boldsymbol{x}=\boldsymbol{B}\boldsymbol{x}+\boldsymbol{f},\ \ \boldsymbol{B}\in\mathbf{R}^{n\times n}$$

and iterative scheme

$$\boldsymbol{x}^{(k+1)}=\boldsymbol{B}\boldsymbol{x}^{(k)}+\boldsymbol{f}$$

If some norm of $\boldsymbol{B}$ satisfies $\|\boldsymbol{B}\|=q<1$, then

(1) The iterative method converges, i.e., for arbitrary $\boldsymbol{x}^{(0)}$, we have

$$\lim_{k\to\infty}\boldsymbol{x}^{(k)}=\boldsymbol{x}^*\ \text{ and }\ \boldsymbol{x}^*=\boldsymbol{B}\boldsymbol{x}^*+\boldsymbol{f}$$

(2) $\|\boldsymbol{x}^*-\boldsymbol{x}^{(k)}\|\leqslant q^k\|\boldsymbol{x}^*-\boldsymbol{x}^{(0)}\|$

(3) $\|\boldsymbol{x}^*-\boldsymbol{x}^{(k)}\|\leqslant\dfrac{q}{1-q}\|\boldsymbol{x}^{(k)}-\boldsymbol{x}^{(k-1)}\|$

(4) $\|\boldsymbol{x}^*-\boldsymbol{x}^{(k)}\|\leqslant\dfrac{q^k}{1-q}\|\boldsymbol{x}^{(1)}-\boldsymbol{x}^{(0)}\|$

Proof (1) Form Theorem 7.3.5, conclusion (1) is clearly.

(2) From

$$\boldsymbol{x}^*-\boldsymbol{x}^{(k+1)}=\boldsymbol{B}(\boldsymbol{x}^*-\boldsymbol{x}^{(k)})$$

and

$$\boldsymbol{x}^{(k+1)}-\boldsymbol{x}^{(k)}=\boldsymbol{B}(\boldsymbol{x}^{(k)}-\boldsymbol{x}^{(k-1)})$$

We have

(a) $\|\boldsymbol{x}^*-\boldsymbol{x}^{(k+1)}\|\leqslant q\|\boldsymbol{x}^*-\boldsymbol{x}^{(k)}\|$

(b) $\|\boldsymbol{x}^{(k+1)}-\boldsymbol{x}^{(k)}\|\leqslant q\|\boldsymbol{x}^{(k)}-\boldsymbol{x}^{(k-1)}\|$

By repeating (a) we can obtain (2).

(3) From (b),we have

$$\begin{aligned}\left\|x^{(k+1)}-x^{(k)}\right\| &= \left\|x^{*}-x^{(k)}-(x^{*}-x^{(k+1)})\right\| \\ &\geqslant \left\|x^{*}-x^{(k)}\right\|-\left\|x^{*}-x^{(k+1)}\right\| \\ &\geqslant (1-q)\left\|x^{(k)}-x^{*}\right\|\end{aligned}$$

$$\left\|x^{*}-x^{(k)}\right\| \leqslant \frac{1}{1-q}\left\|x^{(k+1)}-x^{(k)}\right\| \leqslant \frac{q}{1-q}\left\|x^{(k)}-x^{(k-1)}\right\|$$

By repeating (b) we can obtain (4). □

7.3.2 Some Special Systems of Equations

Definition 7.3.2 Let $A=(a_{ij})_{n\times n}$

(1) If the entries of A satisfies

$$|a_{ii}| > \sum_{\substack{j=1\\ j\neq i}}^{n} |a_{ij}| \quad (i=1,2,\cdots,n)$$

then A is said to be **strictly diagonally dominant matrix** (严格对角占优矩阵).

(2) If the entries of A satisfies

$$|a_{ii}| \geqslant \sum_{\substack{j=1\\ j\neq i}}^{n} |a_{ij}| \quad (i=1,2,\cdots,n)$$

with inequality holding for at least one row i, then A is said to be **weakly diagonally dominant matrix** (弱对角占优矩阵).

Definition 7.3.3 A $n\times n$ matrix $A=(a_{jk})$ is called **reducible** (可约) if there exist two nonempty sets $N, M \subset \{1,\cdots,n\}$ such that

$$N\cap M=\varnothing, \quad N\cup M=\{1,\cdots,n\}$$

and

$$a_{jk}=0, \quad j\in N,\ k\in M$$

Otherwise the matrix is called **irreducible** (不可约).

A reducible matrix A, after a reordering of the rows and columns, can be partitioned into a 2×2 black matrix of the form

$$P^{\mathrm{T}}AP=\begin{pmatrix} A_{11} & A_{12} \\ O & A_{22}\end{pmatrix} \tag{7.3.6}$$

Theorem 7.3.7 If $A=(a_{ij})_{n\times n}$ is a strictly diagonally dominant matrix or an irreducible weakly diagonally dominant matrix, then A is nonsingular.

Proof Let A be a strictly diagonally dominant matrix, using the proof by contradiction. If $\det(A)=0$, then $Ax=0$ has a nonzero solution, denoted $x=(x_1,\cdots,x_n)^{\mathrm{T}}$ and $|x_k|=\max\limits_{1\leqslant i\leqslant n}|x_i|\neq 0$.

Consider the kth equation of the homogeneous system

$$\sum_{j=1}^{n} a_{kj} x_j = 0$$

and

$$|a_{kk}||x_k| = \left|\sum_{\substack{j=1 \\ j\neq k}}^{n} a_{kj} x_j\right| \leqslant \sum_{\substack{j=1 \\ j\neq k}}^{n} |a_{kj}||x_j| \leqslant |x_k| \sum_{\substack{j=1 \\ j\neq k}}^{n} |a_{kj}|$$

that is

$$|a_{kk}| \leqslant \sum_{\substack{j=1 \\ j\neq t}}^{n} |a_{ij}|$$

This is a contradiction with the assumption. Therefore

$$\det(A) \neq 0. \qquad \square$$

Theorem 7.3.8 Let $\boldsymbol{Ax}=\boldsymbol{b}$.

(1) Assume that the matrix $\boldsymbol{A}$ is strictly diagonally dominant. Then the Jacobi and Gauss-Seidel iterations converge to the unique solution of $\boldsymbol{Ax}=\boldsymbol{b}$.

(2) Assume that the matrix $\boldsymbol{A}$ is irreducible and weakly diagonally dominant. Then the Jacobi and Gauss-Seidel iterations converge to the unique solution of $\boldsymbol{Ax}=\boldsymbol{b}$.

Proof (1) For Gauss-Seidel iteration, from the given conditions, $a_{ii} \neq 0 (i = 1,2,\cdots,n)$, Gauss-Seidel matrix $\boldsymbol{G} = (\boldsymbol{D}-\boldsymbol{L})^{-1}\boldsymbol{U}(\boldsymbol{A} = \boldsymbol{D}-\boldsymbol{L}-\boldsymbol{U})$.

Consider the eigenvalues of $\boldsymbol{G}$.

$$\begin{aligned}\det(\lambda\boldsymbol{I}-\boldsymbol{G}) &= \det(\lambda\boldsymbol{I}-(\boldsymbol{D}-\boldsymbol{L})^{-1}\boldsymbol{U}) \\ &= \det((\boldsymbol{D}-\boldsymbol{L})^{-1})\cdot\det(\lambda(\boldsymbol{D}-\boldsymbol{L})-\boldsymbol{U})\end{aligned}$$

Since $\det((\boldsymbol{D}-\boldsymbol{L})^{-1}) \neq 0$, therefore, the eigenvalues of $\boldsymbol{G}$ equal to the roots of the equation $\det(\lambda(\boldsymbol{D}-\boldsymbol{L})-\boldsymbol{U}) = 0$.

Denote

$$\boldsymbol{C} \equiv \lambda(\boldsymbol{D}-\boldsymbol{L})-\boldsymbol{U} = \begin{pmatrix} \lambda a_{11} & a_{12} & \cdots & a_{1n} \\ \lambda a_{21} & \lambda a_{22} & \cdots & a_{2n} \\ \vdots & \vdots & & \vdots \\ \lambda a_{n1} & \lambda a_{n2} & \cdots & \lambda a_{nn} \end{pmatrix}$$

If $|\lambda| \geqslant 1$, since $\boldsymbol{A}$ is strictly diagonally dominant, then

$$\begin{aligned}|c_{ii}| = |\lambda a_{ii}| &> |\lambda|\left(\sum_{j=1}^{i-1}|a_{ij}| + \sum_{j=i+1}^{n}|a_{ij}|\right) \\ &\geqslant \sum_{j=1}^{i-1}|\lambda a_{ij}| + \sum_{j=i+1}^{n}|a_{ij}| = \sum_{\substack{j=1 \\ j\neq i}}^{n}|c_{ij}|\end{aligned}$$

This indicates $\boldsymbol{C}$ is strictly diagonally dominant, from the Theorem 7.3.7 $\det(\boldsymbol{C}) \neq 0$. Hence, $|\lambda| < 1$, i.e., Gauss-Seidel iteration converges.

(2) By the given condition $|a_{ii}| \geqslant \sum_{\substack{j=1\\j\neq i}}^{n} |a_{ij}|$ $(i=1,2,\cdots,n)$, we have that $\|B\|_\infty \leqslant 1$ for the Jacobi matrix $B = D^{-1}(L+U)$. Notice that $\rho(B) \leqslant \|B\|$, it follows that $\rho(B) \leqslant 1$ for the spectral radius.

Now assume that there exists an eigenvalue λ of B with $|\lambda| = 1$. For the associated eigenvector we may assume that $\|x\|_\infty = 1$. Then from $\lambda x = Bx$ we obtain the inequality

$$|\lambda||x_j| \leqslant \sum_{\substack{k=1\\k\neq j}}^{n} \left|\frac{a_{jk}}{a_{jj}}\right| |x_k| \leqslant \sum_{\substack{k=1\\k\neq j}}^{n} \left|\frac{a_{jk}}{a_{jj}}\right| \leqslant 1, \quad j=1,2,\cdots,n \tag{7.3.7}$$

Let $N = \{j: |x_j| = 1\}$. Since $\|x\|_\infty = 1$, we have that $N \neq \varnothing$. For $j \in N$, we have $|\lambda||x_j| = 1$, and therefore equality holds in (7.3.7), i.e.,

$$\sum_{\substack{k=1\\k\neq j}}^{n} \left|\frac{a_{jk}}{a_{jj}}\right| = 1, \quad j \in N$$

From this it follows that

$$M = \{1,\cdots,n\} \setminus N \neq \varnothing$$

Since A is weakly diagonally dominant. Because A is irreducible, there exists $j_0 \in N$ and $k_0 \in M$ such that $a_{j_0 k_0} \neq 0$. Now by using

$$|a_{j_0 k_0}||x_{k_0}| < |a_{j_0 k_0}|$$

we obtain the contradiction

$$1 = |x_{j_0}| = |\lambda||x_{j_0}| \leqslant \sum_{\substack{k=1\\k\neq j}}^{n} \left|\frac{a_{j_0 k}}{a_{j_0 j_0}}\right| |x_k| < \sum_{\substack{k=1\\k\neq j}}^{n} \left|\frac{a_{j_0 k}}{a_{j_0 j_0}}\right| \leqslant 1$$

Therefore, we have $\rho(B) < 1$, and the statement of the theorem follows from Theorem 7.3.5. □

Theorem 7.3.9 A necessary condition for the SOR method to be convergent is that $0 < \omega < 2$.

Proof Assume that the SOR iteration is convergent, then $\rho(L_\omega) < 1$. Let the eigenvalues of L_ω are $\lambda_1, \lambda_2, \cdots, \lambda_n$, then

$$|\det(L_\omega)| = |\lambda_1 \lambda_2 \cdots \lambda_n| \leqslant (\rho(L_\omega))^n$$

or

$$|\det(L_\omega)|^{1/n} \leqslant \rho(L_\omega) < 1$$

In the other hand

$$\det(L_\omega) = \det[(D-\omega L)^{-1}]\det((1-\omega)D + \omega U) = (1-\omega)^n$$

hence $$\left|\det(\boldsymbol{L}_\omega)\right|^{1/n} = |1-\omega| < 1$$

that is $$0 < \omega < 2$$ □

Theorem 7.3.10 Let $\boldsymbol{Ax}=\boldsymbol{b}$, if

(1) $\boldsymbol{A}$ is a symmetric positive definite matrix.

(2) $0 < \omega < 2$.

Then the SOR method converges to the unique solution of $\boldsymbol{Ax}=\boldsymbol{b}$.

Theorem 7.3.11 Let $\boldsymbol{Ax}=\boldsymbol{b}$, if

(1) Matrix $\boldsymbol{A}$ is strictly diagonally dominant (or irreducible and weakly diagonally dominant).

(2) $0 < \omega \leqslant 1$.

Then the SOR method converges to the unique solution of $\boldsymbol{Ax}=\boldsymbol{b}$.

Now given the system

$$\boldsymbol{x} = \boldsymbol{Bx} + \boldsymbol{f}, \quad \boldsymbol{B} \in \mathbf{R}^{n\times n}$$

the iterative method

$$\boldsymbol{x}^{(k+1)} = \boldsymbol{Bx}^{(k)} + \boldsymbol{f} \quad (k = 0,1,\cdots)$$

and the iteration converges, denoted that $\lim\limits_{k\to\infty} \boldsymbol{x}^{(k)} = \boldsymbol{x}^*$, then $\boldsymbol{x}^* = \boldsymbol{Bx}^* + \boldsymbol{f}$.

From the Theorem 7.3.5 we have $0 < \rho(\boldsymbol{B}) < 1$, and the error $\boldsymbol{\varepsilon}^{(k)} = \boldsymbol{x}^{(k)} - \boldsymbol{x}^*$ satisfies

$$\boldsymbol{\varepsilon}^{(k)} = \boldsymbol{B}^k \boldsymbol{\varepsilon}^{(0)}$$

therefore

$$\left\|\boldsymbol{\varepsilon}^{(k)}\right\| \leqslant \left\|\boldsymbol{B}\right\|^k \left\|\boldsymbol{\varepsilon}^{(0)}\right\|$$

Let $\boldsymbol{B}$ be symmetric, from Theorem 3.3.2

$$\left\|\boldsymbol{\varepsilon}^{(k)}\right\|_2 \leqslant \left\|\boldsymbol{B}\right\|_2^k \left\|\boldsymbol{\varepsilon}^{(0)}\right\|_2 = (\rho(\boldsymbol{B}))^k \left\|\boldsymbol{\varepsilon}^{(0)}\right\|_2$$

In order to achieve the initial error reduced 10^{-s}, i.e.,

$$(\rho(\boldsymbol{B}))^k \leqslant 10^{-s}$$

take logarithm, we obtain the least number of iterations it needs

$$k \geqslant \frac{s\ln 10}{-\ln \rho(\boldsymbol{B})} \tag{7.3.8}$$

Definition 7.3.4 $R(\boldsymbol{B}) = -\ln \rho(\boldsymbol{B})$ is called **asymptotic rate of convergence** (渐近收敛速度) for iteration $\boldsymbol{x}^{(k+1)} = \boldsymbol{Bx}^{(k)} + \boldsymbol{f}$, or **convergence rate for iteration method.**

Example 7.3.6 Consider the convergence of Jacobi and G-S methods for solving the following system.

$$\begin{cases} 6x_1 + x_2 - 3x_3 = 12 \\ x_1 + 5x_2 + 3x_3 = 8 \\ -x_1 + 4x_2 + 8x_3 = -6 \end{cases}$$

Solution Since

$$A=\begin{pmatrix} 6 & 1 & -3 \\ 1 & 5 & 3 \\ -1 & 4 & 8 \end{pmatrix}$$

is a strictly diagonally dominant matrix, therefore, both Jacobi and G-S methods converge.

7.4 Computer Experiments

7.4.1 Functions Needed in The Experiments by Mathematica

1．实验环境

Mathematica 5.1.

2．所需系统函数

（1）DiagonalMatrix——对角矩阵生成函数.

格式：`DiagonalMatrix[list]`

功能：生成以表 list 的元素为对角元素的对角矩阵.

（2）Inverse——逆矩阵生成函数.

格式：`Inverse[A]`

功能：生成方阵 $\boldsymbol{A}$ 的逆 $\boldsymbol{A}^{-1}$.

（3）Max——取最大值函数.

格式：`Max[x1,x2,…,xn]`

功能：计算 x_i 中的最大值，x_i 为数或数值表.

（4）Return[ ].

功能：退出函数所有过程和循环,返回值为 Null.

7.4.2 Experiments by Mathematica

1．Jacobi 迭代

（1）算法.

① 输入：变量个数 n,系数矩阵 $\boldsymbol{A}$,常数矩阵 $\boldsymbol{b}$,初值向量 $\boldsymbol{x}_0$,迭代精度 eps,迭代次数的上限 n_{max};

② 作 n 次判断$|a_{ii}|<$ eps?

如果为真，输出“迭代失败”,结束；

否则,执行③；

③ $jm\leftarrow \boldsymbol{E}-\boldsymbol{D}^{-1}\boldsymbol{A}$;

④ $\boldsymbol{f}\leftarrow \boldsymbol{D}^{-1}\boldsymbol{b}$;

⑤ 作 n_{max} 次循环：

$\boldsymbol{x}\leftarrow jm\,\boldsymbol{x}_0+\boldsymbol{f}$;

判断$\| x-x_0 \| < eps$ 否？

如果为真,输出解向量 x，结束；

否则，$x_0 \leftarrow x \quad (k=1,2,\cdots n_{max})$;

⑥ 如果$\|x-x_0\|>eps$,输出“迭代失败”.

（2）程序清单.

```
(*Jacobi Iteration*)
Clear[a,b,x];
nmax=500;
n=Input["n="];
a=Input["A="];
b=Input["b="];
x0=Input["x0="];
r=10^(-6);
eps=Input["eps="];
Do[ If[Abs[a[[k,k]]]<r,
            t1=1;Return[],
            t1=0
    ],
    {k,1,n}
  ];
If[ t1==1,
Print["Jacobi 迭代失败"];
   d=DiagonalMatrix[Table[a[[k,k]],{k,1,n}]];
     d1=Inverse[d];
     jm=IdentityMatrix[n]-d1.a;
     f=d1.b;
     Do[x=jm.x0+f;
            err=Max[Abs[x-x0]];
            Print["x=",x//N,"   ""k=",k,"   ""err=",err//N];
            If[N[err]<eps,Break[],x0=x],
            {k,1,nmax}
        ];
If[err>eps,Print["迭代失败"]]
]
```

（3）变量说明.

nmax：保存迭代次数的上限值；

x：保存迭代过程中的向量 $x^{(k+1)}$；

x0：保存初始向量和迭代过程中的向量 $x^{(k)}$；

err：保存误差$\|x-x_0\|_\infty$；

r：微小值；

Jm：保存 Jacobi 迭代阵.

（4）计算实例.

Example 7.4.1 Use the Jacobi iteration to solve the following linear system

$$\begin{pmatrix} 4 & 1 & 1 & 0 & 1 \\ -1 & -3 & 1 & 1 & 0 \\ 2 & 1 & 5 & -1 & -1 \\ -1 & -1 & -1 & 4 & 0 \\ 0 & 2 & -1 & 1 & 4 \end{pmatrix}\begin{pmatrix} x_1 \\ x_2 \\ x_3 \\ x_4 \\ x_5 \end{pmatrix}=\begin{pmatrix} 8 \\ 8 \\ 8 \\ 8 \\ 8 \end{pmatrix}$$

操作步骤如下：

① 光标定位在要执行的 Cell 中，按小键盘的【Enter】键；

② 在弹出的对话框中按提示分别输入

```
5,
{{4,1,1,0,1},{-1,-3,1,1,0},{2,1,5,-1,-1},{-1,-1,-1,4,0},
{0,2,-1,1,4}},
{8,8,8,8,8},
{0,0,0,0,0}, 0.0001
```

③ 每次输入后单击 OK 命令按钮,输出结果如图 7-1 所示.

```
x={2., -2.66667, 1.6, 2., 2.}    k=1    err=2.66667
x={1.76667, -2.13333, 2.13333, 2.23333, 3.23333}    k=2    err=1.23333
x={1.19167, -1.8, 2.41333, 2.44167, 3.04167}    k=3    err=0.575
x={1.08625, -1.44556, 2.58, 2.45125, 2.89292}    k=4    err=0.354444
x={0.99316, -1.35167, 2.52344, 2.55517, 2.75497}    k=5    err=0.137951
x={1.01831, -1.30485, 2.5351, 2.54123, 2.6679}    k=6    err=0.0870642
x={1.02546, -1.31399, 2.49547, 2.56214, 2.65089}    k=7    err=0.0396264
x={1.04191, -1.32262, 2.49522, 2.55173, 2.64033}    k=8    err=0.0164463
x={1.04677, -1.33165, 2.48617, 2.55363, 2.64718}    k=9    err=0.00904722
x={1.04957, -1.33566, 2.48779, 2.55032, 2.64896}    k=10    err=0.00400409
x={1.04973, -1.33716, 2.48716, 2.55043, 2.65219}    k=11    err=0.00323157
x={1.04945, -1.33738, 2.48806, 2.54993, 2.65276}    k=12    err=0.000906245
x={1.04914, -1.33715, 2.48823, 2.55003, 2.65322}    k=13    err=0.00046259
x={1.04892, -1.33696, 2.48843, 2.55006, 2.65313}    k=14    err=0.000215562
x={1.04885, -1.33681, 2.48846, 2.5501, 2.65307}    k=15    err=0.000143057
x={1.04882, -1.33676, 2.48846, 2.55012, 2.653}    k=16    err=0.000074149
```

图 7-1 Jacobi 迭代计算结果

2. Seidel 迭代

（1）算法.

① 输入：变量个数 n，系数矩阵 $\boldsymbol{A}$，常数矩阵 $\boldsymbol{b}$，初值向量 $\boldsymbol{x}_0$，迭代精度 eps，迭代次数的上限 $n_{\max}$；

② 作 n 次判断 $|a_{ii}|$<eps？

如果为真，输出“迭代失败”，结束；

否则，执行 3)；

③ 作 n_{max} 次循环（i=1,2,⋯,n_{max}）.

a. 作 n 次循环：

$$x_i^{(k+1)}=\frac{b_i-\sum_{j=1}^{i-1}a_{ij}x_j^{(k+1)}-\sum_{j=i+1}^{n}a_{ij}x_j^{(k)}}{a_{ii}};$$

b. 判断 $\|x^{(k+1)}-x^{(k)}\|_\infty<\text{eps}$？

如果为真，输出解向量 x，结束；

否则，$x^{(k)}\leftarrow x^{(k+1)}$；

④ 判断：如果 $\|x-x_0\|_\infty\geqslant \text{eps}$，输出“迭代失败”,结束.

（2）程序清单.

```
(*Seidel Iteration*)
Clear[a,b,x];
nmax=500;
n=Input["n="];
a=Input["A="];
b=Input["b="];
x0=Input["x0="];
r=10^(-6);
eps=Input["eps="];
x=x0;
Do[If[Abs[a[[i,i]]]<r,t1=1; Return[], t1=0],{i,1,n}];
If [ t1==1,
   Print["Seidel迭代失败"],
   Do[ Do [ u1=Sum[a[[i,j]]*x[[j]],{j,1,i-1}];
           u2=Sum[a[[i,j]]*x0[[j]],{j,i+1,n}];
           x[[i]]=(b[[i]]-u1-u2)/a[[i,i]],
           {i,1,n}
       ];
       err=Max[Abs[x-x0]];
       Print["x=",x//N,"   ""k=",k,"   ""err=",err//N];
       If[N[err]<eps,Break[],x0=x],
       {k,1,nmax}
     ];
    If [err>=eps, Print["  迭代失败"]]
]
```

（3）变量说明.

nmax：保存迭代次数的上限值；

x：保存迭代过程中的向量 $x^{(k+1)}$；

x0：保存初始向量和迭代过程中的向量 $x^{(k)}$；

err：保存误差$\|x-x_0\|_\infty$；

r：微小值.

（4）计算实例.

Example 7.4.2 Use the Seidel iteration to solve the following linear system

$$\begin{pmatrix} 4 & 1 & 1 & 0 & 1 \\ -1 & -3 & 1 & 1 & 0 \\ 2 & 1 & 5 & -1 & -1 \\ -1 & -1 & -1 & 4 & 0 \\ 0 & 2 & -1 & 1 & 4 \end{pmatrix} \begin{pmatrix} x \\ x_2 \\ x_3 \\ x_4 \\ x_5 \end{pmatrix} = \begin{pmatrix} 8 \\ 8 \\ 8 \\ 8 \\ 8 \end{pmatrix}$$

操作步骤如下：

① 光标定位在要执行的 Cell 中，按小键盘的【Enter】键；

② 在弹出的对话框中按提示分别输入

```
5,
{{4,1,1,0,1},{-1,-3,1,1,0},{2,1,5,-1,-1},{-1,-1,-1,4,0},
{0,2,-1,1,4}},
{8,8,8,8,8,8},
{0,0,0,0,0}, 0.0001
```

③ 每次输入后单击 OK 命令按钮，输出结果如图 7-2 所示.

```
x={2., -3.33333, 1.46667, 2.03333, 3.525}    k=1    err=3.525
x={1.58542, -2.02847, 2.48319, 2.51003, 3.00753}    k=2    err=1.30486
x={1.13444, -1.3804, 2.52582, 2.56996, 2.67917}    k=3    err=0.648069
x={1.04386, -1.31602, 2.49549, 2.55583, 2.64293}    k=4    err=0.0905829
x={1.0444, -1.33103, 2.4882, 2.55039, 2.64996}    k=5    err=0.0150032
x={1.04822, -1.33654, 2.48809, 2.54994, 2.65281}    k=6    err=0.00551471
x={1.04891, -1.33696, 2.48838, 2.55008, 2.65305}    k=7    err=0.000693271
x={1.04888, -1.33681, 2.48844, 2.55013, 2.65298}    k=8    err=0.000151091
x={1.04885, -1.33676, 2.48843, 2.55013, 2.65296}    k=9    err=0.0000457643
```

图 7-2 Seidel 迭代计算结果

7.4.3 Experiments by Matlab

1. Jacobi 迭代

（1）函数语句.

```
[x,err]=JacobiIte(A,b,x0,delta,l)
```

（2）参数说明.

A：矩阵变量，输入参数，系数矩阵；

b：向量变量，输入参数，常数项（以列向量形式输入）；

x0：向量变量，输入参数，初始向量（以列向量形式输入）；

delta：实变量，输入参数，误差限；

l：实变量，输入参数，迭代次数的上限；

x：向量变量，输出参数，方程组的近似解；

err：实变量，输出参数，误差.

（3）JacobiIte.m 程序.

```
function [x,err]=JacobiIte(A,b,x0,delta,l)
%Jacobi 迭代定义的 Matlab 程序。其中 b,x0 要求以列向量的形式输入。
if nargin<3
   x0=zeros(1,n);
end
if nargin<4
   delta=1e-6;
end
if nargin<5
   l=100;
end
D=diag(diag(A));
L=-triu(A,1);
U=-tril(A,-1);
if det(D)==0
   fprintf('失败信息：因为对角矩阵奇异，所以迭代失败。');
   return;
end
B=D\(L+U);
f=D\b;
for k=1:l
   x=B*x0+f;
   err=max(abs(x-x0));
   if err<delta
     fprintf('经过%3d 次迭代，得到近似解为：',k);
     x
     fprintf('误差为：%4d',err);
     return
   end
   x0=x;
end
 fprintf('迭代了%3d 次，还是没有达规定的误差要求',l);
```

（4）计算实例.

Example 7.4.3 Use the Jacobi iteration to solve the following linear system

$$\begin{pmatrix} 4 & 1 & 1 & 0 & 1 \\ -1 & -3 & 1 & 1 & 0 \\ 2 & 1 & 5 & -1 & -1 \\ -1 & -1 & -1 & 4 & 0 \\ 0 & 2 & -1 & 1 & 4 \end{pmatrix}\begin{pmatrix} x \\ x_2 \\ x_3 \\ x_4 \\ x_5 \end{pmatrix}=\begin{pmatrix} 8 \\ 8 \\ 8 \\ 8 \\ 8 \end{pmatrix}$$

在命令窗口输入：

```
>> A=[4,1,1,0,1;-1,-3,1,1,0;2,1,5,-1,-1;-1,-1,-1,4,0;0,2,-1,1,4];
b=[8;8;8;8;8];x0=[0;0;0;0;0];
```

```
delta=0.0001;JacobiIte(A,b,x0,delta)
```

经过 16 次迭代，得到近似解为：

```
x =
    1.0488
   -1.3368
    2.4885
    2.5501
    2.6530
```

误差为：7.414904e-005

2. Gauss-Seidel 迭代

（1）函数语句.

```
x=GaussSeidelIte(A,b,x0,delta,l)
```

（2）参数说明.

A：矩阵变量，输入参数，系数矩阵；

b：向量变量，输入参数，常数项（以行向量形式输入）；

x0：向量变量，输入参数，初始向量（以行向量形式输入）；

delta：实变量，输入参数，误差限；

l：实变量，输入参数，迭代次数的上限；

x：向量变量，输出参数，方程组的近似解.

（3）GaussSeidelIte.m 程序.

```
function x=GaussSeidelIte(A,b,x0,delta,l)
if nargin<3
    x0=zeros(1,n);
end
if nargin<4
    delta=1e-6;
end
if nargin<5
    l=100;
end
[n,m]=size(A);
for k=1:l
   a=x0;
    for i=1:n
        x0(i)=(b(i)-A(i,1:n)*x0(1:n)'-A(i,i)*x0(i))/A(i,i);
    end
    err=max(abs(a-x0));
    fprintf('k= %4d   error= %14.10f ',k,err);x0
    if err<delta
        x=x0;
      return
    end
end
fprintf('迭代了%3d次，还是没有达规定的误差要求',l);
```

（4）计算实例.

Example 7.4.4 Use the G-S iteration to solve the following linear system

$$\begin{pmatrix} 4 & 1 & 1 & 0 & 1 \\ -1 & -3 & 1 & 1 & 0 \\ 2 & 1 & 5 & -1 & -1 \\ -1 & -1 & -1 & 4 & 0 \\ 0 & 2 & -1 & 1 & 4 \end{pmatrix} \begin{pmatrix} x_1 \\ x_2 \\ x_3 \\ x_4 \\ x_5 \end{pmatrix} = \begin{pmatrix} 8 \\ 8 \\ 8 \\ 8 \\ 8 \end{pmatrix}$$

在命令窗口输入：

```
>> A=[4,1,1,0,1;-1,-3,1,1,0;2,1,5,-1,-1;-1,-1,-1,4,0;0,2,-1,1,4];
b=[8,8,8,8,8];
x0=[0,0,0,0,0]; delta=0.0001;GaussSeidelIte(A,b,x0,delta)

k=   1  error=  3.5250000000
x0 = 2.0000   -3.3333   1.4667   2.0333   3.5250

k=   2  error=  1.3048611111
x0 = 1.5854   -2.0285   2.4832   2.5100   3.0075

k=   3  error=  0.6480692998
x0 = 1.1344   -1.3804   2.5258   2.5700   2.6792

k=   4  error=  0.0905828609
x0 = 1.0439   -1.3160   2.4955   2.5558   2.6429

k=   5  error=  0.0150032092
x0 = 1.0444   -1.3310   2.4882   2.5504   2.6500

k=   6  error=  0.0055147134
x0 = 1.0482   -1.3365   2.4881   2.5499   2.6528

k=   7  error=  0.0006932708
x0 = 1.0489   -1.3370   2.4884   2.5501   2.6531

k=   8  error=  0.0001510906
x0 = 1.0489   -1.3368   2.4884   2.5501   2.6530

k=   9  error=  0.0000457643
x0 = 1.0488   -1.3368   2.4884   2.5501   2.6530
ans =
   1.0488   -1.3368   2.4884   2.5501   2.6530
```

Exercises 7

Questions

1. Find the first two iterations of the Jacobi and G-S methods respectively for the following linear systems, using $\boldsymbol{x}^{(0)}=0$:

(1) $\begin{cases}3x_1-x_2+x_3=1\\3x_1+6x_2+2x_3=0;\\3x_1+3x_2+7x_3=4\end{cases}$ (2) $\begin{cases}4x_1+x_2-x_3+x_4=-2\\x_1+4x_2-x_3-x_4=-1\\-x_1-x_2+5x_3+x_4=0\\x_1-x_2+x_3+3x_4=1\end{cases}$.

2. Find the first two iterations of the SOR method with $\omega=1.1$ for the following linear systems, using $\boldsymbol{x}^{(0)}=0$:

(1) $\begin{cases}10x_1-x_2=9\\-x_1+10x_2-2x_3=7;\\-2x_2+10x_3=6\end{cases}$ (2) $\begin{cases}10x_1+5x_2=6\\5x_1+10x_2-4x_3=25\\-4x_2+8x_3-x_4=-11\\-x_3+5x_4=-11\end{cases}$.

3. Given the system

$$\begin{cases}5x_1+2x_2+x_3=-12\\-x_1+4x_2+2x_3=20\\2x_1-3x_2+10x_3=3\end{cases}$$

Consider the convergence of Jacobi and G-S methods for solving this system, please give the iterative scheme if convergence.

4. Given the system

(1) $\begin{cases}x_1+0.4x_2+0.4x_3=1\\0.4x_1+x_2+0.8x_3=2;\\0.4x_1+0.8x_2+x_3=3\end{cases}$ (2) $\begin{cases}2x_1-x_2+x_3=1\\x_1+x_2+x_3=1\\x_1+x_2-2x_3=1\end{cases}$.

Consider the convergence of Jacobi and G-S methods for solving these systems.

5. Prove that both the Jacobi and G-S iterations must be convergent for solving the following system, and write the iterative schemes.

$$\begin{cases}10x_1-2x_2-2x_3=1\\-2x_1+10x_2-x_3=0.5\\-x_1-2x_2+3x_3=1\end{cases}$$

6. Show that $\lim\limits_{k\to\infty}\boldsymbol{A}_k=\boldsymbol{A}$ if and only if $\lim\limits_{k\to\infty}\boldsymbol{A}_k\boldsymbol{x}=\boldsymbol{A}\boldsymbol{x}$ holds for any vector $\boldsymbol{x}$.

7. Consider the solution of the linear system (take $\omega=0.9$)

$$\begin{cases} 5x_1 + 2x_2 + x_3 = -12 \\ -x_1 + 4x_2 + 2x_3 = 20 \\ 2x_1 - 3x_2 + 10x_3 = 3 \end{cases}$$

with the SOR method, ensure that $\left\| x^{(k+1)} - x^{(k)} \right\|_\infty < 10^{-4}$ when the iteration is halted.

8. Let the system $\boldsymbol{Ax}=\boldsymbol{b}$, where $\boldsymbol{A}$ is a positive definite. Given the iterative formula

$$\boldsymbol{x}^{(k+1)} = \boldsymbol{x}^{(k)} + \omega(\boldsymbol{b} - \boldsymbol{Ax}^{(k)}) \quad (k = 0,1,2,\cdots)$$

prove that the above iteration converges when $0 < \omega < \frac{2}{\beta}$, where $(0 < \alpha \leqslant \lambda(\boldsymbol{A}) \leqslant \beta)$.

9. Prove that the matrix

$$\boldsymbol{A} = \begin{pmatrix} 1 & a & a \\ a & 1 & a \\ a & a & 1 \end{pmatrix}$$

is positive definite for $-\frac{1}{2} < \alpha < 1$, and the Jacobi iteration converges only for $-\frac{1}{2} < \alpha < \frac{1}{2}$.

10. Discuss the convergence of the Jacobi and G-S iterations for solving the system $\boldsymbol{Ax}=\boldsymbol{b}$, if convergence, compare their rate of convergence, where

$$\boldsymbol{A} = \begin{pmatrix} 3 & 0 & -2 \\ 0 & 2 & 1 \\ -2 & 1 & 2 \end{pmatrix}$$

11. Given the system

$$\begin{cases} x_1 - \frac{1}{4}x_3 - \frac{1}{4}x_4 = \frac{1}{2} \\ x_2 - \frac{1}{4}x_3 - \frac{1}{4}x_4 = \frac{1}{2} \\ -\frac{1}{4}x_1 - \frac{1}{4}x_2 + x_3 = \frac{1}{2} \\ -\frac{1}{4}x_1 - \frac{1}{4}x_2 + x_4 = \frac{1}{2} \end{cases}$$

(1) Solve the spectral radius of Jacobi iterative matrix $\boldsymbol{J}$.

(2) Solve the spectral radius of G-S iterative matrix $\boldsymbol{G}$.

(3) Discuss the convergence of the Jacobi and G-S iterations for solving this system.

Computer Questions

1. Program the G-S method and test it on these examples:

$$(1)\begin{cases}3x+y+z=5\\x+3y-z=3\\3x+y-5z=-1\end{cases};\qquad (2)\begin{cases}3x+y+z=5\\3x+y-5z=-1.\\x+3y-z=3\end{cases}$$

Analyze what happens when these systems are solved by simple Gaussian elimination without pivoting.

2. Apply G-S iteration on the system in which

$$\boldsymbol{A}=\begin{pmatrix}0.963\,26 & 0.813\,21\\0.813\,21 & 0.686\,54\end{pmatrix},\quad \boldsymbol{b}=\begin{pmatrix}0.888\,24\\0.749\,88\end{pmatrix}$$

Use $(0.331\,16, 0.700\,00)^{\mathrm{T}}$ as the starting point and explain what happens.

3. The linear system

$$\begin{cases}x_1+2x_2-2x_3=7\\x_1+x_2+x_3=2\\2x_1+2x_2+x_3=5\end{cases}$$

has the solution $(1,2,-1)^{\mathrm{T}}$.

(1) Show that $P(\boldsymbol{J})=0$.

(2) Use the Jacobi method with $\boldsymbol{x}^{(0)}=0$ to approximate the solution to the linear system to within 10^{-5} in the l_∞ norm.

(3) Show that $P(\boldsymbol{G})=2$.

(4) Show that the G-S method applied as in part (2) fails to give a good approximation in 25 iterations.

4. Solve the systems in Problems 3 by Jacobi and G-S methods, and the iteration is halted when $\left\|\boldsymbol{x}^{(k+1)}-\boldsymbol{x}^{(k)}\right\|_\infty<10^{-4}$.

5. Solve the linear system (take $\omega=1.03$, $\omega=1$ and $\omega=1.1$ respectively) with the SOR method, given the exact solution $\boldsymbol{x}^*=\left(\frac{1}{2},1,-\frac{1}{2}\right)^{\mathrm{T}}$. Estimate the number of iterations such that $\left\|\boldsymbol{x}^*-\boldsymbol{x}^{(k)}\right\|_\infty<5\times10^{-6}$ when the iteration is halted

$$\begin{cases}4x_1-x_2=1\\-x_1+4x_2-x_3=4\\-x_2+4x_3=-3\end{cases}$$

8 Numerical Solution of Ordinary Differential Equations

提　要

在自然科学和工程技术中，常常会遇到常微分方程的求解问题. 除了一些简单的、常系数线性微分方程外，要求出复杂的变系数的或非线性问题的解析表达式一般是很困难的，甚至是不可能的. 实际上，只需要获得解在若干点上的近似值，再根据这些近似值构造拟合函数即可解决问题，因此，需要研究常微分方程的数值解法，即求出在离散点上解的近似值的方法.

本章主要介绍求解一阶常微分方程的数值问题：

$$\begin{cases} y' = f(x,y) \\ y(x_0) = y_0 \end{cases} \quad x \in [a,b]$$

的常用算法，包括 Euler 法、线性多步法和 Runge-Kutta 方法等。

词　汇

polygon method	折线法	single-step Methods	单步法
explicit method	显式法	multistep methods	多步法
implicit method	隐式法	variational equation	变分方程
principal term	主项	Taylor-Series	泰勒级数
equidistant grid points	等距节点	modified Euler method	改进的欧拉法
consistent	相容的	analytic solution	解析解
rectangle	矩形	magnitude	大小，量
subscript	下标	partial derivative	偏导数
local truncation error	局部截断误差	global truncation error	全局截断误差

8.1 Introduction

As we all known, many physical phenomena can be modeled by higher-order ordinary differential equations. For example, the motion of a mass acted upon by a force can be modeled by a second-order differential equation using Newton's second law. In almost all cases of practical interest, a higher-order differential equation can be **recast** (重作) as a system of first order differential equations by introducing new variables.

As an example of a system that is nonlinear, consider the **swinging pendulum** (钟摆) show in Figure 8-1. When the mass of the pendulum is small in comparison with the mass m at the end of pendulum, the equation of motion of the pendulum is as follows

$$mR^2\frac{\mathrm{d}^2\theta}{\mathrm{d}t^2}+\mu\frac{\mathrm{d}\theta}{\mathrm{d}t}+mgR\sin\theta=0 \tag{8.1.1}$$

Here θ is the angular position of the pendulum measured relative to vertical (垂直的), m is mass at the end of the pendulum, R is the length of the pendulum, μ is the **coefficient of viscous friction** (黏滞摩擦系数), and g is the **acceleration due to gravity** (重力加速度). Again, this is a second-order equation. If we define the state vector to be $\boldsymbol{x}=[x_1,x_2]^{\mathrm{T}}=[\theta,\mathrm{d}\theta/\mathrm{d}t]^{\mathrm{T}}$, we can get the following first-order system

$$\begin{cases}\dfrac{\mathrm{d}x_1}{\mathrm{d}t}=x_2\\[2ex]\dfrac{\mathrm{d}x_2}{\mathrm{d}t}=-\left(\dfrac{g}{R}\right)\sin x_1-\left(\dfrac{\mu}{mR^2}\right)x_2\end{cases} \tag{8.1.2}$$

This is nonlinear due to the presence of the $\sin x_1$ term.

When you finish this chapter, you will be able to compute numerical solutions to systems of ordinary differential equations. You will know how to convert higher-order differential equations into equivalent systems of first-order differential equations using the notion of a state vector. You will know how to solve initial value problems using single-step methods ranging from Euler's method to the Runge-Kutta method. You will know how and when to apply multi-step solution techniques, including the Adams method. These overall goals will be achieved by mastering the following chapter objectives.

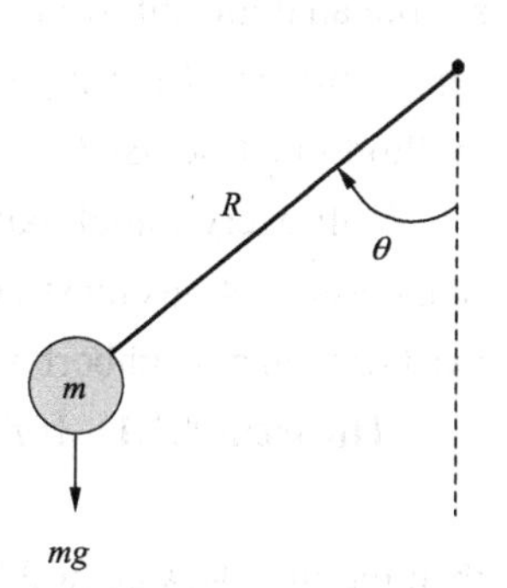

Figure 8-1 A pendulum

Objectives

- Understand the Euler's method.
- Understand the Taylor-Series method.
- Be able to apply the Runge-Kutta single-step solution methods.
- Be able to apply the Adams multi-step solution method.
- Know how to convert a higher-order ordinary differential equation into an equivalent system of first-order equations.

8.2 The Existence and Uniqueness of Solutions

Let us consider an initial-value problem written in the form

$$\begin{cases} y' = f(x,y) \\ y(x_0) = y_0 \end{cases} \tag{8.2.1}$$

Here y is an unknown function of x that we hope to construct from the information given in Equation (8.2.1), where $y'=dy(x)/dx$. The second of the two equations in (8.2.1) specifies (指定,给定) one particular value of the function $y(x)$. The first equation gives the slope of the curve y at any point x. Of course, the function $f(x,y)$ must be specified. For a concrete example, we can take

$$\begin{cases} y' = y\tan(x+2) \\ y(-2) = 1 \end{cases} \tag{8.2.2}$$

We would like to determine $y(x)$ on an interval containing the initial point $x_0=-2$. The **analytic solution** (解析解) of this initial-value problem is $y(x)=\sec(x+2)$, as we can easily verify. Since $\sec x$ approaches **infinite** (无穷) at $x=\pm\pi/2$, our solution is **valid** (有效) only for $-\pi/2 < x+2 < \pi/2$. The example in (8.2.2) is exceptional because it has a simple analytic solution from which numerical values are readily calculated. Typically, for problems of the type in (8.2.1), analytic solutions are not available and numerical method must be used.

Will every initial-value problem of the form in Equations(8.2.1)have a solution? No. Some assumptions must be made about $f(x,y)$, and even then we can expect the solution to exist only in a neighborhood of $x=x_0$.

Theorem 8.2.1 If $f(x,y)$ is continuous in a **rectangle** (矩形) R centered at (x_0,y_0), say

$$R = \{(x,y): |x-x_0| \leqslant \alpha,\ |y-y_0| \leqslant \beta\} \tag{8.2.3}$$

then the initial-value problem (8.2.1) has a solution $y(x)$ for $|x-x_0| \leqslant \min(\alpha, \beta/M)$, where $M = \max\limits_{(x,y)\in R} |f(x,y)|$.

Example 8.2.1 Prove that the initial-value problem

$$\begin{cases} y' = (x+\sin y)^2 \\ y(0) = 3 \end{cases}$$

has a solution on the interval $-1 \leqslant x \leqslant 1$.

Solution In this example, $f(x,y)=(x+\sin y)^2$ and $(x_0,y_0)=(0,3)$. In the rectangle

$$R = \{(x,y): |x| \leqslant \alpha,\ |y-3| \leqslant \beta\}$$

the **magnitude** (大小,量) of f is bounded by $|f(x,y)| \leqslant (\alpha+1)^2 \equiv M$. We want $\min(\alpha, \beta/M) \geqslant 1$, and so we can let $\alpha=1$. Then $M=4$, and our objective is met by letting $\beta \geqslant 4$. The Theorem 8.2.1 **asserts** (断定) that a solution of the initial-value problem exists on the interval $|x_0| \leqslant$

$\min(\alpha,\beta/M)=1$.

Theorem 8.2.2 If $f(x,y)$ and $\partial f/\partial y$ are continuous in the rectangle: $R=\{(x,y): |x-x_0|\leqslant\alpha, |y-y_0|\leqslant\beta\}$, then the initial-value problem (8.2.1) has a unique solution in the interval $|x-x_0|\leqslant\min(\alpha,\beta/M)$.

In both Theorems (8.2.1) and (8.2.2), the interval on the x-axis in which the solution is asserted to exist may be smaller than the base of the rectangle in which we have defined $f(x,y)$. The next theorem is of a different type that allows us to infer the existence and uniqueness of a solution on a prescribed interval $[a,b]$.

Theorem 8.2.3 If $f(x,y)$ is continuous in the **strip** (带) $a\leqslant x\leqslant b$, $-\infty<y<\infty$, and satisfies Lipschitz condition:

$$|f(x,y_1)-f(x,y_2)|\leqslant L|y_1-y_2| \tag{8.2.4}$$

then the initial-value problem (8.2.1) has a unique solution in the interval $[a,b]$.

For a function of one variable, Lipschitz condition would assert simply

$$|f(x_1)-f(x_2)|\leqslant L|x_1-x_2| \tag{8.2.5}$$

We see immediately that this condition is stronger than continuity because if x_2 approaches x_1, then the right-hand side in (8.2.5) approaches 0, and this forces $f(x_2)$ to approach $f(x_1)$. The condition (8.2.5) is weaker than having a bounded derivative. Indeed, if $f'(x)$ exists everywhere and does not exceed L in modulus, then by the Mean-value theorem,

$$|f(x_1)-f(x_2)|=|f'(\xi)||x_1-x_2|\leqslant L|x_1-x_2|$$

Example 8.2.2 Prove that the function $f(x)=\sum_{i=1}^{n}a_i|x-b_i|$ satisfies a Lipschitz condition with the constant $L=\sum_{i=1}^{n}|a_i|$.

$$\begin{aligned}|f(x_1)-f(x_2)|&=\left|\sum_{i=1}^{n}a_i|x_1-b_i|-\sum_{i=1}^{n}a_i|x_2-b_i|\right|\\&=\left|\sum_{i=1}^{n}a_i\left(|x_1-b_i|-|x_2-b_i|\right)\right|\\&\leqslant\sum_{i=1}^{n}|a_i|\left||x_1-b_i|-|x_2-b_i|\right|\\&\leqslant\sum_{i=1}^{n}|a_i||x_1-x_2|\\&=L|x_1-x_2|\end{aligned}$$

Example 8.2.3 Consider the initial-value problem

$$\begin{cases}y'=x^2+y^2\\y(0)=0\end{cases}$$

on $R=\{(x,y):-0.5<x<0.5,-0.5<y<0.5\}$. For $f(x,y)=x^2+y^2$, we have

$$|f(x,y)| \leqslant 0.5$$

on R. Hence M=0.5, we can estimate

$$|f(x,y_1)-f(x,y_2)| = |y_1^2 - y_2^2| = |(y_1+y_2)(y_1-y_2)| \leqslant |y_1 - y_2|$$

fall all $(x,y_1),(x,y_2)\in G$, i.e., f satisfies Lipschitz condition with Lipschitz constant L=1. Therefore, the given problem has a unique solution in the interval [−0.5,0.5].

8.3 Taylor-Series Method

In the numerical solution of differential equations, we rarely expect to obtain the solution directly as a formula giving $y(x)$ as a function of x. Instead, we usually construct a table of function values of the form, see Table 8-1.

Table 8-1 Table of function values

x_i	x_0	x_1	x_2	x_3	$\cdots$	x_n
y_i	y_0	y_1	y_2	y_3	$\cdots$	y_n

Here, y_i is the computed approximate value of $y(x_i)$, our notation for the exact solution at x_i.

From Table 8-1, a spline function or other approximating function can be constructed. However, most numerical methods for solving ordinary differential equations produce such a table first.

Let us consider again the initial-value problem

$$\begin{cases} y' = f(x,y) \\ y(x_0) = y_0 \end{cases} \tag{8.3.1}$$

where $f(x,y)$ is a prescribed function of two variables, and (x_0, y_0) is a single given point through which the solution curve passes. A solution of (8.3.1) is a function $y(x)$ such that $\mathrm{d}y(x)/\mathrm{d}x = f(x, y(x))$ for all x in some neighborhood of x_0, and $y(x_0) = y_0$.

Assume that various partial derivatives of f exist. To illustrate the method, we take a concrete example:

$$\begin{cases} y' = \cos x - \sin y + x^2 \\ y(-1) = 3 \end{cases} \tag{8.3.2}$$

At the heart of the procedure is the Taylor series for y, which we write as

$$y(x+h) = y(x) + hy'(x) + \frac{h^2}{2!}y''(x) + \frac{h^3}{3!}y'''(x) + \frac{h^4}{4!}y^{(4)}(x) + \cdots \tag{8.3.3}$$

The derivatives appearing here can be obtained from the differential equation in (8.3.2). They are

$$y'' = -\sin x - y'\cos y + 2x$$

$$y''' = -\cos x - y''\cos y + (y')^2 \sin y + 2$$

$$y^{(4)} = \sin x - y'''\cos y + 3y'y''\sin y + (y')^3 \cos y$$

Now, we decide to use only terms up to and including h^4 in formula (8.3.3). The terms that we have not included start with a term in h^5, and they constitute collectively the truncation error inherent in the produce. The resulting numerical method is said to be of order 4. (The order of the Taylor-series method is n if terms up to and including $h^n y^{(n)}(x)/n!$ are used)

Here is an algorithm to solve the initial-value problem (8.3.3), starting at $x_0=-1$ and step size $h=0.01$. We desire a solution in the x-interval [−1, 1], and thus we must take 200 steps.

input $M \leftarrow 200;\ h \leftarrow 0.01;\ x \leftarrow -1.0;\ y \leftarrow 3.0$

output $0, x, y$

for $k=1$ to M do

$y' \leftarrow \cos x - \sin y + x^2$

$y'' \leftarrow -\sin x - y'\cos y + 2x$

$y''' \leftarrow -\cos x - y''\cos y + (y')^2 \sin y + 2$

$y^{(4)} \leftarrow \sin x + ((y')^3 - y''')\cos y + 3y'y''\sin y$

$$y \leftarrow y + h\left(y' + \frac{h}{2}\left(y'' + \frac{h}{3}\left(y''' + \frac{h}{4}y^{(4)}\right)\right)\right)$$

$x \leftarrow x + h$

output k, x, y

end do

Here is a sample of the output from that computer program:

k	x	y
0	−1.000 00	3.000 00
1	−0.990 00	3.014 00
2	−0.980 00	3.028 03
3	−0.970 00	3.042 09
4	−0.960 00	3.056 17
5	−0.950 00	3.070 28
6	−0.940 00	3.084 43
7	−0.930 00	3.098 61
⋮	⋮	⋮
196	0.960 00	6.365 66
197	0.970 00	6.379 77
198	0.980 00	6.393 86
199	0.990 00	6.407 91
200	1.000 00	6.421 94

8.4 Euler's Method

In the sequel we **confine** (限制) our expression to the initial value problem for a differential equation of the first order. The generalization to systems and henceforth to equations of higher order is straightforward. We shall always tacitly (默许) assume that the assumptions of the Theorem 8.2.3 are satisfied.

The following simple method for the numerical solution of the initial value problem

$$\begin{cases} y' = f(x,y) \\ y(x_0) = y_0 \end{cases} \tag{8.4.1}$$

was first used by Euler. Given a step size $h>0$, it consists in replacing the derivative $y' = f(x,y)$ throughout the interval $[x_0, x_0+h]$ by the derivative $y_0' = f(x_0, y_0)$ at the initial point, i.e., geometrically speaking, by replacing the solution by its tangent line at the initial point x_0. This leads to the approximation

$$y_1 = y_0 + hf(x_0, y_0) \tag{8.4.2}$$

for the value $y(x_1)$ of the exact solution at the point $x_1=x_0+h$. Repeating this procedure leads to the Euler method as described in the following definition.

See Figure 8-2, for obvious reason, this method is also known as the **polygon method** (折线法).

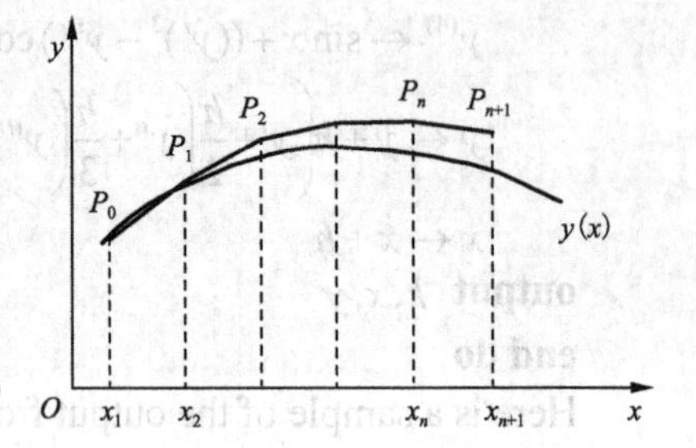

Figure 8-2 Euler's method

Definition 8.4.1 The Euler method for the numerical solution of the initial-value problem (8.4.1) constructs approximations y_n to the exact solution $y(x_n)$ at the **equidistant grid points** (等距节点)

$$x_n=x_0+nh, \quad n=1,2,\cdots$$

with step size h by

$$y_{n+1}=y_n+hf(x_n,y_n), \quad n=0,1,\cdots$$

Example 8.4.1 Solve the initial-value problem

$$\begin{cases} y' = y - \dfrac{2x}{y} \quad (0<x<1) \\ y(0)=1 \end{cases}$$

Solution The Euler method is of the form

$$y_{n+1} = y_n + h\left(y_n - \frac{2x_n}{y_n}\right), \quad n=0,1,\cdots$$

Take the step size $h=0.1$, some results are given in Table 8-2.

Table 8-2 Some results by using Euler method

x_n	y_n	$y(x_n)$	x_n	y_n	$y(x_n)$
0.1	1.100 0	1.095 4	0.6	1.509 0	1.483 2
0.2	1.191 8	1.183 2	0.7	1.580 3	1.549 2
0.3	1.277 4	1.264 9	0.8	1.649 8	1.612 5
0.4	1.358 2	1.341 6	0.9	1.717 8	1.673 3
0.5	1.435 1	1.414 2	1.0	1.784 8	1.732 1

There are three different interpretations of the approximation formula of Euler's method.

(1) Replace the derivative by the difference quotient

$$\frac{y(x_1)-y(x_0)}{h} \approx y'(x_0)=f(x_0,y_0)$$

and solve for $y(x_1)$.

(2) Transform the initial-value problem equivalently into the integral equation

$$y(x)=y(x_0)+\int_{x_0}^{x} f(t,y(t))\mathrm{d}t$$

then

$$y(x_1)=y_0+\int_{x_0}^{x_1} f(t,y(t))\mathrm{d}t$$

approximately by the rectangular rule

$$\int_{x_0}^{x_1} f(t,y(t))\mathrm{d}t \approx hf(x_0,y_0)$$

(3) Use Taylor's formula

$$y(x_1)=y(x_0)+hy'(x_0)+\frac{h^2}{2}y''(x_0+\theta h)$$

with $0<\theta<1$ and neglect the remainder term, i.e., approximate

$$y(x_1)\approx y(x_0)+hy'(x_0)$$

Each of these three interpretations opens up possibilities for improvements of Euler's method. For example, instead of the rectangular ruler we can use the more accurate trapezoidal ruler

$$\int_{x_0}^{x_1} f(t,y(t))\mathrm{d}t \approx \frac{h}{2}(f(x_0,y(x_0))+f(x_1,y(x_1)))$$

which yields trapezoidal method

$$y_1=y_0+\frac{h}{2}(f(x_0,y_0)+f(x_1,y_1)) \tag{8.4.3}$$

Repeating this procedure leads to the following method.

Definition 8.4.2 The implicit Euler method for the numerical solution of the initial value problem (8.4.1) constructs approximations y_n to the exact solution $y(x_n)$ at the equidistant grid points

$$x_n=x_0+nh,\quad n=1,2,\cdots$$

with step size h by

$$y_{n+1}=y_n+\frac{h}{2}(f(x_n,y_n)+f(x_{n+1},y_{n+1})),\quad n=0,1,\cdots \tag{8.4.4}$$

This method is called an **implicit method** (隐式法). Since determining y_{n+1} requires the solution of an equation that in general is nonlinear. In contrast, the Euler method of Definition 8.4.1 is an **explicit method** (显式法), since it provides an explicit expression for the computation of y_{n+1}.

Remark 8.4.1 The nonlinear equations of the implicit Euler method can be solved by successive approximations, provided that the Lipschitz constant L for f and the step size h satisfy $Lh<2$.

Proof We have to solve equation (8.4.3) for y_1. Setting

$$g(y)=y_0+\frac{h}{2}(f(x_0,y_0)+f(x_1,y))$$

one can rewrite (8.4.3) as the fixed point equation $y_1=g(y_1)$. The function g is a contraction, since

$$|g(u)-g(v)|=\frac{h}{2}|f(x_1,u)-f(x_1,v)|\leqslant\frac{hL}{2}|u-v|$$

and therefore the assertion follows. □

Because the solution of the nonlinear Equation (8.4.3) will deliver only an approximation to the solution of the initial value problem, there is no need to solve (8.4.3) with high accuracy. Using the approximate value from the explicit Euler method as a starting point and carrying out only one iteration, we arrive at the following method.

Definition 8.4.3 The predictor corrector method for the Euler method for the numerical solution of the initial value problem (8.4.1), also known as the improved (modified) Euler method, constructs approximations y_n to the exact solution $y(x_n)$ at the equidistant grid points

$$x_n=x_0+nh,\quad n=1,2,\cdots$$

by

$$y_{n+1}=y_n+\frac{h}{2}(f(x_n,y_n)+f(x_{n+1},y_n+hf(x_n,y_n))),\quad n=0,1,\cdots$$

or written in the form

$$\begin{cases} y_{n+1}=y_n+\frac{h}{2}(K_1+K_2) \\ K_1=f(x_n,y_n) \\ K_2=f(x_n+h,y_n+hK_1) \end{cases}$$

Example 8.4.2 Consider again the initial value problem from Example 8.4.1. Table 8-3 gives the difference between the exact solution and the approximate solution obtained by the improved Euler method for step sizes $h=0.1$.

Table 8-3 The difference between the exact solution and the approximate solution

x_n	y_n	$y(x_n)$	x_n	y_n	$y(x_n)$
0.1	1.095 9	1.095 4	0.6	1.486 0	1.483 2
0.2	1.184 1	1.183 2	0.7	1.552 5	1.549 2
0.3	1.266 2	1.264 9	0.8	1.615 3	1.612 5
0.4	1.343 4	1.341 6	0.9	1.678 2	1.673 3
0.5	1.416 4	1.414 2	1.0	1.737 9	1.732 1

8.5 Single-step Methods

8.5.1 Single-step Methods

We generalize the Euler methods into more general single-step methods by the following definition.

Definition 8.5.1 Explicit single-step methods for the approximate solution of the initial value problem

$$\begin{cases} y' = f(x,y) \\ y(x_0) = y_0 \end{cases}$$

construct approximations y_n to the exact solution $y(x_n)$ at the equidistant grid points

$$x_n = x_0 + nh, \quad n = 1,2,\cdots$$

with step size h by

$$y_{n+1} = y_n + h\varphi(x_n, y_n, h), \quad n = 0,1,\cdots \tag{8.5.1}$$

Example 8.5.1 The Euler method and the improved Euler method are single-step methods with

$$\varphi(x,y,h) = f(x,y)$$

and

$$\varphi(x,y,h) = \frac{1}{2}(f(x,y) + f(x+h, y+hf(x,y)))$$

respectively.

The function φ describes how the differential equation

$$y' = f(x,y)$$

is approximated by the difference equation

$$\frac{1}{h}(y(x+h) - y(x)) = \varphi(x,y,h)$$

From a reasonable approximation we expect that the exact solution to the initial value problem approximately satisfies the difference equation. Hence,

$$\frac{1}{h}(y(x+h) - y(x)) - \varphi(x,y,h) \to 0, \quad h \to 0$$

must be fulfilled for the exact solution y. We also expect that the order of this convergence will influence the accuracy of the approximate solution.

8.5.2 Local Truncation Error

Definition 8.5.2 Let $y(x)$ be the exact solution to the initial value problem (8.2.1), then

$$T_{n+1} = y(x_{n+1}) - y(x_n) - h\varphi(x_n, y(x_n), h) \tag{8.5.2}$$

is called the **local truncation error** of the single-step method (8.5.1).

Assume that there is no error before x_n. When $y_n = y(x_n)$, we have

$$
\begin{aligned}
y(x_{n+1})-y_{n+1} &= y(x_{n+1})-(y(x_n)+h\varphi(x_n,y(x_n),h)) \\
&= y(x_{n+1})-y(x_n)-h\varphi(x_n,y(x_n),h)=T_{n+1}
\end{aligned}
$$

From the Definition 8.5.2, one obtain the local truncation error of Euler method

$$
\begin{aligned}
T_{n+1} &= y(x_{n+1})-y(x_n)-hf(x_n,y(x_n)) \\
&= y(x_n+h)-y(x_n)-hy'(x_n) \\
&= \frac{h^2}{2}y''(x_n)+O(h^3)
\end{aligned}
\tag{8.5.3}
$$

Definition 8.5.3 Let $y(x)$ be the exact solution to the initial value problem (8.2.1) , if there exists the maximum integer p such that the local truncation error of the single-step method (8.5.1) satisfies:

$$
T_{n+1}=y(x+h)-y(x)-h\varphi(x,y,h)=O(h^{p+1}) \tag{8.5.4}
$$

then the method (8.5.1) has accuracy of order p.

If rewrite (8.5.4) in the form

$$
T_{n+1}=\psi(x_n,y(x_n))h^{p+1}+O(h^{p+2})
$$

then $\psi(x_n,y(x_n))h^{p+1}$ is called **the principal term of local truncation error**.

For (8.5.3),p=1, the accuracy is order one, the principal term of local truncation error is $\frac{h^2}{2}y''(x_n)$.

For implicit Euler method (8.4.4)

$$
\begin{aligned}
T_{n+1} &= y(x_{n+1})-y(x_n)-\frac{h}{2}(y'(x_n)+y'(x_{n+1})) \\
&= hy'(x_n)+\frac{h^2}{2}y''(x_n)+\frac{h^3}{3!}y'''(x_n)- \\
&\quad \frac{h}{2}(y'(x_n)+y'(x_n)+hy''(x_n)+\frac{h^2}{2}y'''(x_n))+O(h^4) \\
&= -\frac{h^3}{12}y'''(x_n)+O(h^4)
\end{aligned}
$$

where p=2, the accuracy is order two, the principal term of local truncation error is $-\frac{h^3}{2}y'''(x_n)$.

8.6 Runge-Kutta Methods

8.6.1 Second-Order Runge-Kutta Method

We begin with the Taylor series for $y(x+h)$:

$$
y(x+h)=y(x)+hy'(x)+\frac{h^2}{2!}y''(x)+\frac{h^3}{3!}y'''(x)+\cdots \tag{8.6.1}
$$

From the differential equation, we obtain

$$y'(x) = f$$
$$y''(x) = f_x + f_y y' = f_x + f_y f$$
$$y'''(x) = f_{xx} + f_{xy} f + (f_{yx} + f_{yy} f) f + f_y (f_x + f_y f)$$
$$\cdots\cdots$$

Here **subscripts** (下标) denote **partial derivatives** (偏导数), and the chain rule of differentiation is used repeatedly. The first three terms in Equation (8.6.1) can be written now in the form

$$\begin{aligned} y(x+h) &= y + hf + \frac{1}{2}h^2(f_x + ff_y) + O(h^3) \\ &= y + \frac{1}{2}hf + \frac{1}{2}h(f + hf_x + hff_y) + O(h^3) \end{aligned} \tag{8.6.2}$$

where y means $y(x)$, f means $f(x,y)$, and so on. We are able to eliminate the partial derivatives with the aid of the first few terms in the Taylor series in two variables.

$$f(x+h, y+hf) = f + hf_x + hff_y + O(h^2)$$

Equation (8.6.2) can be rewritten as

$$y(x+h) = y + \frac{1}{2}hf + \frac{1}{2}hf(x+h, y+hf) + O(h^3)$$

Hence, the formula for advancing the solution is

$$y(x+h) = y(x) + \frac{1}{2}hf(x,y) + \frac{1}{2}hf(x+h, y+hf(x,y))$$

or equivalently

$$y(x+h) = y(x) + \frac{h}{2}(K_1 + K_2) \tag{8.6.3}$$

where

$$\begin{cases} K_1 = f(x,y) \\ K_2 = f(x+h, y+hK_1) \end{cases} \tag{8.6.4}$$

This formula is just improved Euler method, it is called a **second-order Runge-Kutta method**.

In general, second-order Runge-Kutta formulas are of the following form

$$y(x+h) = y + c_1 hf + c_2 hf(x+\alpha h, y+\beta hf) + O(h^3) \tag{8.6.5}$$

where c_1,c_2,α and β are parameters at our disposal. Equation (8.6.5) can be rewritten with the aid of the Taylor series in two variables as

$$y(x+h) = y + c_1 hf + c_2 h(f + \alpha hf_x + \beta hff_y) + O(h^3) \tag{8.6.6}$$

Comparing Equations (8.6.2) and (8.6.6), we see that we should impose these conditions:

$$\begin{cases} c_1 + c_2 = 1 \\ c_2\alpha = \dfrac{1}{2} \\ c_2\beta = \dfrac{1}{2} \end{cases} \tag{8.6.7}$$

One solution is $c_1 = c_2 = \dfrac{1}{2}, \alpha = \beta = 1$, which is the one corresponding to improved Euler method in Equation (8.6.3). The system of Equations (8.6.7) has solutions other than this one, such as the one obtained by letting $c_1 = 0, c_2 = 1, \alpha = \beta = \dfrac{1}{2}$. The resulting formula from (8.6.5) is called the **mid-point method** (中点法):

$$y(x+h) = y + hK_2 \tag{8.6.8}$$

where

$$\begin{cases} K_1 = f(x, y) \\ K_2 = f\left(x + \dfrac{h}{2}, y + \dfrac{h}{2}K_1\right) \end{cases}$$

or rewritten as:

$$\begin{cases} y_{n+1} = y_n + hK_2 \\ K_1 = f(x_n, y_n) \\ K_2 = f\left(x_n + \dfrac{h}{2}, y_n + \dfrac{h}{2}K_1\right) \end{cases} \tag{8.6.9}$$

8.6.2 Fourth-Order Runge-Kutta Method

The higher-order Runge-Kutta formulas are very **tedious** (厌烦的) to derive, and we shall not do so. The formulas are rather **elegant** (漂亮的), however, and are easily programmed once they have been derived. Here are the formulas for the classical fourth-order Runge-Kutta method:

$$y(x+h) = y(x) + \frac{h}{6}(K_1 + 2K_2 + 2K_3 + K_4) \tag{8.6.10}$$

where

$$\begin{cases} K_1 = f(x, y) \\ K_2 = f\left(x + \dfrac{h}{2}, y + \dfrac{h}{2}K_1\right) \\ K_3 = f\left(x + \dfrac{h}{2}, y + \dfrac{h}{2}K_2\right) \\ K_4 = f(x + h, y + hK_3) \end{cases}$$

or rewritten as

$$\begin{cases} y_{n+1} = y_n + \dfrac{h}{6}(K_1 + 2K_2 + 2K_3 + K_4) \\ K_1 = f(x_n, y_n) \\ K_2 = f\left(x_n + \dfrac{h}{2}, y_n + \dfrac{h}{2}K_1\right) \\ K_3 = f\left(x_n + \dfrac{h}{2}, y_n + \dfrac{h}{2}K_2\right) \\ K_4 = f(x_n + h, y_n + hK_3) \end{cases} \tag{8.6.11}$$

This is called a fourth-order method because it reproduces the terms in the Taylor series up to and including the one involving h^4. The error is therefore $O(h^5)$.

Example 8.6.1 Solve the following initial-value problem by using the Runge-Kutta method of order 4

$$\begin{cases} y' = y - \dfrac{2x}{y} \quad (0 < x < 1) \\ y(0) = 1 \end{cases} \tag{8.6.12}$$

on the interval [0,1], using steps of h=0.2.

Solution The fourth-order Runge-Kutta method gives

$$\begin{cases} y_{n+1} = y_n + \dfrac{h}{6}(K_1 + 2K_2 + 2K_3 + K_4) \\ K_1 = y_n - \dfrac{2x_n}{y_n} \\ K_2 = y_n + \dfrac{h}{2}K_1 - \dfrac{2x_n + h}{y_n + \dfrac{h}{2}K_1} \\ K_3 = y_n + \dfrac{h}{2}K_2 - \dfrac{2x_n + h}{y_n + \dfrac{h}{2}K_2} \\ K_4 = y_n + hK_3 - \dfrac{2(x_n + h)}{y_n + hK_3} \end{cases}$$

Some results are given in Table 8-4.

Table 8-4 The results of computation

x_n	y_n	$y(x_n)$
0.2	1.1832	1.1832
0.4	1.3417	1.3416
0.6	1.4833	1.4832
0.8	1.6125	1.6125
1.0	1.7321	1.7321

8.7 Multistep Methods（多步法）

The methods discussed to this point in the chapter are single-step methods because they don't use any knowledge of prior values of $y(x)$ when the solution is advanced from x to $x+h$. If $x_0, x_1, x_2, \cdots, x_i$ are steps along the x-axis, then y_{i+1} (the approximate value of $y(x_{i+1})$) depends only on y_i, and knowledge of the approximate values $y_{i-1}, y_{i-2}, \cdots, y_0$ is not used.

8.7.1 General Formulas of Multistep Methods

A method is called **Multistep method** if the values $y_{n+i}(i=0,1,\cdots,k-1)$ is used in computing y_{n+k}. General Formula of Multistep method can be represented in the form

$$y_{n+k}=\sum_{i=0}^{k-1}\alpha_i y_{n+i}+h\sum_{i=0}^{k}\beta_i f_{n+i} \tag{8.7.1}$$

where y_{n+i} is approximate to $y(x_{n+i})$, $f_{n+i}=f(x_{n+i},y_{n+i})$, $x_{n+i}=x_n+ih$, α_i,β_i are constants, α_0 or β_0 is not zero at least, then (8.7.1) is called **linear *k*-step method.** If $\beta_k=0$ then (8.7.1) is called **explicit linear *k*-step method**, if $\beta_k\neq 0$, then (8.7.1) is called **implicit linear *k*-step method.**

Definition 8.7.1 Let $y(x)$ be the exact solution of the initial problem (8.2.1), the local truncation of the multistep method is

$$\begin{aligned}T_{n+k}&=L(y(x_n),h)\\&=y(x_{n+k})-\sum_{i=0}^{k-1}\alpha_i y(x_{n+i})-h\sum_{i=0}^{k}\beta_i y'(x_{n+i})\end{aligned} \tag{8.7.2}$$

if $T_{n+k}=O(h^{p+1})$, then method (8.7.1) is order p. If $p\geqslant 1$ then Equations (8.7.1) and (8.2.1) are **consistent**(相容的).

Using Taylor expansion for T_{n+k} at the point x_n

$$y(x_n+ih)=y(x_n)+ihy'(x_n)+\frac{(ih)^2}{2!}y''(x_n)+\frac{(ih)^3}{3!}y'''(x_n)+\cdots$$

$$y'(x_n+ih)=y'(x_n)+ihy''(x_n)+\frac{(ih)^2}{2!}y'''(x_n)+\cdots$$

consider (8.7.2), one obtain

$$T_{n+k}=c_0y(x_n)+c_1hy'(x_n)+c_2h^2y''(x_n)+\cdots+c_ph^py^{(p)}(x_n)+\cdots \tag{8.7.3}$$

where

$$c_0=1-(\alpha_0+\cdots+\alpha_{k-1})$$

$$c_1=k-(\alpha_1+2\alpha_2+\cdots+(k-1)\alpha_{k-1})-(\beta_0+\beta_1+\cdots+\beta_k)$$

$$c_m=\frac{1}{m!}(k^m-(\alpha_1+2^m\alpha_2+\cdots+(k-1)^m\alpha_{k-1}))-\frac{1}{(m-1)!}(\beta_1+2^{m-1}\beta_2+\cdots+k^{m-1}\beta_k)$$

$$m=2,3,\cdots \tag{8.7.4}$$

If we chose the coefficients α_i and β_i, such that

$$c_0=c_1=\cdots=c_p=0,\quad c_{p+1}\neq 0$$

From the Definition 8.7.1 we obtain that the multistep method is degree of accuracy p, and

$$T_{n+k}=c_{p+1}h^{p+1}y^{(p+1)}(x_n)+O(h^{p+2}) \tag{8.7.5}$$

The first term of the right-hand side is called **principal term of the local truncation error** (局部截断误差主项), and c_{p+1} is called **error constant**.

From the Definition 8.7.1, $p \geqslant 1$, i.e., $c_0=c_1=0$, and from (8.7.4) one obtain

$$\begin{cases} \alpha_0 + \alpha_1 + \cdots + \alpha_{k-1} = 1 \\ \sum_{i=1}^{k-1} i\alpha_i + \sum_{i=0}^{k} \beta_i = k \end{cases} \tag{8.7.6}$$

For $k=1$, if $\beta_1=0$, then

$$\alpha_0=1, \quad \beta_0=1$$

Now, the Formula (8.7.1) is of the form

$$y_{n+1} = y_n + hf_n$$

that is the Euler method. From (8.7.4) one obtain $c_2=1/2 \neq 0$, therefore the method is degree of accuracy one, and the truncation error is

$$T_{n+1} = \frac{1}{2}h^2 y''(x_n) + O(h^3)$$

For $k=1$, if $\beta_1 \neq 0$,the formula is implicit, from $c_0=c_1=c_2=0$ we obtain $\alpha_0=1, \beta_0=\beta_1=1/2$. Therefore

$$y_{n+1} = y_n + \frac{h}{2}(f_n + f_{n+1})$$

From (8.7.4), $c_3=-1/12$, therefore $p=2$, so trapezoidal method is of order two.

8.7.2 Adams Explicit and Implicit Formulas

Consider

$$y_{n+k} = y_{n+k-1} + h\sum_{i=0}^{k} \beta_i f_{n+i} \tag{8.7.7}$$

It is called **Adams method**. If $\beta_k = 0$, then it is called **Adams explicit method** or **Adams-Bashforth formula**. If $\beta_k \neq 0$, then it is called **Adams implicit method** or **Adams-Monlton formula.** These formulas can be obtained directly by integrating the Equation (8.2.1) (from x_{n+k-1} to x_{n+k}). In the following, one can deduce by using (8.7.4) from $c_1=\cdots=c_p=0$, comparing (8.7.7) and (8.7.1) we can obtain $\alpha_0=\alpha_1=\cdots=\alpha_{k-2}=0$, $\alpha_{k-1}=1$. Obviously, $c_0=0$. Now we need to determine the coefficients $\beta_0,\beta_1,\cdots\beta_k$. For this reason let $c_1=\cdots= c_{k+1}=0$, then $\beta_0,\beta_1,\cdots, \beta_k$ can be solved (if $\beta_k=0$, then let $c_0=\cdots=c_k=0$and solve $\beta_0, \beta_1, \cdots, \beta_{k-1}$). Take $k=3$ as example, from $c_1 = c_2 = c_3 = c_4 =0$, from (8.7.4), we have

$$\begin{cases} \beta_0 + \beta_1 + \beta_2 + \beta_3 = 1 \\ 2(\beta_1 + 2\beta_2 + 3\beta_3) = 5 \\ 3(\beta_1 + 4\beta_2 + 9\beta_3) = 19 \\ 4(\beta_1 + 8\beta_2 + 27\beta_3) = 65 \end{cases}$$

If $\beta_3=0$, then from the previous three equations one obtain

$$\beta_0 = \frac{5}{12}, \quad \beta_1 = -\frac{16}{12}, \quad \beta_2 = \frac{23}{12}$$

and Adams explicit formula for $k = 3$ is obtained

$$y_{n+3} = y_{n+2} + \frac{h}{12}(23f_{n+2} - 16f_{n+1} + 5f_n) \tag{8.7.8}$$

From (8.7.4) we have c_4=3/8, therefore (8.7.8) is of order three, and its local truncation error is

$$T_{n+3} = \frac{3}{8}h^4 y^{(4)}(x_n) + O(h^5)$$

If $\beta_3 \neq 0$, then

$$\beta_0 = \frac{1}{24},\quad \beta_1 = -\frac{5}{24},\quad \beta_2 = \frac{19}{24},\quad \beta_3 = \frac{3}{8}$$

Therefore, Adams implicit formula for $k = 3$ is obtained

$$y_{n+3} = y_{n+2} + \frac{h}{24}(9f_{n+3} + 19f_{n+2} - 5f_{n+1} + f_n) \tag{8.7.9}$$

It is of order four with local truncation error

$$T_{n+3} = -\frac{19}{720}h^5 y^{(5)}(x_n) + O(h^6) \tag{8.7.10}$$

Analogously, we can obtain some formulas listed in Table 8-5 and Table 8-6, where k is the number of steps, p is the degree of accuracy, c_{p+1} is the error constant.

Table 8-5 Adams explicit formulas

k	p	formulas	c_{p+1}
1	1	$y_{n+1} = y_n + hf_n$	$\frac{1}{2}$
2	2	$y_{n+2} = y_{n+1} + \frac{h}{2}(3f_{n+1} - f_n)$	$\frac{5}{12}$
3	3	$y_{n+3} = y_{n+2} + \frac{h}{12}(23f_{n+2} - 16f_{n+1} + 5f_n)$	$\frac{3}{8}$
4	4	$y_{n+4} = y_{n+3} + \frac{h}{24}(55f_{n+3} - 59f_{n+2} + 37f_{n+1} - 9f_n)$	$\frac{251}{720}$

Table 8-6 Adams implicit formulas

k	p	formulas	c_{p+1}
1	2	$y_{n+1} = y_n + \frac{h}{2}(f_{n+1} + f_n)$	$-\frac{1}{12}$
2	3	$y_{n+2} = y_{n+1} + \frac{h}{12}(5f_{n+2} + 8f_{n+1} - f_n)$	$-\frac{1}{24}$
3	4	$y_{n+3} = y_{n+2} + \frac{h}{24}(9f_{n+3} + 19f_{n+2} - 5f_{n+1} + f_n)$	$-\frac{19}{720}$
4	5	$y_{n+4} = y_{n+3} + \frac{h}{720}(251f_{n+4} + 646f_{n+3} - 264f_{n+2} + 106f_{n+1} - 19f_n)$	$-\frac{3}{160}$

Example 8.7.1 Solve the initial-value problem with Adams explicit and implicit methods of order four.

$$\begin{cases} y' = -y + x + 1 \\ y(0) = 1 \end{cases}$$

with step size h=0.1.

Solution $f_n = -y_n + x_n + 1,\ x_n = nh = 0.1n$. From Adams explicit formula one obtain

$$\begin{aligned} y_{n+4} &= y_{n+3} + \frac{h}{24}(55f_{n+3} - 59f_{n+2} + 37f_{n+1} - 9f_n) \\ &= \frac{1}{24}(18.5y_{n+3} + 5.9y_{n+2} - 3.7y_{n+1} + 0.9y_n + 0.24n + 3.24) \end{aligned}$$

From Adams implicit formula one obtain

$$\begin{aligned} y_{n+3} &= y_{n+2} + \frac{h}{24}(9f_{n+3} + 19f_{n+2} - 5f_{n+1} + f_n) \\ &= \frac{1}{24}(-0.9y_{n+3} + 22.1y_{n+2} + 0.5y_{n+1} - 0.1y_n + 0.24n + 3) \end{aligned}$$

From this, we obtain

$$y_{n+3} = \frac{1}{24.9}(22.1y_{n+2} + 0.5y_{n+1} - 0.1y_n + 0.24n + 3)$$

The results are given in Table 8-7, where y_0, y_1, y_2, y_3 in the explicit formula and y_0, y_1, y_2 in the implicit formula are obtained from the exact solution $y(x) = e^{-x} + x$.

Table 8-7 The results of computation

x_n	the exact solution	Adams explicit formula		Adams implicit formula	
	$y(x_n) = e^{-x_n} + x_n$	y_n	$\lvert y(x_n)-y_n \rvert$	y_n	$\lvert y(x_n)-y_n \rvert$
0.3	1.040 818 22			1.040 818 01	2.1×10^{-7}
0.4	0.070 320 05	1.070 322 92	2.87×10^{-6}	1.070 319 66	3.9×10^{-7}
0.5	1.106 530 66	1.106 535 48	4.82×10^{-6}	1.106 530 14	5.2×10^{-7}
0.6	1.148 811 64	1.148 818 41	6.77×10^{-6}	1.148 811 01	6.3×10^{-7}
0.7	1.196 585 30	1.196 593 40	8.10×10^{-6}	1.196 584 59	7.1×10^{-7}
0.8	1.249 328 96	1.249 338 16	9.20×10^{-6}	1.249 328 19	7.7×10^{-7}
0.9	1.306 569 66	1.306 579 62	9.96×10^{-6}	1.306 568 84	8.2×10^{-7}
1.0	1.367 879 44	1.367 889 96	1.05×10^{-5}	1.367 878 59	8.5×10^{-7}

8.8 Systems and Higher-Order Differential Equations

The standard form for a system of first-order differential equations is

$$\begin{cases} y_1' = f_1(x, y_1, y_2, \cdots, y_N) \\ y_2' = f_2(x, y_1, y_2, \cdots, y_N) \\ \cdots\cdots \\ y_N' = f_N(x, y_1, y_2, \cdots, y_N) \end{cases} \tag{8.8.1}$$

In this system, N unknown functions, $y_1, y_2, \cdots, y_N$ are to be determined. They are functions of the single independent variable, x. For a concrete example, consider this system, in which x and y have been written in place of y_1 and y_2:

$$\begin{cases} x' = x + 4y - e^t \\ y' = x + y + 2e^t \end{cases} \tag{8.8.2}$$

The general solution of System (8.8.2) is

$$\begin{cases} x = 2ae^{3t} - 2be^{-t} - 2e^t \\ y = ae^{3t} + be^{-t} + \dfrac{1}{4}e^t \end{cases} \tag{8.8.3}$$

where a and b are arbitrary constants. The solution can be verified, of course, by differentiation in System (8.8.2). Notice that the example is a linear system in the unknown functions x and y.

If System (8.8.2) were accompanied by initial conditions

$$x(0) = 4, \ y(0) = \frac{5}{4}$$

Then the solution would be

$$\begin{cases} x = 4e^{3t} + 2e^{-t} - 2e^t \\ y = 2e^{3t} - e^{-t} + \dfrac{1}{4}e^t \end{cases} \tag{8.8.4}$$

The initial-value problem for the general System (8.8.1) consists of the N differential equations together with a prescribed initial value for x, say $x=x_0$, and a specification of the value of each function y_i at x_0.

8.8.1 Vector Notation

A convenient vector notation can be used to rewrite System (8.8.1). We let Y denote the column vector whose components are $y_1, y_2, \cdots y_N$. These components are functions of x, hence, Y is a mapping of $\mathbf{R}$ (or an interval in $\mathbf{R}$) to $\mathbf{R}^N$, i.e., $Y=(y_1, y_2, \cdots, y_N)^{\mathrm{T}}$. Similarly, let F denote the column vector with components $f_1, f_2, \cdots, f_N$, each of this is a function on $\mathbf{R}^{N+1}$ (or a subset), $f_i=f_i(x, y_1, y_2, \cdots, y_N)$, and so F is a mapping of $\mathbf{R}^{N+1}$ to $\mathbf{R}^N$, i.e., $F=(f_1, f_2, \cdots, f_N)^{\mathrm{T}}$. System (8.8.1) can now be written as

$$Y' = F(x, Y) \tag{8.8.5}$$

An initial-value problem for System (8.8.5) would also include numerical values for the vector $Y(x_0)=(y_1^0, y_2^0, \cdots, y_N^0)^{\mathrm{T}}$, where x_0 is the initial value of x. A system in the form of (8.8.5) is said to be autonomous.

A differential equation of high order can be converted to a system of first-order equations. Suppose that a single differential equation is given in the form

$$y^{(n)}=f(x, y, y', y'', \cdots, y^{(n-1)})$$

Here, of course, all derivatives are with respect to x: $y^{(i)}=\mathrm{d}^i y/\mathrm{d}x^i$. Next, we introduce new variables $y_1, y_2, \cdots, y_n$ according to these definitions:

$$y_1=y, \quad y_2=y', \quad y_3=y'', \quad \cdots, \quad y_n=y^{(n-1)}$$

The new variables satisfy the following system of the first-order differential equations:

$$\begin{cases} y_1'=y_2 \\ y_2'=y_3 \\ y_3'=y_4 \\ \cdots\cdots \\ y_n'=f(x, y_1, y_2, y_3, \cdots, y_n) \end{cases}$$

This is a system of the form presented in the Equation (8.8.5).

To solve differential equation using widely available software, it is always necessary to convert the problem into a system such as (8.8.5). We illustrate this process with two examples.

Example 8.8.1 Convert the initial-value problem

$$\begin{cases} y'''-(\sin y''+\mathrm{e}^x y')^2+\cos y=0 \\ y(0)=3 \\ y'(0)=4 \\ y''(0)=5 \end{cases} \tag{8.8.6}$$

into a system of first-order differential equations with initial values.

Solution We first introduce new variables y_1, y_2 and y_3 as follows:

$$y_1=y, \quad y_2=y', \quad \text{and} \quad y_3=y''$$

The system of equations governing $Y=(y_1, y_2, y_3)^{\mathrm{T}}$ is

$$\begin{cases} y_1'=y_2 \\ y_2'=y_3 \\ y_3'=(\sin y_3+\mathrm{e}^x y_2)^2-\cos y_1 \end{cases} \tag{8.8.7}$$

The initial conditions at x=0 are $Y=(3,4,5)^{\mathrm{T}}$.

System of higher-order equations can be treated in a similar manner, as illustrated in the next example.

Example 8.8.2 Convert the system shown into a system of first-order differential equations:

$$\begin{cases} x'''-5tx''y'+\ln(x')z=0 \\ y''-\sin(ty)+7tx''=0 \\ z'+16ty'-e^t zx'=0 \end{cases} \tag{8.8.8}$$

Solution After introducing the new variables:

$$x_1=x,\ x_2=x',\ x_3=x'',\ x_4=y,\ x_5=y',\ \text{and}\ x_6=z$$

the problem can be written as

$$\begin{cases} x_1'=x_2 \\ x_2'=x_3 \\ x_3'=5tx_3x_5-\ln(x_2)x_6 \\ x_4'=x_5 \\ x_5'=\sin(tx_4)-7tx_3 \\ x_6'=-16tx_5+e^t x_6x_2 \end{cases} \tag{8.8.9}$$

8.8.2 Taylor-Series Method for Systems

From Taylor series for $y_i(x+h)$:

$$y_i(x+h)=y_i(x)+hy_i'(x)+\frac{h^2}{2!}y_i''(x)+\frac{h^3}{3!}y_i'''(x)+\cdots+\frac{h^n}{n!}y_i^{(n)}(x)$$

or in vector notation:

$$Y(x+h)=Y(x)+hY'(x)+\frac{h^2}{2!}Y''(x)+\frac{h^3}{3!}Y'''(x)+\cdots+\frac{h^n}{n!}Y^{(n)}(x) \tag{8.8.10}$$

The derivatives appearing here can be obtained from the differential equation. Usually these derivatives must be computed in a particular order when used in a computer program. We must be sure that quantities needed at one step are available as the results from prior steps.

Example 8.8.3 Write the Taylor-series algorithm of order 3 for the following initial-value problem. Use $|h|=0.1$, and compute the solution on the interval $-2\leqslant t\leqslant 1$.

$$\begin{cases} x'=x+y^2-t^3 \\ y'=y+x^3+\cos t \\ x(1)=3 \\ y(1)=1 \end{cases} \tag{8.8.11}$$

Solution The higher derivatives required are

$$x''=x'+2yy'-3t^2$$

$$y''=y'+3x^2x'-\sin t$$

$$x'''=x''+2yy''+2(y')^2-6t$$

$$y'''=y''+6x(x')^2+3x^2x''-\cos t$$

A suitable algorithm to carry out the computation follows:

input $t \leftarrow 1; x \leftarrow 3; y \leftarrow 1; h \leftarrow -0.1; M \leftarrow 30$

output $0, t, x, y$

for $k=1$ to M do

$$x' \leftarrow x + y^2 - t^3$$

$$y' \leftarrow y + x^3 + \cos t$$

$$x'' \leftarrow x' + 2yy' - 3t^2$$

$$y'' \leftarrow y' + 3x^2 x' - \sin t$$

$$x''' \leftarrow x'' + 2yy'' + 2(y')^2 - 6t$$

$$y''' \leftarrow y'' + 6x(x')^2 + 3x^2 x'' - \cos t$$

$$x \leftarrow x + h\left(x' + \frac{1}{2}h\left(x'' + \frac{1}{3}h(x''')\right)\right)$$

$$y \leftarrow y + h\left(y' + \frac{1}{2}h\left(y'' + \frac{1}{3}h(y''')\right)\right)$$

$$t \leftarrow t + h$$

output k, t, x, y

end do

8.8.3 Fourth-Order Runge-Kutta Formula for Systems

The Runge-Kutta formula for Systems of first-order equations is most easily written down in the case when our system is autonomous, that is, when it has the form of Equation (8.8.5). The classical fourth-order Runge-Kutta formulas, in vector form, are

$$\begin{cases} Y_{n+1} = Y_n + \frac{h}{6}(k_1 + 2k_2 + 2k_3 + k_4) \\ k_1 = F(x_n, Y_n) \\ k_2 = F\left(x_n + \frac{h}{2}, Y_n + \frac{h}{2}k_1\right) \\ k_3 = F\left(x_n + \frac{h}{2}, Y_n + \frac{h}{2}k_2\right) \\ k_4 = F(x_n + h, Y_n + hk_3) \end{cases} \tag{8.8.12}$$

or presented in the form

$$y_{i,n+1} = y_{in} + \frac{h}{6}(K_{i1} + 2K_{i2} + 2K_{i3} + K_{i4}) \qquad (i = 1, 2, \cdots, N)$$

where

$$K_{i1} = f_i(x_n, y_{1n}, y_{2n}, \cdots, y_{Nn})$$

$$K_{i2} = f_i\left(x_n + \frac{h}{2}, y_{1n} + \frac{h}{2}K_{11}, y_{2n} + \frac{h}{2}K_{21}, \cdots, y_{Nn} + \frac{h}{2}K_{N1}\right)$$

$$K_{i3} = f_i\left(x_n + \frac{h}{2}, y_{1n} + \frac{h}{2}K_{12}, y_{2n} + \frac{h}{2}K_{22}, \cdots, y_{Nn} + \frac{h}{2}K_{N2}\right)$$

$$K_{i4} = f_i(x_n + h, y_{1n} + hK_{13}, y_{2n} + hK_{23}, \cdots, y_{Nn} + hK_{N3})$$

where y_{in} is approximate to $y_i(x)$ at the point $x_n = x_0 + nh$.

Let us consider the special case with

$$\begin{cases} y' = f(x,y,z) \\ z' = g(x,y,z) \\ y(x_0) = y_0 \\ z(x_0) = z_0 \end{cases}$$

now the fourth-order Runge-Kutta formula is

$$\begin{cases} y_{n+1} = y_n + \dfrac{h}{6}(K_1 + 2K_2 + 2K_3 + K_4) \\ z_{n+1} = z_n + \dfrac{h}{6}(L_1 + 2L_2 + 2L_3 + L_4) \end{cases}$$

where

$$\begin{cases} K_1 = f(x_n, y_n, z_n) \\ K_2 = f\left(x_n + \dfrac{h}{2}, y_n + \dfrac{h}{2}K_1, z_n + \dfrac{h}{2}L_1\right) \\ K_3 = f\left(x_n + \dfrac{h}{2}, y_n + \dfrac{h}{2}K_2, z_n + \dfrac{h}{2}L_2\right) \\ K_4 = f(x_n + h, y_n + hK_3, z_n + hL_3) \\ L_1 = g(x_n, y_n, z_n) \\ L_2 = g\left(x_n + \dfrac{h}{2}, y_n + \dfrac{h}{2}K_1, z_n + \dfrac{h}{2}L_1\right) \\ L_3 = g\left(x_n + \dfrac{h}{2}, y_n + \dfrac{h}{2}K_2, z_n + \dfrac{h}{2}L_2\right) \\ L_4 = g(x_n + h, y_n + hK_3, z_n + hL_3) \end{cases}$$

For the second-order equations with initial-value problem

$$\begin{cases} y'' = f(x,y,y') \\ y(x_0) = y_0 \\ y'(x_0) = y_0' \end{cases}$$

Introduce new variable $z = y'$, it can be converted into the first-order initial problem:

$$\begin{cases} y' = z \\ z' = f(x,y,z) \\ y(x_0) = y_0 \\ z(x_0) = y_0' \end{cases}$$

The fourth-order Runge-Kutta formula is

$$K_1 = z_n,\quad L_1 = f(x_n, y_n, z_n)$$
$$K_2 = z_n + \frac{h}{2}L_1,\quad L_2 = f\left(x_n + \frac{h}{2}, y_n + \frac{h}{2}K_1, z_n + \frac{h}{2}L_1\right)$$
$$K_3 = z_n + \frac{h}{2}L_2,\quad L_3 = f\left(x_n + \frac{h}{2}, y_n + \frac{h}{2}K_2, z_n + \frac{h}{2}L_2\right)$$
$$K_4 = z_n + hL_3,\quad L_4 = f(x_n + h, y_n + hK_3, z_n + hL_3)$$

eliminate K_1, K_2, K_3, K_4, and obtain

$$\begin{cases} y_{n+1} = y_n + hz_n + \dfrac{h^2}{6}(L_1 + L_2 + L_3) \\ z_{n+1} = z_n + \dfrac{h}{6}(L_1 + 2L_2 + 2L_3 + L_4) \end{cases}$$

where

$$L_1 = f(x_n, y_n, z_n)$$
$$L_2 = f\left(x_n + \frac{h}{2}, y_n + \frac{h}{2}z_n, z_n + \frac{h}{2}L_1\right)$$
$$L_3 = f\left(x_n + \frac{h}{2}, y_n + \frac{h}{2}z_n + \frac{h^2}{4}L_1, z_n + \frac{h}{2}L_2\right)$$
$$L_4 = f\left(x_n + h, y_n + hz_n + \frac{h^2}{2}L_2, z_n + hL_3\right)$$

8.9 Computer Experiments

8.9.1 Functions Needed in the Experiments by Mathematica

格式：`f[x_,y_ ]=Input["提示"]`

功能：定义二元函数 $f(x, y)$，并由键盘输入函数表达式.

8.9.2 Experiments by Mathematica

1．欧拉法

（1）算法.

① 输入：函数 $f(x, y)$，自变量区间端点 a，b，步长 h；

② 计算节点数：$n=(b-a)/h$；

③ 置计数变量 $i=0$，作循环(当 $i<n$ 时)：

计算：$y_{n+1} = y_n + hf(x_n, y_n)$；

$i=i+1$；

④ 输出数值解.

（2）程序清单.

```
(*Euler Method*)
```

```
Clear[x,y,f]
f[x_,y_]=Input["f(x, y)="]
y0=Input["y0="];
a=Input["a="];
b=Input["b="];
h=Input["h="];
n=(b-a)/h;
i=0;
Print["区间[a,b]=","[",a,",",b,"]","  ""初值 y0=",y0,"  ""步长=",h];
Print["常微分方程 f(x,y)=",f[x,y],"初值问题在各节点处的数值解为: "];
While[i<n,
   xk=a+i*h;
   y1=y0+h*f[xk, y0];
   Print["y(", xk+h//N, ")=", y1//N];
   y0=y1;
   i=i+1
]
```

（3）变量说明.

f[x_, y_]：保存函数 $f(x, y)$；

y0：保存初值及前次计算结果；

a：保存自变量取值区间左端点；

b：保存自变量取值区间右端点；

n：保存节点个数；

h：保存节点步长；

xk：保存节点值 x_k；

y1：保存数值解.

（4）计算实例.

Example 8.9.1 Solve the initial-value problem

$$\begin{cases} y'=\dfrac{0.9y}{1+2x} \quad (0\leqslant x\leqslant 0.1) \\ y(0)=1 \end{cases}$$

by using Euler method with step size h=0.01.

操作步骤如下：

① 将光标定位在要执行的 Cell 中，按小键盘的【Enter】键；

② 在弹出的对话框中按提示分别输入：

```
0.9y/(1+2x),1,0,0.1,0.01
```

③ 每次输入后单击 OK 命令按钮，系统输出结果如图 8-3 所示.

```
Out[13]= 0.9 y / (1 + 2 x)

区间[a,b]=[0,0.1]   初值y0=1   步长=0.01

常微分方程f(x,y)= 0.9 y / (1 + 2 x) 初值问题在各节点处的数值解为:

y(0.01)=1.009
y(0.02)=1.0179
y(0.03)=1.02671
y(0.04)=1.03543
y(0.05)=1.04406
y(0.06)=1.0526
y(0.07)=1.06106
y(0.08)=1.06944
y(0.09)=1.07773
y(0.1)=1.08595
```

图 8-3 欧拉法计算结果

2. 改进欧拉法

（1）算法.

① 输入：函数 $f(x, y)$，初值 y_0 自变量区间端点 a，b，步长 h;

② 计算节点数：$n=(b-a)/h$;

③ 置计数变量 $i=0$，作循环(当 $i<n$ 时)：

计算：$\overline{y}_{n+1}=y_n+hf(x_n,y_n)$；

$$y_{n+1}=y_n+\frac{h}{2}(f(x_n,y_n)+f(x_n,\overline{y}_{n+1}))；$$

$i=i+1$;

④ 输出数值解.

（2）程序清单.

```
(*Improved Euler Method*)
Clear[x,y,f]
f[x_,y_]=Input["f(x,y)="]
y0=Input["y0="];
a=Input["a="];
b=Input["b="];
h=Input["h="];
n=(b-a)/h;
i=0;
Print["区间[a,b]=","[",a,",",b,"]","  ""初值 y0=",y0,"  ""步长=",h];
Print["常微分方程 f (x,y)=",f[x,y],"初值问题在各节点处的数值解为: "];
While[i<n,
   xk=a+i*h;
   y1=y0+h*f[xk, y0];
   y2=y0+h/2*(f[xk, y0]+f[xk+h, y1]);
   Print["y(",xk+h//N,")=",y2//N];
```

```
  y0=y2;
  i=i+1;
]
```

（3）变量说明.

f[x_, y_]：保存函数 $f(x, y)$；

y0：保存初值及前次计算结果；

a：保存自变量取值区间左端点；

b：保存自变量取值区间右端点；

n：保存节点个数；

h：保存节点步长；

xk：保存节点值 x_k；

y1：保存数值解.

（4）计算实例.

Example 8.9.2 Solve the initial-value problem

$$\begin{cases} y' = \dfrac{0.9y}{1+2x} \\ y(0)=1 \end{cases} \quad (0 \leqslant x \leqslant 0.1)$$

by using improved Euler method with step size h=0.01.

操作步骤如下：

① 将光标定位在要执行的 Cell 中，按小键盘的【Enter】键；

② 在弹出的对话框中按提示分别输入：

```
0.9y/(1+2x),1,0,0.1,0.01
```

③ 每次输入后单击 OK 命令按钮，系统输出结果如图 8-4 所示.

Out[168]= $\dfrac{0.9\,y}{1+2\,x}$

区间[a,b]=[0,0.1]　初值y0=1　步长=0.01

常微分方程f(x,y)= $\dfrac{0.9\,y}{1+2\,x}$ 初值问题在各节点处的数值解为:

```
y(0.01)=1.00895
y(0.02)=1.01781
y(0.03)=1.02657
y(0.04)=1.03524
y(0.05)=1.04382
y(0.06)=1.05232
y(0.07)=1.06074
y(0.08)=1.06907
y(0.09)=1.07733
y(0.1)=1.08551
```

图 8-4 改进欧拉法计算结果

3. Fourth-Order Runge-Kutta 方法

（1）算法.

① 输入：函数 $f(x,y)$，初值 y_0 自变量区间端点 a，b，步长 h;

② 计算节点数：$n=(b-a)/h$;

③ 置计数变量 $i=0$，作循环（当 $i<n$ 时）：

计算：y_{k+1}，K_1，K_2，K_3，K_4;

$i=i+1$;

④ 输出数值解.

（2）程序清单.

```
(*Fourth-Order Runge-Kutta Method*)
Clear[x,y,f]
f[x_,y_]=Input["f(x,y)="]
y0=Input["y0="];
a=Input["a="];
b=Input["b="];
h=Input["h="];
n=(b-a)/h;
i=0;
Print["区间[a,b]=","[",a,",",b,"]","  ""初值 y0=",y0,"  ""步长=",h];
Print["常微分方程 f(x,y)=",f[x,y],"初值问题在各节点处的数值解为: "];
While[i<n,
   xk=a+i*h;
   k1=f[xk,y0];
   k2=f[xk+h/2,y0+k1*h/2];
   k3=f[xk+h/2,y0+k2*h/2];
   k4=f[xk+h,y0+k3*h];
   y1=y0+h*(k1+2k2+2k3+k4)/6;
   Print["y(",xk+h//N,")=",y1//N];
   y0=y1;
   i=i+1;
]
```

（3）变量说明.

f[x_, y_]：保存常微分方程 $f(x,y)$;

y0：保存初值及前次计算结果；

a：保存自变量取值区间左端点；

b：保存自变量取值区间右端点；

n：保存节点个数；

h：保存节点步长；

xk：保存节点值 x_k；

y1：保存数值解.

（4）计算实例.

Example 8.9.3　Solve the initial-value problem

$$\begin{cases} y' = \dfrac{0.9y}{1+2x} \quad (0 \leqslant x \leqslant 0.1) \\ y(0) = 1 \end{cases}$$

by using classical fourth-order Runge-Kutta method with step size h=0.01.

操作步骤如下：

① 光标定位在要执行的 Cell 中，按小键盘的【Enter】键；

② 在弹出的对话框中按提示分别输入：

```
0.9y/(1+2x),1,0,0.1,0.01
```

③ 每次输入后单击 OK 命令按钮，系统输出结果如图 8-5 所示.

Out[46]= $\dfrac{0.9\,y}{1+2\,x}$

区间[a,b]=[0,0.1]　初值y0=1　步长=0.01

常微分方程f(x,y) = $\dfrac{0.9\,y}{1+2\,x}$ 初值问题在各节点处的数值解为：

```
y(0.01)=1.00895
y(0.02)=1.01781
y(0.03)=1.02657
y(0.04)=1.03524
y(0.05)=1.04382
y(0.06)=1.05232
y(0.07)=1.06074
y(0.08)=1.06907
y(0.09)=1.07733
y(0.1)=1.0855
```

图 8-5　Fourth-Order Runge-Kutta 方法计算结果

8.9.3　Experiments by Matlab

1．欧拉法

（1）函数语句.

```
[x,y]=euler(f,x0,y0,b,n)
```

（2）参数说明.

f：符号变量，输入参数，常微分方程 $y'=f(x,y)$等式右边的函数；

x0：实变量，输入参数，自变量取值区间的左端点；

b：实变量，输入参数，自变量取值区间的右端点；

y0：实变量，输入参数，初始条件 y(x0)=y0；

n：正整数，输入参数，将区间[x0,b]n 等分；
x：向量变量，输出参数，等距节点；
y：向量变量，输出参数，数值解.

（3）euler.m 程序.

```
function [x,y]=euler(f,x0,y0,b,n)
if nargin<5
   n=10;
end
h=(b-x0)/n;
f=inline(f);
x(1)=x0;y(1)=y0;
for k=1:n
   y(k+1)=y(k)+h*feval(f,x(k),y(k));
   x(k+1)=x(k)+h;
end
flag=input('是否绘制欧拉法得到的数据以及解函数的图形（输入‘1’表示绘制，输入‘0’表示不绘制。）？');
if flag==1
   g=input('请输入方程精确解的函数解析式：');
   x1=x0:(b-x0)/100:b;
   y1=subs(g,'x',x1);
   plot(x,y,'*',x1,y1,'r');
end
```

（4）计算实例.

Example 8.9.4 Solve the initial-value problem

$$\begin{cases} y' = \dfrac{0.9y}{1+2x} & (0 \leqslant x \leqslant 0.1) \\ y(0) = 1 \end{cases}$$

by using Euler method with step size h=0.01.

在命令窗口输入：

```
f='0.9*y/(1+2*x)';x0=0;y0=1;b=0.1;n=10;[x,y]=euler(f,x0,y0,b,n)
是否绘制欧拉法得到的数据以及解函数的图形(输入‘1’表示绘制,输入‘0’表示不绘制。)?
0
x =   0     0.0100     0.0200     0.0300     0.0400     0.0500     0.0600
0.0700    0.0800    0.0900    0.1000
y =  1.0000    1.0090    1.0179    1.0267    1.0354    1.0441    1.0526
1.0611    1.0694    1.0777    1.0860
```

2. 改进欧拉法

（1）函数语句.

```
[x,y]=EulerModi(f,x0,y0,b,n)
```

（2）参数说明.

f：符号变量，输入参数，常微分方程 y'=f(x,y)等式右边的函数；

x0：实变量，输入参数，自变量取值区间的左端点；

b：实变量，输入参数，自变量取值区间的右端点；

y0：实变量，输入参数，初始条件 y(x0)=y0；

n：正整数，输入参数，将区间[x0，b]n 等分；

x：向量变量，输出参数，等距节点；

y：向量变量，输出参数，数值解.

（3）EulerModi.m 程序.

```
function [x,y]=EulerModi(f,x0,y0,b,n)
if nargin<5
    n=10;
end
h=(b-x0)/n;
f=inline(f);
x(1)=x0;y(1)=y0;
for k=1:n
    x(k+1)=x(k)+h;
    yy=y(k)+h*feval(f,x(k),y(k));
    y(k+1)=y(k)+h/2*(feval(f,x(k),y(k))+feval(f,x(k+1),yy));
end
flag=input('是否绘制欧拉法得到的数据以及解函数的图形（输入‘1’表示绘制，输入‘0’表示不绘制。）？');
if flag==1
    g=input('请输入方程精确解的函数解析式：');
    x1=x0:(b-x0)/100:b;
    y1=subs(g,'x',x1);
    plot(x,y,'*',x1,y1,'r');
end
```

（4）计算实例.

Example 8.9.5 Solve the initial-value problem

$$\begin{cases} y' = \dfrac{0.9y}{1+2x} & (0 \leqslant x \leqslant 0.1) \\ y(0) = 1 \end{cases}$$

by using Euler method with step size h=0.01.

在命令窗口输入：

```
>> clear
>> f='0.9*y/(1+2*x)';x0=0;y0=1;b=0.1;n=10;[x,y]=EulerModi(f,x0,y0,b,n)
是否绘制欧拉法得到的数据以及解函数的图形(输入‘1’表示绘制,输入‘0’表示不绘制。)?
0
x =   0      0.0100      0.0200      0.0300      0.0400      0.0500      0.0600
0.0700      0.0800      0.0900      0.1000
```

```
y =  1.0000    1.0090    1.0178    1.0266    1.0352    1.0438    1.0523
1.0607    1.0691    1.0773    1.0855
```

3. Fourth-Order Runge-Kutta 方法

（1）函数语句.

```
[x,y]=RungeKutta4(f,x0,y0,b,n,C)
```

（2）参数说明.

f：符号变量，输入参数，常微分方程 $y'=f(x,y)$等式右边的函数；

x0：实变量，输入参数，自变量取值区间的左端点；

b：实变量，输入参数，自变量取值区间的右端点；

y0：实变量，输入参数，初始条件 y(x0)=y0；

n：正整数，输入参数，将区间[x0，b]n 等分；

C：向量变量，输入参数，参数组；

x：向量变量，输出参数，等距节点；

y：向量变量，输出参数，数值解.

（3）RungeKutta4.m 程序.

```
function [x,y]=RungeKutta4(f,x0,y0,b,n,C)
if nargin<5
    n=10;
end
if nargin<6
   C=[1/6,1/3,1/3,1/6,1/2,1/2,1/2,0,1/2,1,0,0,1];  %默认为中点法
end
if length(C)~=13
    fprintf('参数的个数不等于13。\n');
    return
end
h=(b-x0)/n;
f=inline(f);
x(1)=x0;y(1)=y0;
for k=1:n
    K1=feval(f,x(k),y(k));
    K2=feval(f,x(k)+C(5)*h,y(k)+C(6)*h*K1);
    K3=feval(f,x(k)+C(7)*h,y(k)+h*(C(8)*K1+C(9)*K2));
    K4=feval(f,x(k)+C(10)*h,y(k)+h*(C(11)*K1+C(12)*K2+C(13)*K3));
    y(k+1)=y(k)+h*(C(1)*K1+C(2)*K2+C(3)*K3+C(4)*K4);
    x(k+1)=x(k)+h;
end
flag=input('是否绘制欧拉法得到的数据以及解函数的图形（输入‘1’表示绘制，输
入‘0’表示不绘制。）？');
if flag==1
    g=input('请输入方程精确解的函数解析式：');
```

```
    x1=x0:(b-x0)/100:b;
    y1=subs(g,'x',x1)
    plot(x,y,'*',x1,y1,'r');
end
```

（4）计算实例.

Example 8.9.6 Solve the initial-value problem

$$\begin{cases} y' = \dfrac{0.9y}{1+2x} & (0 \leqslant x \leqslant 0.1) \\ y(0) = 1 \end{cases}$$

by using fourth-order Runge-Kutta method with step size h=0.01.

在命令窗口输入：

```
>> clear
>>
f='0.9*y/(1+2*x)';x0=0;y0=1;b=0.1;n=10;[x,y]=RungeKutta4(f,x0,y0,b,n)
是否绘制欧拉法得到的数据以及解函数的图形(输入‘1’表示绘制,输入‘0’表示不绘制。)?
0
x =  0     0.0100    0.0200    0.0300    0.0400    0.0500    0.0600    0.0700
0.0800    0.0900    0.1000
y =  1.0000    1.0090    1.0178    1.0266    1.0352    1.0438    1.0523
1.0607    1.0691    1.0773    1.0855
```

Exercises 8

Questions

1. Solve the initial-value problem with Euler method

$$\begin{cases} y' = -y - xy^2 & (0 \leqslant x \leqslant 0.6) \\ y(0) = 1 \end{cases}$$

take the step size h=0.2.

2. Solve the initial-value problem with Euler method

$$\begin{cases} y' = -\dfrac{0.9}{1+2x} y & (0 < x < 1) \\ y(0) = 1 \end{cases}$$

with the numerical solutions when x=0,0.02,0.04,⋯,0.10.

3. Solve the initial-value problem with improved Euler method

$$\begin{cases} y' = \dfrac{1}{x} y - \dfrac{1}{x} y^2 & (1 \leqslant x \leqslant 1.5) \\ y(1) = 0.5 \end{cases}$$

take the step size h=0.1, and compare the result with the exact solution $y(x)=\frac{x}{1+x}$.

4. Solve the initial-value problem with the fourth-order Runge-Kutta method

$$\begin{cases} y'=x+y \\ y(0)=1 \end{cases} \quad (0<x<1)$$

take the step size h=0.2.

5. Solve the initial-value problem with Adams explicit formula

$$\begin{cases} y'=3x-2y \\ y(0)=1 \end{cases} \quad (0\leqslant x\leqslant 0.5)$$

take the step size h=0.1.

6. Solve the approximate solution at x=1.5 of the initial-value problem with the second-order Taylor-series method (take the step size h=0.25)

$$\begin{cases} y'=x^2+y^2 \\ y(1)=1 \end{cases}$$

7. Solve the initial-value problem with the fourth-order Runge-Kutta method

$$\begin{cases} y'=8-3y \\ y(0)=2 \end{cases}$$

take the step size h=0.2, and compute the approximate value of $y(0.4)$ with four decimal digits.

8. Write an autonomous system of first-order equations equivalent to

$$\begin{cases} x'''-\sin(x'')+\mathrm{e}^t x'+2t\cos x=25 \\ x(0)=5 \\ x'(0)=3 \\ x''(0)=7 \end{cases}$$

9. Write the third-order differential equation

$$\begin{cases} x'''+2x''-x'-2x=\mathrm{e}^t \\ x(8)=3 \\ x'(8)=2 \\ x''(8)=1 \end{cases}$$

as an autonomous system of first-order equations.

10. Convert the system of second-order ordinary differential equations

$$\begin{cases} x''' - x'y = 3y'x\log t \\ y'' - 2xy' = 5x'y\sin t \end{cases}$$

into a system of first-order equations in which t does not appear explicitly.

Computer Questions

1. Use computer to solve the Problems 1-4 and 6, 7.

2. Write a computer program to solve this initial-value problem using the Taylor-series method. Include terms in h, h^2, and h^3, and continue the solution to t=1. Let h=0.01.

$$\begin{cases} x_1' = t + x_1^2 + x_2, & x_1(-1) = 0.43 \\ x_2' = t - x_1 + x_2^2, & x_2(-1) = -0.69 \end{cases}$$

3. Write a computer program to solve this initial-value problem using the Taylor-series method. Include terms in h, h^2, and h^3, and continue the solution to t=1. Let h=0.01.

$$\begin{cases} x_1' = \sin x_1 + \cos(tx_2), & x_1(-1) = 2.37 \\ x_2' = t^{-1}\sin(tx_1), & x_2(-1) = -3.48 \end{cases}$$

The programming of x_2', x_2'' and x_2''' must be very carefully carried out because of the singularity at t=0.

Appendix

Appendix A Mathematica Basic Operations

Mathematica 软件是一个集成化的软件系统．它由美国伊利诺大学复杂系统研究中心主任、物理学、数学和计算机科学教授 Stephen Wolfram 负责研制，能够完成符号运算、数值计算、数学图形绘制、动画制作等多种操作．Mathematica 系统是目前世界上应用最广泛的符号计算系统之一．

A.1 Mathematica 的基本操作

1．启动 Mathematica

双击桌面图标，或单击“开始”按钮，选择“程序”→“Mathematica 5.1”→“Mathematica 5.1”命令，启动 Mathematica 5.1，集成工作界面如图 A-1 所示．

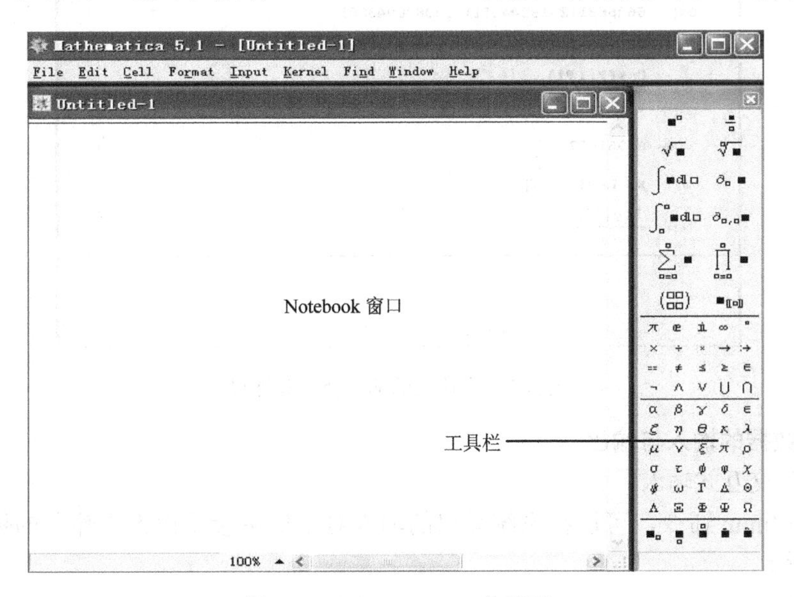

图 A-1 Mathematica 工作界面

在 Notebook 窗口内，可以用交互方式完成各种运算，如函数作图、求极限、解方程等，也可以编写结构化程序．

2．Notebook 窗口的交互方式操作

基本操作步骤：

（1）在 Notebook 窗口的工作区内单击，定位输入点；

（2）输入一个表达式或表达式序列；

（3）按【Shift+Enter】键或数字键盘的【Enter】键，系统开始计算并输出计算结果.

每次执行运算命令后，系统会显示输入和输出信息，并自动附上带有次序编号 n 的标识 In[n]和 Out[n]. 相应的输入和输出，都被称为“细胞”（Cell），系统自动加上“]”标识.

我们进行如下运算：

（1）(-12.8+33)/24　　四则运算

（2）39 ^20　　方幂运算

（3）Sin[4.5]、Cos[2.1Pi]　　求函数值

（4）x=Log[3, 20]　　为变量赋值

各次计算结果如图 A-2 所示.

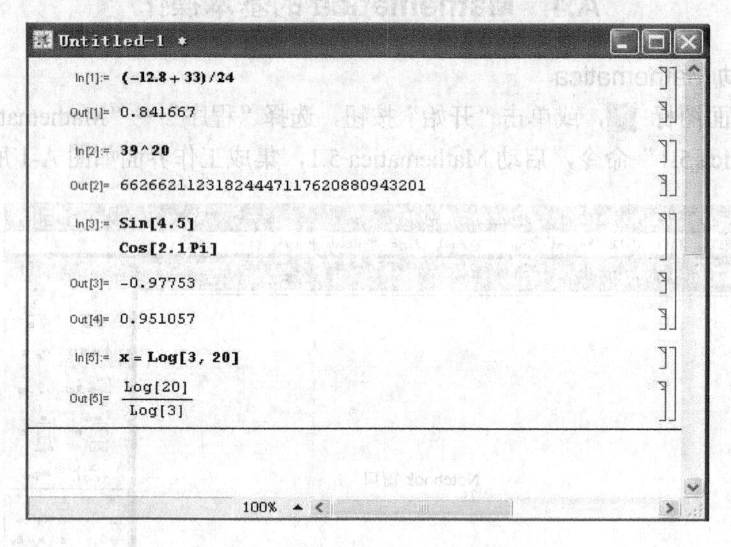

图 A-2　表达式的输入及计算结果

3. 数据的输入与输出

（1）交互式输入.

使用 Input 命令，可以在系统弹出的输入对话框中接受由键盘输入的数据，如表 A-1 所示.

表 A-1　Input 命令格式

命　令	功　能
Input[ ]	产生显示？号的输入对话框并等待输入
Input["提示"]	产生显示“提示信息”的输入对话框并等待输入
InputString[　]	产生显示？号的输入对话框并等待输入字符串
InputString["提示"]	产生显示“提示信息”的输入对话框并等待输入字符串

例如：输入一个数组 x_i={1, 3, 5, 7, 9}，如图 A-3 所示.

（2）输出命令.

格式 1：`Print[expr1,expr2,…]`

功能：依次输出表达式 expr1，expr2，….

格式 2：`Print["string",expr1,expr2,…]`

功能：依次输出字符串 string（通常为提示信息）、表达式 expr1，expr2，….

例如：展开多项式 $(1+x)^{20}$，如图 A-4 所示.

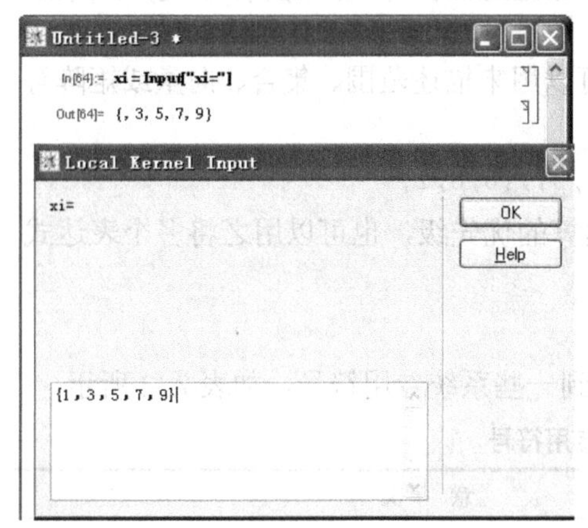

图 A-3　数据输入界面

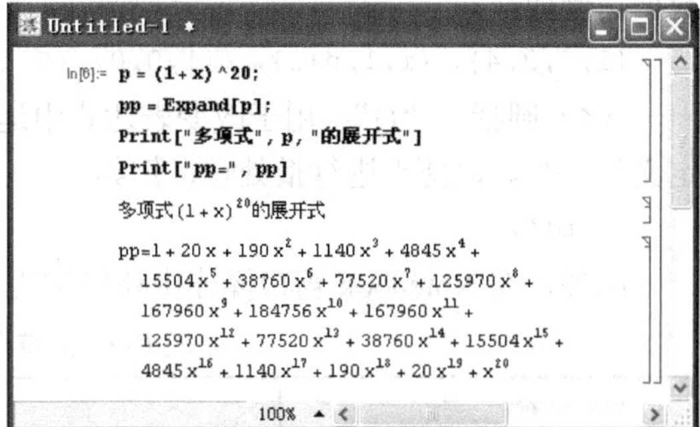

图 A-4　多项式展开

4．文件的保存与使用

选择“File”→“Save”命令（或者选择“Save As”命令），可以将之前的输入、输出内容保存成文件，文件以“.nb”作为扩展名，称为 Notebook 文件.

如果想使用某个 Notebook 文件，可以通过选择“File”→“Open”命令读入，也可以直接运行 Notebook 文件，系统会首先启动 Mathematica，再将其打开.

5．退出 Mathematica

单击程序窗口右上方的“关闭”按钮或选择“File”→“Exit”命令.

A.2　Mathematica 的语法规则

1．表述格式

Mathematic 提供了两种数学表达式的格式，即一维格式和二维格式.

以行式写出的称为一维格式，如 $f[x\_]=x\verb|^|3/(2x+3)+y\verb|^|(x-2)$，可以直接在 Notebook 窗口的工作区内输入.

形如 $f[x\_]=\dfrac{x^3}{2x+3}+y^{x-2}$ 的称为二维格式，输入时先单击工具栏中的相应公式模板，系统自动生成公式框架，然后在框架的占位符中填入参数而成，如图 A-5 所示.

2. 定界符号

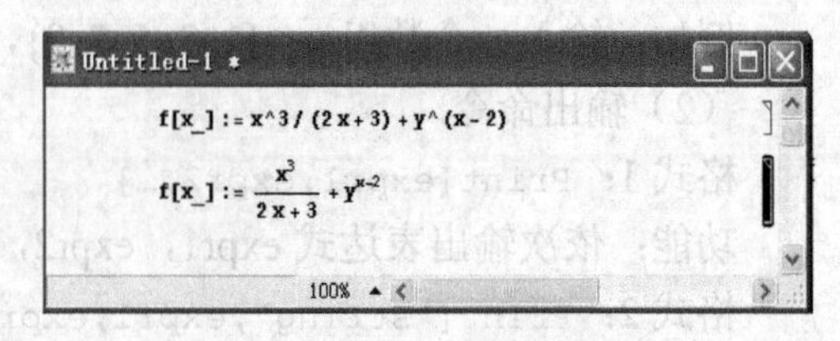

图 A-5　函数的输入

用于 Mathematic 中起定界作用的有如下四种符号：

（1）方括号“[　]”：用于描述函数的自变量相关参数．例如：

```
Random[Real,{0,1},30]   %求 0 到 1 之间精度为 30 位的随机数
```

（2）双方括号“[[　]]”：用于表示表（Mathematica 的一种数据结构）的元素．例如：

```
a[[n]]          %表 a 的第 n 个元素
```

（3）花括号“{ }”：用于表示一个表，可以用来描述范围、集合、向量或矩阵等数据．例如：

```
{1,2,3,4}，{i,1,n+1}，{{1,0,0},{0,1,0},{0,0,1}}
```

（4）圆括号“()”：用于改变表达式中运算的优先级，也可以用之将多个表达式定义为一个，以便于进行批处理．例如：

```
(a+b)^2
```

此外，在 Notebook 窗口操作中还经常用到一些系统专用符号，如表 A-2 所示．

表 A-2　系统专用符号

符　号	意　义
%	重复倒数第一次输出的内容
%%	重复倒数第二次输出的内容
%n	重复第 Out[n]次输出的内容
?	显示其后命令的简单帮助信息，如：?Clear
??	显示其后命令的详细帮助信息
;	禁止显示其前面表达式的运算结果

3. 注释与续行

注释一般有两种作用，一种是区分程序中的标记或解释性的文字；另一种是调试程序时暂时禁止某段程序运行．要想使程序中的某个部分成为注释内容，只需要在其两端分别加上“(*”和“*)”就可以了．

当一条语句不能在一行内写完需换行时，则要加“\”，然后按【Enter】键，再在下一行继续输入．

如果输入的语句和表达式过长不能在一行显示完时，系统会自动分断，以多行的形式显示，如图 A-6 所示．

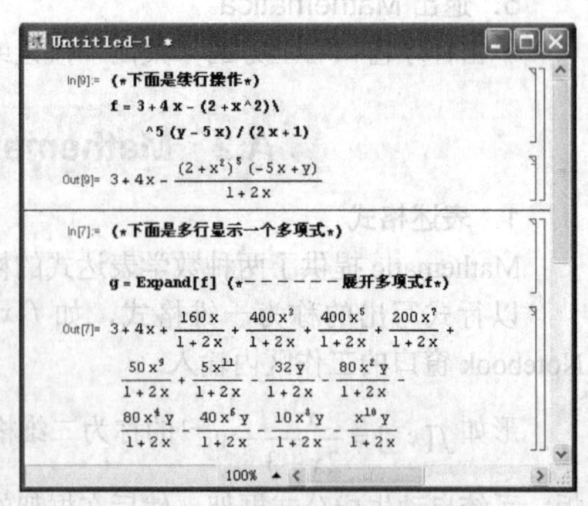

图 A-6　注释语句与分段输出

A.3 Mathematica 的数据

1. 常数

Mathematica 的常数分为两大类：一类是我们平常写出的数，叫普通数；另一类是系统内部的常数，有固定的写法.

Mathematica 中的普通数有整数、有理数、实数、复数四种类型，如表 A-3 所示.

表 A-3 Mathematica 中的普通数

类　型	描　述	意　义
整数	Integer	任意长度的精确数
有理数	Rational	化简过的分数
实数	Real	任意精确度的近似数
复数	Complex	实部、虚部可为整数、有理数、实数

Mathematica 的系统内部常数是指用特定的字符串表示的数学常数，这些常数都是精确数．需要特别注意的是这些内部常数书写时必须以大写字母开头，如表 A-4 所示.

2. 运算符

（1）算术运算符.

+、−、*、/、^分别表示加、减、乘、除、乘方的运算，其中在不引起混淆的情况下乘法运算符“*”也可省略不写，另外开方可以表示成分数指数，上述运算的优先顺序同数学运算完全一致.

（2）关系运算符（见表 A-5）.

表 A-4 Mathematica 系统内部常数

符　号	意　义
Pi	无理数π=3.141 59…
E	自然对数的底 e=2.718 28…
Degree	1°，π/180
I	虚数单位
Infinity	无穷∞
-Infinity	负无穷-∞
GoldenRatio	黄金分割数1.618 3…

表 A-5 关系运算符

运 算 符	意　义	实　例
==	相等	3x= =y−1, x= =y= =z
!=	不相等	x!=y , x!=y!=z
>	大于	x>y, x>y>z
>=	大于或等于	x>=y
<	小于	x<y
<=	小于等于	x<=y

（3）逻辑运算符（见表 A-6）.

表 A-6 逻辑运算符

运 算 符	意　义	实　例
!	非．若A为真，则!A为假	!A
&&	与．只有A，B均为真时，A&&B才为真	A&&B
||	或．只要A，B中有一为真，则A||B为真	A||B

3．变量

在 Mathematica 中，变量即取即用，不需先说明变量的类型后再使用．在 Mathematica 中变量不仅可存放一个整数或复数，还可存放一个多项式或复杂的算式．

（1）变量的命名．

变量名以小写字母开头，可以是单个字母，也可以是包含任意多字母或数字（不能作为开头）的字符串，但不能有空格和标点符号，例如：abc、g2、aSDF 均是合法的变量名．

如果两个变量名仅仅是大小写的不同，系统也视为不同的变量，如：asdf 和 aSDF 是不同的．

（2）变量的赋值（见表 A-7）．

表 A-7　变量的赋值格式

格　式	意　义
变量名=表达式	为一个变量赋值
变量名 1=变量名 2=表达式	为多个变量赋同一值
{变量 1,变量 2,…}= {表达式 1, 表达式 2,…}	分别为多个变量赋不同值

例如：
```
a=3*5^2
x=y=2*x^2-1
{n,dx0,dxn}={11,x^2,x+2y}
```

（3）变量替换．

代数式中的变量可以用另一个变量（或代数式）替换，例如“把上例中变量 y 中的 x 用 Pi-x 替换”，可表述为

```
y=2*x^2-1;
y/.x->Pi-x
```

x->Pi-x 中的“->”是由键盘上的减号及大于号组成的（以下同）．

（4）变量的清除．

当一个变量无用时，可以用命令 Clear 加以清除，以免影响后面计算的结果．

格式 1：`Clear[变量 1,变量 2,…]`

格式 2：变量= ．

4．函数

（1）函数的命名规则．

① 函数名的首字母必须大写，有时一个函数名是由几个单词构成，则每个单词的首写字母也必须大写，例如：

```
FindMinimum[f[x],{x,x0}]                %求局部极小值函数
```

② 函数名和自变量之间的分隔符是方括号“[ ]”，而不是一般数学书上用的圆括号“()”．自变量可以是数值，也可以是算术表达式．

③ 计算三角函数时，要注意使用弧度制，如果要使用角度制，要把角度制先乘以 Degree 常数（Degree=π/180），转换为弧度制．

④ 如果函数的定义中同时包含上、下标，则下标优先.

（2）内部函数.

Mathematica 中定义了许多功能强大的函数，称为系统函数或内部函数．可以将系统函数分为两类：

① 数学意义上的函数（见表 A-8）.

表 A-8 数学函数

函　数	功　能	函　数	功　能
N[x,k]	求出表达式的近似值，其中*k*为可选项，它指有效数字的位数	Sqrt[x]	求平方根
Round[x]	舍入取整	Exp[x]	以e为底的指数函数
Abs[x]	取绝对值	Log[a,x]	以*a*为底的对数函数
Max[x1,x2,…]	取*x*1,*x*2,…中的最大值	Log[x]	以e为底的对数函数
Min[x1,x2,…]	取*x*1,*x*2,…中的最小值	Sin[x]	正弦函数
x+Iy	复数*x*+i*y*	Cos[x]	余弦函数
Re[z]	复数*z*的实部	Tan[x]	正切函数
Im[z]	复数*z*的虚部	Cot[x]	余切函数
Abs[z]	复数*z*的模	Sec[x]	正割函数
Arg[z]	复数*z*的辐角	Csc[x]	余割函数
PrimeQ[n]	*n*为素数时为真，否则为假	ArcSin[x]	反正弦函数
Mod[m,n]	*m*被*n*除的正余数	ArcCos[x]	反余弦函数
GCD[n1,n2,…]	*n*1,*n*2…的最大公约数	ArcTan[x]	反正切函数
LCM[n1,n2,…]	*n*1,*n*2…的最小公倍数	ArcCot[x]	反余切函数

② 命令意义上的函数，例如：

```
Plot[f[x],{x,xmin,xmax}]          %作函数图形的函数
Solve[eqn,x]                      %解方程函数
```

（3）自定义函数.

当一个计算需要多次操作时，可以通过自定义函数提高使用效率.

格式 1：函数名[自变量_]:=表达式

功能：定义一个一元函数.

格式 2：函数名[自变量 1_ ,自变量 2 _,…]:=表达式

功能：定义一个多元函数.

例：定义函数 $f(x)=x^3+2\sqrt{x}+\sin x$，先分别求 $x=5.1$，$\dfrac{\pi}{2}$ 时的函数值，再求 $f(x^2)$.

输入及输出结果如图 A-7 所示.

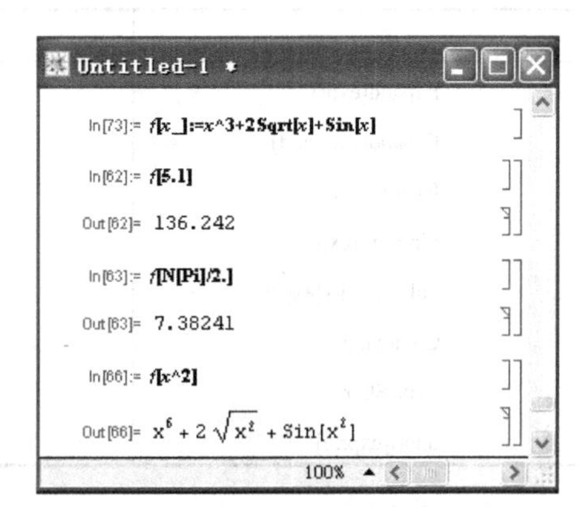

图 A-7 自定义函数及求值运算

由于系统不知道变量 x 的符号，所以没有对 $\sqrt{x^2}$ 进行开方运算.

（4）带附加条件的自定义函数.

在使用“f[x_]:=表达式”定义规则时，可以给规则附加条件，附加条件放在定义规则表达式后面，通过“/;”与表达式连接．规则的附加条件形式为：

```
f[x_ ]:=表达式/;条件
```

（5）函数求值.

求函数在某处的函数值有两种方法：一种是直接在函数中把自变量用一个值或式子代替，如 Sin[135*Pi/180]、Sqrt[a+b]；另一种为变量替换.

格式 1：函数/.变量->数值（或代数式）

格式 2：函数/.{变量1->数值（或代数式）1,变量2->数值（或代数式）2,…}

5．表达式

（1）定义.

表达式是以变量、常量、运算符构成的式子、表、甚至是图形，例如 3*x^3-2*x+5 和 x<=0 分别是算术表达式和关系表达式.

写表达式时，要注意以下几点：

① 所有表达式必须以线性形式写出．因此分子、分母、指数、下标等都必须写在同一行上.

② 只能使用合法的标识符（字符或字符串）.

③ 为了指定运算的次序可以利用括号．括号必须成对出现，且只有一种括号“(”与“)”，除了特定符号外不得使用方括号“[”与“]”及花括号“{”与“}”.

（2）表达式的表示形式.

在显示表达式时，有时需要表达式的展开形式，有时又需要其因子乘积的形式．在计算过程中可能得到很复杂的表达式，这时又需要对它们进行化简．常用的处理这种情况的函数如表 A-9 所示.

表 A-9　表达式操作函数

函　数	功　能
Expand[expr]	对代数式 expr 进行展开
Expand[expr,part]	对代数式 expr 进行展开，并保留 part 因式不展开
Factor[expr]	对 expr 进行因式分解
Simplify[expr]	将 expr 进行化简
FullSimplity[expr]	将 expr 化成最简的形式
Command	执行命令 Command，屏幕上不显示结果
Expr/Short	显示表达式的一行形式
Short[expr,n]	显示表达式的 n 行形式

例：将多项式 $2\,401+4\,116y+2\,646y^2+156y^3+81y^4$ 进行因式分解．将表达式(x+y)^4 (x+y^2)展开.

输入及输出结果如图 A-8 所示.

（3）关系表达式与逻辑表达式.

关系表达式是最简单的逻辑表达式，我们常用关系表达式表示一个判别条件. 关系表达式的一般形式是：

表达式 + 关系运算符 + 表达式

其中表达式可为数字表达式、字符表达式或意义更广泛的表达式，如一个图形表达式等. 关系运算的结果是一个逻辑值（True 或 False），如图 A-9 所示.

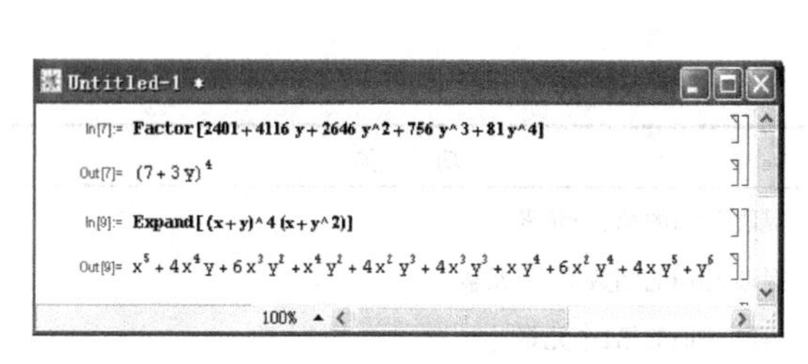

图 A-8　多项式因式分解及展开

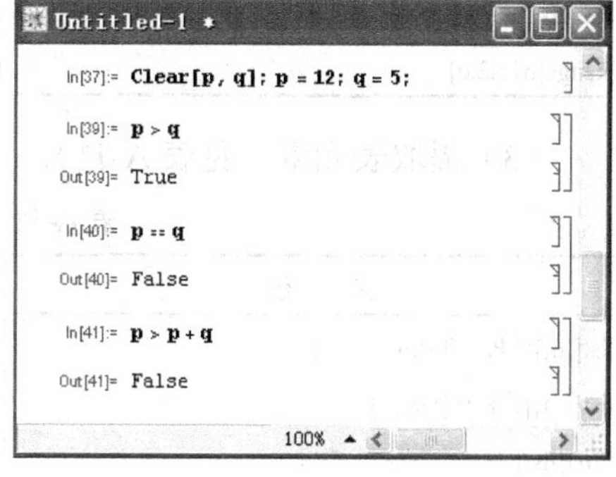

图 A-9　关系运算

用一个关系式只能表示一个判定条件，要表示几个判定条件组合，必须用逻辑运算符将关系表达式组织在一起，称表示判定条件的表达式为逻辑表达式.

下面的例子说明了关系表达式与逻辑表达式的应用.

```
Ln[4]:=3x^2<Y+1&&3^2==y
Out[4]=false
Ln[5]:=3x^2+1||3^2==y
Out[5]=True
```

6. 表

表是 Mathematica 中最重要的数据类型，它可以表示数组和矩阵等多元素的数据. 表的生成方法很灵活，主要有直接构造和利用建表函数生成两类.

（1）直接构造表.

将多个表达式（可以是常量、变量或函数）放在一对花括号内，中间用逗号分隔开，就构成了一个表，表中的每一项称为元素，没有任何元素的表称为空表，表的元素也可以是表.

格式 1：变量={表达式 1,表达式 2,…, 表达式 n}

功能：生成 n 个元素的一维表.

格式 2：变量={{序列 1},{序列 2},…,{序列 n}}

功能：生成 n 个元素的二维表.

（2）利用建表函数（见表 A-10）生成表.

表 A-10　建表函数

函　数	功　能
Table[f,{i,min,max,step}]	生成f的数值表，i从min变到max，步长为step
Table[f,{i,min,max}]	生成f的数值表，i从min变到max，步长为1
Table[f,{n}]	生成n个f的表
Table[f,{i,imin,imax},{j,jmin,jmax},…]	生成一个多维表
TableForm[list]	以表列格式显示一个表list
Rang[n]	生成数列{1,2,…,n}
Range[n1,n2,d]	生成数列{n1, n1+d, n1+2d,…, n2}

（3）提取表元素（见表 A-11）.

表 A-11　提取表元素函数

函　数	功　能
list[[n]]或Part [list,n]	提取表list的第n个元素
list[[-n]]或Part [list,-n]	提取表list的倒数第n个元素
First[list]	提取表list的第1个元素
Last[list]	提取表list的最后1个元素
list[[{n1,n2,…}]]	提取表list的第n1,n2,…个元素生成子表
list[[m,n]]	提取表list的第m个子表中的第n个元素
Take[list,n]	提取表list的前n个元素
Take[list,-n]	提取表list的后n个元素
Take[list,{m,n}]	提取表list的从第m个到第n个元素

（4）表的操作（见表 A-12）.

表 A-12　表的操作函数

函　数	功　能
Insert[list,expr,n]	生成在表 list 的第 n 个位置插入 expr 后的表
Insert[list,expr,-n]	生成在表 list 的倒数第 n 个位置插入 expr 后的表
Prepend[list,expr]	生成在表 list 的首部插入 expr 后的表
Append[list,expr]	生成在表 list 的尾部插入 expr 后的表
AppendTo[list,expr]	将元素 expr 插在表 list 的尾部
Rest[list]	删除表 list 的第 1 个元素
Drop[list,n]	删除表 list 的第 n 个元素
Drop[list,-n]	删除表 list 的倒数第 n 个元素
Drop[list,{m,n}]	删除表 list 的从第 m 到第 n 个元素
ReplacePart[list,expr,n]	将表 list 的第 n 个元素替换为 expr
Length[list]	计算表 list 中的元素个数

A.4 Mathematica 的数学函数

Mathematica 系统提供了丰富的数学计算函数，包括极限、积分、微分、最值、极值、统计、规划等各个领域，复杂的数学问题简化为对函数的调用，极大地提高了解决问题的效率．下面简单例举几类．

1．极限（见表 A-13）

2．微商（导数）

在 Mathematica 中能方便地计算任何函数表达式的任意阶微商（导数）．如果 f 是一元函数，D[f,x]表示 $\frac{df(x)}{dx}$；如果 f 是多元函数，D[f,x]表示 $\frac{\partial}{\partial x}f$（见表 A-14）．

表 A-13　求极限函数

函　数	功　能
Limit[expr,x->x0]	x->x0 时函数的极限
Limit[expr,x->x0, Direction->-1]	x->x0+时函数的极限
Limit[expr,x->x0, Direction->1]	x-> x0−时函数的极限

表 A-14　求微商函数

函　数	功　能
D[f,x]	计算偏导数
D[f,x1,x2,…]	计算多重导数
D[f,{x,n}]	计算n阶导数

3．不定积分和定积分

（1）不定积分（见表 A-15）．

（2）定积分．

计算定积分的命令和计算不定积分是同一个 Integrate 函数．在计算定积分时，除了要给出变量外还要给出积分的上下限．当定积分算不出准确结果时，用 N[%]命令总能得到其数值解．Nintegrate 也是计算定积分的函数，其使用方法及形式和 Integrate 函数相同．

用 Integrate 函数计算定积分得到的是准确解，用 Nintegrate 函数计算定积分得到的是近似数值解.计算多重积分时，第一个自变量相应于最外层积分放在最后计算（见表 A-16）．

表 A-15　求不定积分函数

函　数	功　能
Integrate[f,x]	计算不定积分
Integrate[f,x,y]	计算不定积分
Integrate[f,x,y,z]	计算不定积分

表 A-16　求定积分函数

函　数	功　能
Integrate[f,{x,a,b}]	计算定积分
NIntegrate[f,{x,a,b}]	计算定积分
Integrate[f,{x,a,b},{y,c,d}]	计算定积分
NIntegrate[f,{x,a,b},{y,c,d}]	计算定积分

4．幂级数

幂级数用 Series 展开后，展开项中含有截断误差（见表 A-17）．

表 A-17　幂级数展开函数

函　数	功　能
Series[expr,{x,x0,n}]	将 expr 在 x=x0 点展开到 n 阶的级数
Series[expr,{x,x0,n},{y,y0,m}]	先对 y 展开到 m 阶再对 x 展开 n 阶幂级数

5．解方程与方程组

格式：`Solve[eqns,vars]`

其中 eqns 可以是单个方程，也可以是方程组，单个方程用 exp==0 的形式（其中 exp 为关于未知元的表达式）；方程组写成用花括号括起来的中间逗号分割的若干单个方程的集合；vars 为未知元表，其形式为{*x*1,*x*2,…,*xn*}.

6．常微分方程（见表 A-18）

表 A-18 微分方程求解函数

函 数	功 能
Dsolve[eqns,y[x],x]	解 *y*(*x*)的微分方程或方程组 eqns, *x* 为变量
Dsolve[eqns,y,x]	在纯函数的形式下求解
NDsolve[eqns,y[x],x,{xmin,xmax}]	在区间{*x*min,*x*max}上求解变量 *x* 的数的形式下求解常微分方程和常微分方程组 eqns 的数值解

A.5 绘制图形

1．作图函数格式

在 Mathematica 中，用函数 Plot 可以很方便地作出一元函数的静态图像.

格式：`Plot[{f1,f2,…},{x,xmin,xmax},可选参数]`

功能：绘制函数 *f*1,*f*2,…的图像，*fi*（*i*=1,2,3,…）是函数名，表{*x*,*x*min,*x*max}中 *x* 为函数 *fi*(*i*=1,2,3,…)的自变量，*x*min 和 *x*max 是自变量取值区间的左端点和右端点.

例：作 $y=(x-1)$^2-3 在[-4,6]内的图像和 $y=\lg x$ 在[0.1,4]内的图像，其输入和输出如图 A-10 所示.

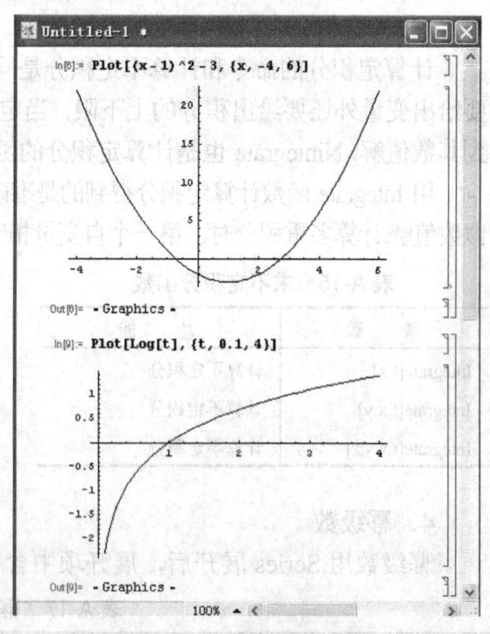

图 A-10 绘制函数图像

2．作图函数的可选参数

（1）参数 AspectRatio（面貌比）.

在 Mathematica 中，根据美学原理系统默认两个坐标轴的单位长度之比为 1:0.618，将参数 AspectRatio 设置为 Automatic 时可使纵横比变为通常的 1:1.

格式：`AspectRatio->Automatic`

例：①作 $y=\sin x$ 和 $y=\cos x$ 在[0,2π]内的图像，且两坐标轴上的单位比为 0.618.

② 作 $y=\sin x$ 和 $y=\cos x$ 在[0,2π]内的图像，且两坐标轴上的单位比为 1:1.

输入和输出结果如图 A-11 所示.

（2）参数 PlotStyle（画图风格）.

PlotStyle 的值是一个表，它决定画线的虚实、宽度、色彩等属性.

格式 1：`PlotStyle->RGBColor[red,green,blue]`

功能：决定画线的色彩. red，green，blue 分别表示红、绿、蓝色的掺入量，取值范围为 0～1.

例：作 $y=\sin x$ 在$[0, 2\pi]$内的图像，线条用红色.

输入：`Plot[Sin[x], {x,0,2Pi}, PlotStyle->RGBColor[1,0,0]]`

RGBColor[1, 0, 0]表示图像的颜色中没有绿色和蓝色，因此画出的是红色的图像.

格式 2：`PlotStyle->Thickness[t]`

功能：决定画线的宽度. t 的取值范围为 0～1，且远远小于 1（系统认为整个图形的宽度为 1）.

例：作 $y=\sin x$ 在$[0, 2\pi]$内的图像，线条宽度 $t=0.02$.

输入：`Plot[Sin[x], {x,0,2Pi},  PlotStyle->Thickness[0.02]]`

输入和输出结果如图 A-12 所示.

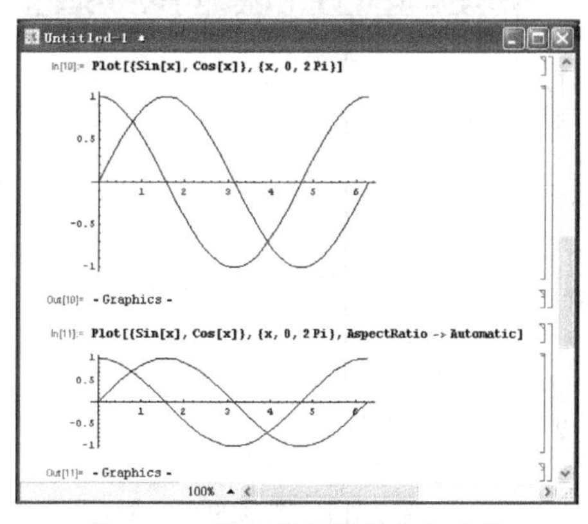

图 A-11 设置坐标轴单位长度比

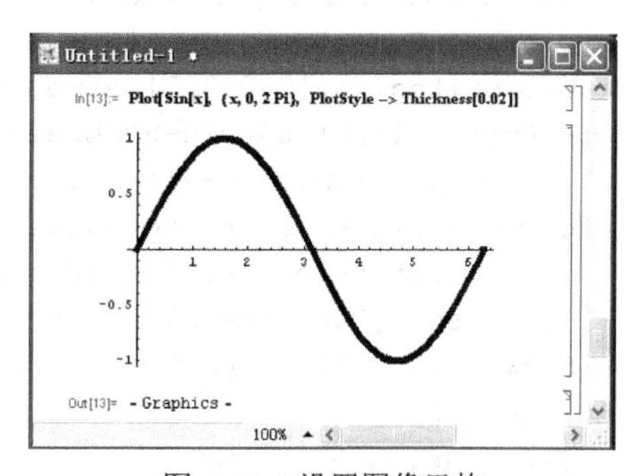

图 A-12 设置图像风格

格式 3：`PlotStyle->Dashing[{d1,d2,…}]`

功能：决定画线的虚实与各分段的长度，其中表$\{d1,d2,\cdots\}$为依次画出的各实、虚分段的线长，每一个 di（$i=1,2,\cdots$）的取值介于 0～1 之间.

例：作 $y=3^x$ 在$[0, 2\pi]$内的图像，线条用虚线，虚实比为 2:3.

输入：`Plot[3^x, {x, 0.1, 5}, PlotStyle->Dashing[{0.03, 0.02}]]`

输出如图 A-13 所示.

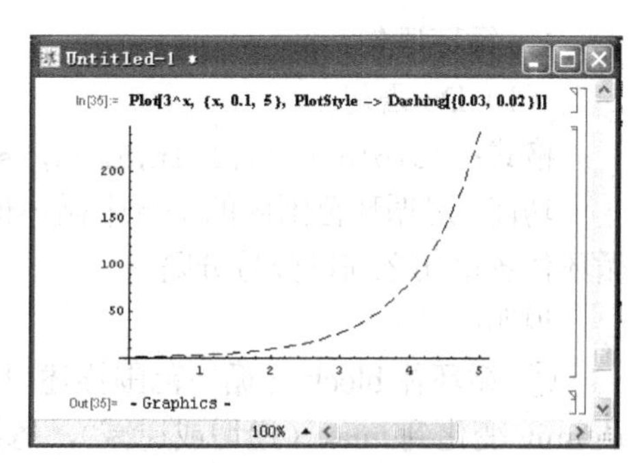

图 A-13 设置图像线型

（3）参数 DisplayFunction（显示函数）.

用于决定图形的显示与否.

格式 1：`DisplayFunction->Identity`

功能：图形不显示.

格式 2：`DisplayFunction-> $DisplayFunction`

功能：恢复图形显示.

3．图形的组合显示函数 Show

Plot 的作用可以同时在同一坐标系的同一区间内作出不同函数的图像，但有时需要在同一坐标系的不同区间作出不同函数的图像，或者在同一坐标系作一个函数而要求函数的各个部分具有不同的形态（如分段函数），这时就需要使用 Show 函数.

格式 1：`Show[fig]`

功能：重绘变量 fig 中保存的图形.

格式 2：`Show[fig1, fig2,…, fign]`

功能：组合绘制多个图形.

例：在同一坐标系中作函数 $y=e^x$ 和 $y=\ln x$ 的图像，并以虚线画出对称轴 $y=x$，从而说明它们的图像关于直线 $y=x$ 对称.

输入：

```
a=Plot[Exp[x],{x,-2,2},AspectRatio->Automatic, PlotStyle->RGBColor [0,1,0],
Display Function->Identity]
b=Plot[Log[x],{x,0.3,3},AspectRatio->Automatic,PlotStyle->RGBColor[1,0
,0],DisplayFunction->Identity]
c=Plot[x,{x,-2,2},AspectRatio->Automatic,PlotStyle->Dashing[{0.09,0.04
}], DisplayFunction->Identity]
Show[a, b, c, DisplayFunction->$Display
Function]
```

输出结果如图 A-14 所示.

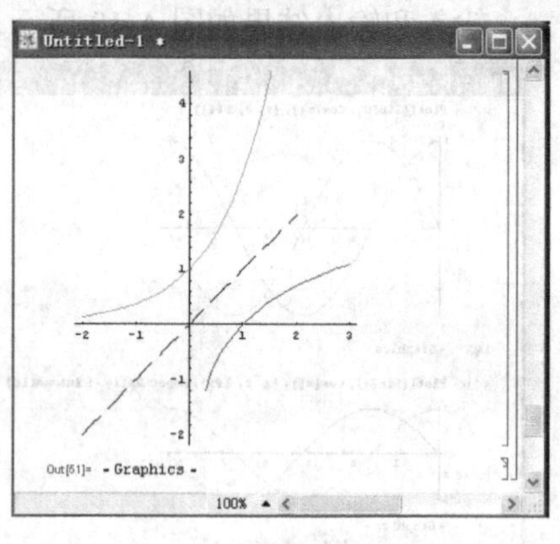

图 A-14　图形的组合显示

A.6　程序控制语句

1．循环语句

（1）Do 语句.

格式：`Do[block,{i,imin,imax,istep}]`

功能：对循环范围内的每一个循环值，执行一次循环体 block（表达式或语句组），循环体各语句之间以分号分隔.

说明：

① 循环体 block 与循环范围描述 {*i*,*i*min,*i*max,*i*step}之间用逗号分隔，循环变量 *i* 从 *i*min 变化到 *i*max（递增或递减），每次改变 *i*step；

② 当初值 *i*min 为 1 或步长 *i*step 为 1 时可省略；

③ 特别地，Do[block, {n}]为执行 *n* 次循环体 block.

（2）While 语句.

格式：`While[test,block]`

功能：当条件 test 满足时，则执行语句组 block，否则跳过（或跳出）循环结构.

（3）For 语句.

格式：`For[start, test, incr, block]`

`[赋,循环条件,增量语句,语句块]`

功能：以 start 为初值，循环计算增量语句 incr 和循环体 block，直到不满足条件 test 为止.

（4）循环控制语句.

```
Return[expr]          %退出函数所有过程和循环，返回 expr 值
Continue[ ]           %转向本层循环的开始处，继续下一次循环
Break[ ]              %跳出本层循环
```

此外，在 Mathematica 的循环结构中，常使用一些特殊的表达式，以达到简洁、方便的目的，如表 A-19 所示.

表 A-19　循环结构中一些特殊的表达式

表达式	意义
i++	$i=i+1$
i--	$i=i-1$
++i	变量 i 先加 1
--i	变量 i 先减 1
i+=k	$i=i+k$
i*=k	$i=i\times k$
{x, y}={y, x}	交换 x 和 y 的值

2. 条件语句

（1）If 语句.

格式 1：`If[条件,语句组]`

功能：若条件（通常为关系表达式）成立，则执行语句组.

格式 2：`If[条件,语句组 1,语句组 2]`

功能：若条件成立，则执行语句组 1，否则执行语句组 2.

格式 3：`If[条件,语句组 1,语句组 2,语句组 3]`

功能：若条件为真，则执行语句组 1；若条件为假，则执行语句组 2；若条件为非真非假，则执行语句组 3.

（2）Which 语句.

格式 1：`Which [条件 1,语句组 1,条件 2,语句组 2,…,条件 n,语句组 n]`

功能：依次判断，若条件 *k* 成立，则执行语句组 *k*，*k*=1，2，…，n.

格式 2：`Which [条件 1, 语句组 1, … , 条件 n, 语句组 n,True, "string"]`

功能：依次判断，若条件 *k* 成立，则执行语句组 *k*，*k*=1，2，…，*n*，若所有条件都不成立，则返回字符串 string.

（3）Swhich 语句.

格式：`Swhich [expr, 模式 1, 表达式 1, … , 模式 n, 表达式 n]`

依次将表达式 expr 的值与模式 1，…，模式 *n* 比较，给出第一个与 expr 匹配的模

式 i 对应的表达式 i 的值，若无匹配模式则整个结构的值为 Null.

3．其他控制及操作语句

（1）使用帮助，Mathematica 的帮助文件提供了 Mathematica 的基本用法的说明，十分详细，可以参照学习.

（2）使用“? 符号名”或“??符号名”来获得关于该符号（函数名或其他）的粗略或详细介绍，符号名中可以使用通配符．例如“?T*”，则系统将给出所有以 T 开头的关键词和函数名，再如“??For”将会得到关于 For 语句的格式和用法的详细情况.

（3）在 Notebook 窗口中运行语句或程序时，如果计算时间太久，可以按【Alt+.】组合键中止.

（4）对函数名不确定的，可先输入前面几个字母（开头一定要大写），然后按【Ctrl+K】组合键，系统会自动补全该函数名.

（5）对程序调试过程中暂不需要显示或执行的语句，可以用一对“(*”“*)”将其括起来.

Appendix B Matlab Basic Operations

Matlab 是"Matrix Laboratory"的缩写，意为"矩阵实验室"，是当今很流行的科学计算软件．它给人们提供了一个方便的数值计算平台．

Matlab 的基本运算单元是不需指定维数的矩阵，系统提供了大量的矩阵及其他运算函数，可以方便地进行一些很复杂的计算，而且运算效率极高．Matlab 命令与数学中的符号、公式非常接近，可读性强，容易掌握，还可以利用它所提供的编程语言进行编程，完成特定的工作．除基本部分外，Matlab 还根据各专门领域中的特殊要求提供了许多可选的工具箱，在很多时候能给予用户极大的帮助．

B.1 Matlab 的基本操作

1．启动 Matlab

双击桌面图标，或单击"开始"按钮，选择"程序"→"Matlab 6.5"→"Matlab 6.5"命令，启动 Matlab 6.5，主工作界面如图 B-1 所示．

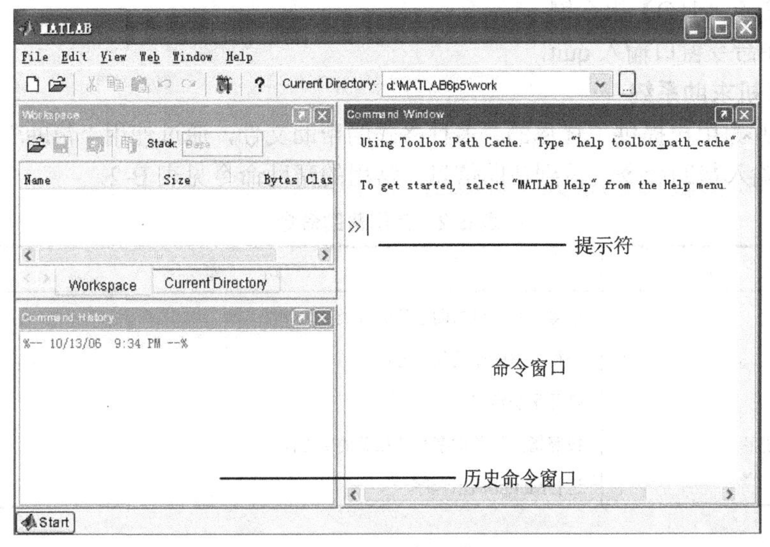

图 B-1 Matlab 主工作界面

2．Matlab 的交互式操作

启动 Matlab 后，在命令窗口（Command Windows）显示一段提示信息后，出现系统提示符">>"，这时就可以输入命令了．Matlab 是一个交互式的系统，输入命令后，系统会马上解释和执行输入的命令并输出结果．常用的控制命令如表 B-1

所示.

表 B-1 常用控制命令

命　令	功　能	命　令	功　能
clear	清除内存中所有的或指定的变量和函数	home	向上滚动显示内容，并将光标定位在命令窗口的左上角
Clc	清除命令窗口中显示的所有内容	Pack	收回内存碎片，释放内存空间
Dir	显示当前目录或指定目录中的内容	Exit	退出Matlab
Cd	显示或改变当前工作目录	Quit	关闭并退出Matlab

如果输入的命令有语法错误，系统会给出提示信息. 在当前提示符下，可以通过上、下箭头或双击历史命令窗口中列出的历史命令，调出并执行以前输入的命令，从而减少重新输入命令的麻烦.

用滚动条可以查看以前的命令及其输出信息. 如果不想看到命令语句的输出结果（特别是运算结果很长时，输出时会长时间的翻屏），可以在命令语句的最后加上分号“;”，表明不输出当前命令的结果.

3．退出 Matlab

退出 Matlab 的方法很多，除单击程序窗口右上方的“关闭”按钮、选择“文件”→“Exit Matlab”命令外，还可以通过如下两种方法退出：

(1) 按【Ctrl+Q】组合键；

(2) 在命令窗口输入 quit.

4．联机求助系统

Matlab 为用户提供了详尽的、多种格式的帮助文件，通过使用“帮助”菜单或在命令窗口输入帮助命令，得到帮助信息. 常用的帮助命令见表 B-2.

表 B-2 常用帮助命令

命　令	功　能
help	获得所有联机求助信息的分类列表
help<命令>	获得具体命令的使用方法
helpwin	打开帮助窗口
lookfor <关键字>	按照指定的关键字查找相关的M文件
demo	运行Matlab演示程序

B.2 Matlab 的语法规则

（1）在命令窗口的提示符后输入 Matlab 命令，按【Enter】键确认并执行；

（2）Matlab 的变量名必须由字母开头，且严格区分大小写，命令中所使用的标点均为英文标点；

（3）输入命令后直接按【Enter】键，则立即显示计算结果；如果不想立即显示计算结果，则需要在命令行后加上分号“;”；

（4）语句行中的百分号“%”为注释符号，其后的命令不被执行；

（5）当表达式过长，在一行输入不便时，输入三个英文句点“...”（续行符），然后回车换行，继续输入.

B.3 Matlab 的数据

1. Matlab 的运算符

（1）算术运算符.

+、−、*、/、\、^分别表示加、减、乘、除、左除（主要用于矩阵的除法）、乘方运算。运算顺序从左到右，优先级依次为乘方、乘除、加减，括号可以改变运算次序和优先级，由最内层的括号向外执行.

（2）关系运算符.

小于	<	小于等于	<=
大于	>	大于等于	>=
等于	==	不等于	~=

关系运算符常被用来构造控制程序流程的条件表达式.

（3）逻辑运算符.

与 & 或 | 非 ~ 异或 XOR

2. 变量和表达式

（1）变量的定义.

Matlab 命令的通常形式为：

```
变量 = 表达式
```

表达式由操作符或其他特殊字符、函数和变量名组成. Matlab 执行表达式并将执行结果显示于命令后，同时存在变量中以留用. 如果变量名和“=”省略，即不指定返回变量，则系统将自动建立名为 ans 的变量. 例如：

键入命令 A=[1.2 3.4 5.6 sin(2.)]，系统将产生 4 维向量 $\boldsymbol{A}$，键入 3*pi/4，系统将自动匹配变量 ans=并显示计算结果.

Matlab 保留本次运行中建立的所有变量的信息. 如果要了解系统当前变量的信息可以输入 whos 命令，屏幕将显示当前系统中所有变量的详细信息，如图 B-2 所示.

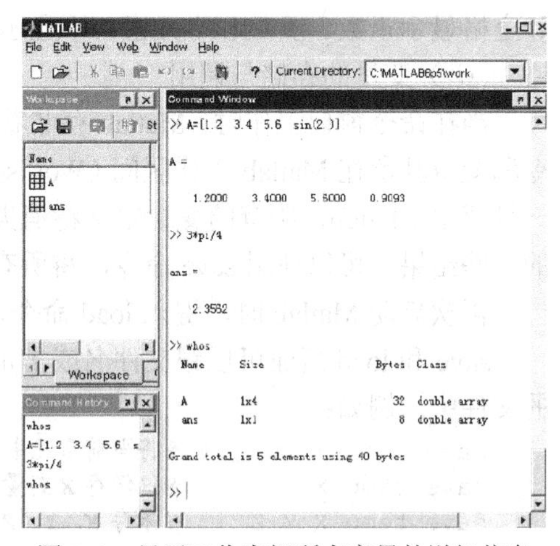

图 B-2 显示工作空间所有变量的详细信息

（2）预定义变量．

除了自定义变量外，系统还有几个预定义的特殊变量，如表 B-3 所示．

表 B-3 系统预定义变量

特殊变量	意 义	特殊变量	意 义
pi	圆周率 π	i和j	虚数单位，用于表示复数，如3＋4i，2-3j
eps	计算机的最小正数	inf	无穷大，由分母为0的运算产生
flops	符点运算次数，用于统计计算量	nan	不定值，由inf/inf或0/0运算产生

需要注意的是，inf 和 nan 能够安全地进行计算和传递．即如果在初始值或中间结果中出现了 inf 和 nan，inf 和 nan 会遵循一定的计算规则进行正确的计算并得到正确的结果．例如，无穷大加上一个有限实数的结果是无穷大．

（3）显示格式．

在默认的状态下，Matlab 以双精度执行所有的运算与存储，以短格式（short 格式）显示计算结果．可以使用 Matlab 命令窗口中的“Option”→“Numerical Format”菜单改变数字显示格式，也可以通过 format 命令改变（见表 B-4）．

表 B-4 改变数字显示格式函数

命 令	功 能	命 令	功 能
format short	5位定点显示	format rat	近似的有理数形式显示
format short e	5位浮点显示	format hex	十六进制显示
format long	15位定点显示	format bank	元角分（美制）定点显示
format long e	15位浮点显示	format compact	压缩变量间空行
format short g	系统自动选择更优的5位显示	format loose	变量间有空行
format long g	系统自动选择更优的15位显示		

另外，还有 Plus 格式（“+”格式），这是显示大矩阵的一种紧凑方法，“+”“-”和空格显示正数、负数和零元素，不显示具体的数值．

（4）变量的存储与调用．

当在命令窗口工作时，Matlab 存储着输入的命令和所有创建的变量的值，这些命令和变量驻留在 Matlab 工作区间（Workspace）中，可以在任何需要的时候被调用．但一旦退出 Matlab，所有的变量定义将消失．如果需要在退出 Matlab 前保持本次计算的一些结果，可以使用 save 命令，将所有变量作为文件存入磁盘 matlab.mat 中．

再次启动 Matlab 时，键入 load 命令，则将变量从 matlab.mat 中重新调出．

save 和 load 后面可以跟文件名或指定的变量名作为参数，将指定的变量存入指定的文件中．例如：

```
save                 %将当前系统中所有变量存入缺省文件 matlab.mat 中
save temp X          %仅保存 X 变量的值到 temp.mat 文件中
save temp X Y Z      %保存 X、Y、Z 三个变量的值到 temp.mat 文件中
load temp            %从 temp.mat 文件中取回所保存的变量
```

3. 数学函数

Matlab 支持所有的常用的数学函数，表 B-5 是基本数学函数的一个简单列表.

表 B-5 常用数学函数

类　型	函　数	类　型	函　数
基本三角函数	sin,cos,tan,sec,csc,cot	指数和对数	exp,log,log10,sqrt
反三角函数	asin,acos,atan,asec,acsc,acot	复数运算	abs,angle,conj,real,imag
双曲函数	sinh,cosh,tanh,sech,asch,coth	数值函数	fix,floor,ceil,round,rem
反双曲函数	asinh,acosh,atanh,asech,acsch,acoth	整数函数	lcm,gcd

其中：

log(x)求 x 的自然对数，log10(x)求 x 以 10 为底的对数.

angle(x)求复数 x 的幅角，conj(x)求 x 的共轭复数，real(x)和 imag(x)分别求复数 x 的实部和虚部.

fix(x)求实数 x 最接近 0 的整数值，floor(x)求 x 最接近负无穷的整数值，ceil(x)求 x 最接近正无穷的整数值，round(x)求 x 最接近 x 的整数值，

rem(x,y)求 x 除以 y 的余数.

lcm(a,b)返回正整数 a 和 b 的最小公倍数，gcd(a,b)返回正整数 a 和 b 的最大公约数.

例如：

```
>>x=sqrt (2) /2
x=
   0.7071
>>y=asin(x)
y=
   0.7854
>>y=sqrt(3^2+4^2)
y=
  5
```

B.4 矩阵

1. 矩阵的生成

（1）基本输入方法.

输入一个小矩阵最简单的方法是直接列出矩阵元素. 矩阵用方括号括起来，元素之间用空格或逗号分隔，矩阵行与行之间用分号隔开. 例如：

```
>>A= [1 2 3 4 5]
A=
  1  2  3  4  5
>>A=[1  2  3;  4  5  6;7  8  0]
A=
  1  2  3
  4  5  6
```

```
7 8 0
```

大的矩阵可以分行输入，用回车键代替分号，例如

```
>>A=[ 1  2  3
      4  5  6
      7  8  0 ]
```

结果和上式一样.

Matlab 的矩阵元素可以是任何数值表达式. 例如:

```
X=[-1.3   sqrt(3)   (1+2+3)*4/5]
x=
  -1.3000    1.7321    4.8000
```

矩阵元素也可以是复数. 输入方法类似. 例如:

```
A=[1+5i   2+6i;  3+7i  4+8i]
```

但当复数作为矩阵的元素输入时，需注意不要留有任何空格. 对 1+5i，如在“+”号左右留有空格，就会被认为是两个分开的数. 所以，有时用以下方法 `A=[1  2; 3  4]+i*[5  6; 7  8]`，这个表达式和前一种输入方法具有相等的结果. 它实际上利用了矩阵的乘法和加法运算.

（2）矩阵元素的存取.

在括号中加注系数可取出单独的矩阵元素. 如

```
>>x=[1 2 3 4 5];
>>x(3)
ans =
   3
```

对于二维矩阵，则需指明行和列，否则，优先按列然后再按行进行元素排序，例如:

```
>>x=[1 4 7; 2 5 8; 3 6 9];
>>x (3,2 )
ans =
   6
>>x (8)
ans=
   8
```

Matlab 中的矩阵可以自动扩充维数. 例如:

```
>>x=[-1.3  sqrt(3)  (1+2+3)*4/5]
x=
   -1.3000   1.7321   4.8000
```

如果直接对 $x(5)$赋值，系统会自动扩展向量的长度，并对未赋值的元素赋初值 0.

```
>>x(5)=abs(x(1))
x=
   -1.3000  1.7321  4.8000   0  1.3000
```

对矩阵也有类似的结果.

（3）“:” 运算符的用法.

在 Matlab 中,“:”是一个很重要的运算符. 例如: 可以产生具有一定规律的向量. 灵活地使用 “:” 是掌握好 Matlab 的一个很重要的技巧. 例如

```
>>Z=1:5;                %生成一个 1~5（单位增量的）的行向量
>>x=pi:-pi/4:0;         %生成一个由 pi~0 的行向量，单位增量是-pi/4=-0.7854
>>x=(0.0:0.2:1.0)';     %生成列向量，“'” 运算符表示取转置矩阵
>>y=sin(x);             %计算函数值
>>[x  y]                %组合出矩阵
ans=
0          0
0.2000     0.1987
0.4000     0.3894
0.6000     0.5646
0.8000     0.7174
1.0000     0.8415
```

（4）生成特殊矩阵的函数.

Matlab 提供了一些用于生成特殊向量和矩阵的函数，如表 B-6 所示.

表 B-6　生成特殊向量和矩阵函数

函　数	功　能
zeros	生成一个零矩阵
ones	生成 “1” 矩阵数
eye	生成单位矩阵
rand	生成随机向量
randn	生成随机矩阵
linspace	生成均匀和对数级数
logspace	生成均匀和对数级数
linspace	生成网络

它们的基本用法是在参数中指定产生的向量或矩阵的维数. 如果只有一个数值参数 n，表示生成 $n\times n$ 的矩阵；如果有两个参数或者一个长度为 2 的向量$[m,n]$，表示生成一个 $m\times n$ 的矩阵.

2. 矩阵操作

Matlab 提供了丰富的矩阵操作函数，可以非常方便地对矩阵进行旋转、变形、扩充等操作.

（1）矩阵转置.

矩阵的转置用符号 “'” 表示. 如

```
>>A=[1  2  3; 4  5  6;7  8  0];
>>B=A'
B=
   1  4  7
   2  5  8
   3  6  0
```

如果 $\boldsymbol{Z}$ 为复矩阵，则 $\boldsymbol{Z}'$为它的复数共轭转置，非共轭转置可以使用 conj(Z')命令求得.

（2）子矩阵.

矩阵的建立和取值不仅可以一个一个元素地进行，也可以成批进行. 首先，大的矩阵可把小的矩阵作为其元素来完成，如：

```
>>A=[1 2 3; 4 5 6; 7 8 0];
>>A=[A;[10 11 12]]
A=
     1   2    3
     4   5    6
     7   8    0
    10  11   12
```

其次，小矩阵可用“:”号从大矩阵中抽取出来，通过指定取值的范围，如从第几行至第几列，从第几列至第几列，可以取出大矩阵选定的行列（见表 B-7).

表 B-7 “:”号的作用及使用

命令	意义
A(:)	取 A 的所有元素
A(:,J)	取 A 的第 J 列
A(J:K,:)	取 A 的第 J 至 K 行组成子阵
A(:,J:K)	取 A 的第 J 至 K 列组成子阵

y=***x*** (2 : 6)表示取出向量 ***x*** 的第 2～6 个元素.

A=***A*** (1 : 3,)表示取出矩阵 ***A*** 的第 1～3 行及所有的列，并重新组成原来的 ***A***.

（3）对角矩阵.

diag 函数用来生成一个对角矩阵或者提取一个矩阵的对角元素. 例如，以下命令通过指定的对角线元素的值生成一对角矩阵.

```
>>diag(2:4)
Ans=
      2  0  0
      0  3  0
      0  0  4
```

（4）上/下三角形矩阵.

tril 函数用于取出矩阵的下三角矩阵；triu 函数用于取出矩阵的上三角形矩阵. 例如：

```
>>a=[11  12  13; 21  22  23; 31  32  33 ];
>>tril(a)
ans=
     11   0   0
     21  22   0
     31  32  33
>>triu(a)
ans=
     11   12   13
     0   22   23
      0    0   33
```

3. 矩阵运算

（1）加法和减法.

如矩阵 $\boldsymbol{A}$ 和 $\boldsymbol{B}$ 的维数相同，则 $\boldsymbol{A}+\boldsymbol{B}$ 和 $\boldsymbol{A}-\boldsymbol{B}$ 分别表示矩阵 $\boldsymbol{A}$ 和 $\boldsymbol{B}$ 的和与差. 如果矩阵 $\boldsymbol{A}$ 和 $\boldsymbol{B}$ 的维数不匹配，Matlab 会给出相应的错误提示信息. 例如：

```
>>A=[1  2  3; 4  5  6; 7  8  0] ;
>>B=[1  4  7;2  5  8; 3  6  0];
>>C=A+B
C=
   2   6  10
   6  10  14
  10  14   0
```

如果运算对象是个标量（即 1*1 矩阵），则可以和其他矩阵进行逐个元素的加减运算.

（2）乘法.

矩阵乘法用“*”符号表示，当 $\boldsymbol{A}$ 矩阵的列数与 $\boldsymbol{B}$ 矩阵的行数相等时，二者可以使用 A*B 命令进行乘法运算，否则是错误的. 例如：

```
>>A=[1  2 ; 3  4]; B=[5  5; 7  8];
>>C=A*B
C=
   19   21
   43   47
```

如果 A（或 B）是标量，则 A*B 返回标量 A（或 B）乘以矩阵 $\boldsymbol{B}$（或 $\boldsymbol{A}$）的每一个元素所得的矩阵.

（3）除法.

在 Matlab 中有两种矩阵除法符号“\”“/”，分别表示左除和右除. 在运行一般的标量运算时，$a/b=b\backslash a$；在进行矩阵运算时，$\boldsymbol{A}\backslash\boldsymbol{B}$ 是 $\boldsymbol{A}$ 的逆矩阵乘以矩阵 $\boldsymbol{B}$，即 $\boldsymbol{A}^{-1}\boldsymbol{B}$，相当于 Matlab 命令 inv(A)*B. $\boldsymbol{A}/\boldsymbol{B}$ 是 $\boldsymbol{A}$ 乘以矩阵 $\boldsymbol{B}$ 的逆矩阵，即 $\boldsymbol{A}\boldsymbol{B}^{-1}$，相当于 Matlab 命令 A*inv(B). 例如：

```
>>A=[1  2  3];
>>B=[1  2  -3;-2  5  6;7  2  1];
>>A/B         % 求 AB^-1，结果是一 3 阶行向量. 注意无法求 B/A.
ans=
    -0.1818     0.3636     0.2727
>>B\A         % 求 B^-1A'，结果是一 3 阶列向量.注意矩阵 A 需要转置.
ans=
    0.2929
    0.4444
    0.0606
```

（4）乘方.

A^P 意思是 A 的 P 次方. 对 A 和 P 分别是整数、实数、向量或矩阵，A^P 具有不同的含义. 最简单的情形是，A 是一个方阵，P 是大于 1 的整数，则 A^P 表示 A 的

P 次幂即 A 自乘 P 次. 例如：

```
>>[1  2  3;4  5  6;7  8  9]^3
ans=
    468    576   684
    1062  1305  1260
    1656  2034  2412
```

（5）数组运算.

数组运算是指向量或矩阵间元素对元素间的运算，由运算符“*”“/”“\”“^”前加一点“.”来表示. 对于矩阵的加和减运算，由于其运算方法就是对应元素的运算，所以没有“.+”“.-”运算符.

数组的乘用符号“.*”表示，如果 $\boldsymbol{A}$ 与 $\boldsymbol{B}$ 矩阵具有相同的阶数，则 **A.*B** 表示 $\boldsymbol{A}$ 和 $\boldsymbol{B}$ 单个元素之间的对应相乘. 例如

```
>>x=[1 2 3]; y=[4 5 6];
>>z=x.*y
z=
   4  10  18
```

数组乘方用符号“.^”表示. 例如：

```
>>x=[1 2 3 ]; y=[4 5 6];
>>z=x.^y
z=
   [1  32   729]
>> 2.^[x y]
ans =
   2   4   8  16  32  64
>> [x y]
ans =
   1   2   3   4   5   6
```

数组除法有操作符“./”和“.\”，为对应元素的“/”和“\”运算. 例如：

```
>>[1 2;3 4]./[2 2;2 2]
ans=
    0.5000   1.0000
    1.5000   2.0000
>>[1 2 ; 3 4].\[2 2 ; 2 2]
ans=
    2.0000   1.0000
    0.6667   0.5000
```

B.5 绘图

1. 二维图形

（1）描点绘图.

plot 命令根据给定的 x-y 点的坐标绘制平面坐标图形. 如果 $\boldsymbol{x}$，$\boldsymbol{y}$ 均是长度为 n 的

实向量，plot(x,y)将绘制点，(x1,y1)，(x2,y2)，…，(xn,yn)的图形．如果没有指定 *x* 坐标，plot(y)函数将按照 ***y*** 的下标绘制一个 ***y*** 中元素的线性图．例如：

```
>>y=[0,1.85,0.88,1,0.24,3.15];
>>plot(y)
```

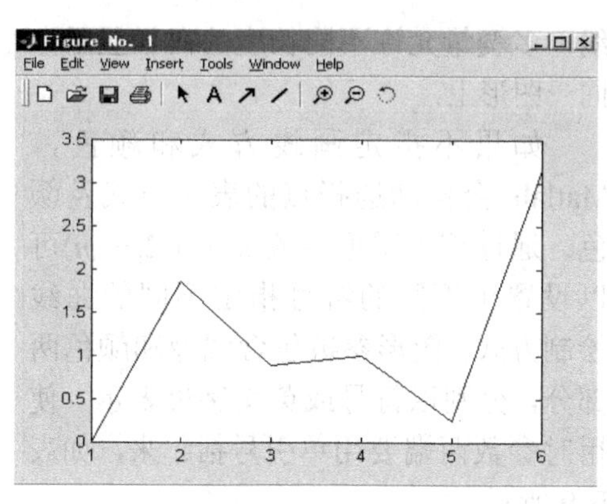

图 B-3　图形窗口

Matlab 会产生一个图形窗口（Figure），显示结果如图 B-3 所示．需要注意的是，*x* 和 *y* 轴的坐标是由计算机自动绘制出的．

如果需要给图形加上标题、*x* 轴、*y* 轴等标注，可以使用 Matlab 图形命令产生相关项，如表 B-8 所示．

例如：

```
t=0:05:4*pi;y=sin(t);
plot(t,y)
grid
title('y=sin(t)曲线图')
xlabel('t=0:0.05:4pi')
```

结果如图 B-4 所示．

表 B-8　图形标注与窗口控制函数

函　　数	功　　能
title	图形标题
xlabel	*X* 坐标轴标注
ylabel	*Y* 坐标轴标注
text	标注数据点
grid	给图形加上网格
hold	保持图形窗口的图像

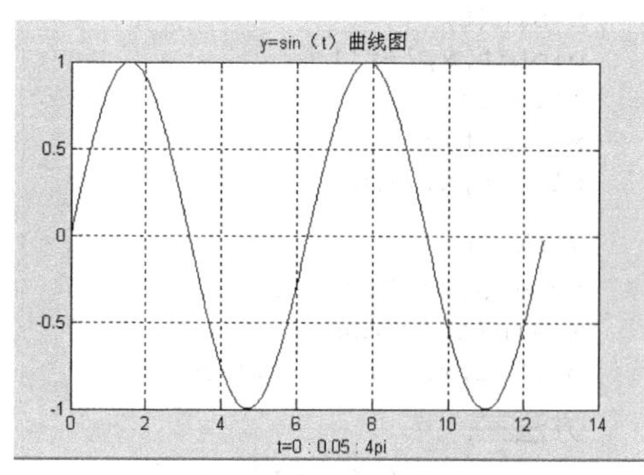

图 B-4　标注后的图形

（2）复线图的绘制．

在一个图形上绘制多条曲线有几种办法．

第一种方法是使用 hold 命令，它的意义是决定在绘制下一幅图形时是否将图形保留在绘图窗口中．一般可以在绘制一个图形后使用 hold on 命令保留绘图窗口，然后继续绘制下一图形直至完成，最后使用 hold off 命令．

第二种方法是利用 plot 函数的多变量方式绘制，它可以接受多个参数．

格式：`plot(x1,y1,s1,x2,y2,s2,…,xn,yn,sn)`

其中，xi，yi(i=1,2,…,n)是成对的向量，根据每一对 ***x***，***y*** 的值，在图上描出一条线．多变量允许不同长度的向量显示在同一图形上．

如果不指定画线方式和颜色，Matlab 会自动选择点的表示方式及颜色．通过使用图形参数 si(i=1,2,⋯,n)可以设置用不同的符号指定不同的曲线绘制方式．图形参数包含线型和颜色两部分，分别以符号或英文字母表示，使用时参数两端要用单引号括起来，如表 B-9 所示．

表 B-9　图形风格设置参数

线型参数	意　义	颜色参数	意　义
.	点	y	黄
o	小圆圈	m	棕色
x	x标记	c	青色
+	加号	r	红色
-	实线	g	绿色
*	星号	b	蓝色
:	虚线	w	白色
-.	点画线	k	黑色
--	间断线		

例如：

```
>>plot(x,y,'x')                    用'x'作为点绘制的图形
plot(x1,y1,':m',x2,y2,'+c')        用':'画第一条棕色线，用'+'画第二条青色线
```

第三种方法是使用矩阵参数绘制多曲线图形，对 plot(x，y)命令来说，如果 $\boldsymbol{y}$ 是矩阵，$\boldsymbol{x}$ 是向量，plot(x，y)用不同的画线形式绘出 $\boldsymbol{y}$ 的各行对应于 $\boldsymbol{x}$ 向量的图形．

例如，绘制[$-\pi,\pi$]区间内 sin(x)的曲线，可以使用如下命令：

方法一：

```
t= - 3.14:0.2:3.14;                %取弧度角
x=sin(t):                          %计算正弦值 sin(t)
plot(t,x,'+r')                     %绘制 sin(t)的图形
hold on                            %保持图形
x=cos(t):                          %计算余弦值 cos(t)
plot(t,x,/-b/)                     %绘制 cos(t)的图形
```

方法二：

```
t= - 3.14:0.2:3.14;
x=sin(t);y=cos(t);
plot(t,x,'+r',t,y,'-b')            %多重图的绘制
```

方法三：

```
t= - 3.14:0.2:3.14;
x(1,:)=sin(t);                     %计算正弦值,保存于矩阵的第一行
x(2,:)=cos(t);                     %计算余弦值,保存于矩阵的第二行
plot(t,x)                          %绘制图形
```

三种命令得到的图形是基本一致的（第三种方式无法指定详细线型），如图 B-5 所示．

（3）根据函数绘图．

```
fplot(fname,lims)          %绘制 fname 指定的函数的图形
```

fplot 函数的绘制区域为 lims=[xmin,xmax,ymin,ymax]，指定 y 轴的区域，函数表达式可以是一个函数名，如 sin, tan 等；也可以是带上参数 x 的函数表达式，如 sin(2x)；也可以是一个用方括号括起的函数组，如[sin,cos]．例如：

```
fplot ('sin(x)',[0,4*pi]           %绘制 sin(x)在[0,4π]间的图形
```

如果在函数说明中使用向量的形式给出多个函数，可以绘制多重图形，例如：

```
fplot('[tan(x),sin(x),cos(x)]',[-2*pi,2*pi,-2*pi,2*pi])
```

显示结果如图 B-6 所示.

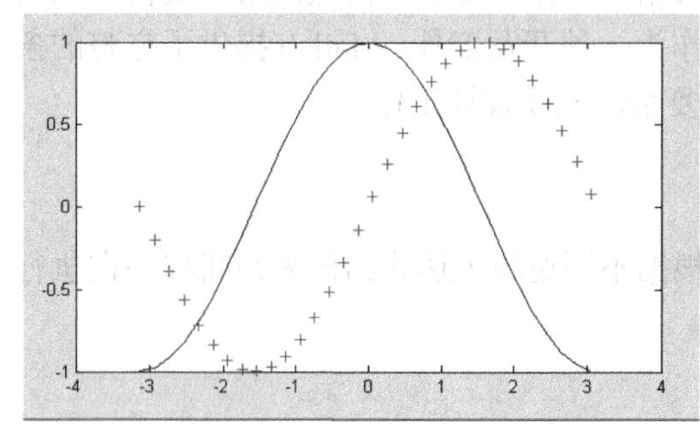

图 B-5　使用不同线型参数绘制的图形

图 B-6　多重图形的绘制

2. 三维图形

Mesh(Z)语句可以给出矩阵 ***Z*** 元素的三维消隐图，网络表面由矩阵 ***Z*** 在 *x-y* 坐标平面上的值所决定，图形由邻近的点连接而成. 例如，绘制 sin(*r*)/*r* 函数的图形（见图 B-7）. 实现方法如下：

```
x=-8:0.5:8;                %建立行向量 x
y=x';                      %建立列向量 y
x=ones(size(y))*x;         %按向量的长度建立 1-矩阵
y=y*ones(size(y))';        %用向量乘法产生的 1-矩阵，生成网格矩阵，它们的值对应
                           %于 x-y 坐标平面
R=sqrt(x.^2+y.^2)+eps;     %计算各网格点的半径
Z=sin(R)./R;               %计算函数值矩阵 Z
Mesh(Z)                    %用 Mesh 函数得到三维图形
```

其他产生三维图形的函数还有 contour，surf，plot3 等，详细用法见 Matlab 相关帮助内容.

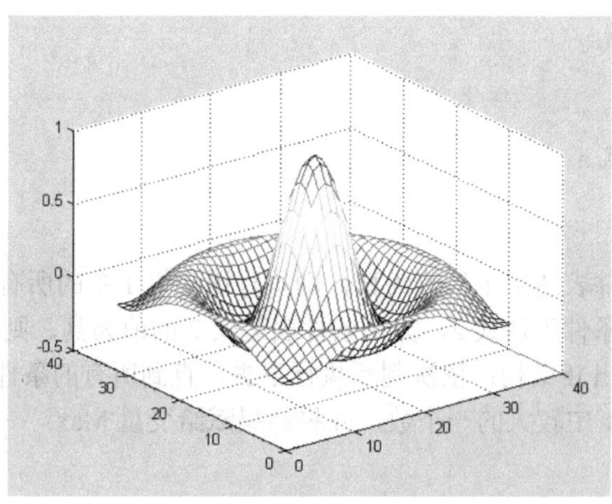

图 B-7　绘制三维图形

B.6 程序设计基础

虽然 Matlab 提供了大量的函数和工具箱，但在实际中，仍然经常需要自己编写 Matlab 程序或函数，以完成特定的功能．作为一种开发工具，Matlab 提供了完整的条件判断和各种控制流语句，可以进行非常复杂的程序设计工作．

1．分支结构

（1）if 分支．

if 语句是一种条件选择结构语句，它根据不同逻辑表达式的值来判断程序在执行过程中需要执行哪些语句．

格式 1：

```
if 条件表达式
   语句组
end
```

功能：如果条件表达式的值为真，则执行语句组中的所有语句；反之，则跳过语句组，执行 end 之后的语句．

格式 2：

```
if 条件表达式
   语句组 1
else
   语句组 2
end
```

功能：如果条件表达式的值为真，则执行语句组 1 中的所有语句；反之，执行语句组 2 中的所有语句．

格式 3：

```
if 条件表达式 1
   语句组 1
else 条件表达式 2
   语句组 2
…
else 条件表达式 n
   语句组 n
end
```

功能：如果条件表达式 1 的值为真，则执行语句组 1 中的所有语句，然后跳出 if 结构；否则，判断条件表达式 2，如果条件表达式 2 的值为真，则执行语句组 2 中的所有语句，然后跳出 if 结构；依次判断执行下去，直到所有的条件都判断完毕．

例如，求 x 和 y 中较大的一个数，并将结果赋给变量 Max：

```
if  x>y
   Max=x;
else
```

```
  Max=y;
end
```

（2）switch 分支.

switch 结构可以根据一个变量或表达式的值执行待定的语句.

格式：

```
switch 表达式 0
   case 常量或表达式 1
      语句组 1
   case 常量或表达式 2
      语句组 2
   …
   otherwise 条件表达式 n
      语句组 n
  end
```

功能：先计算表达式 0 的值，然后依次检查 case 子句后面的值是否与之相等，如果存在相等的值，则执行该 case 子句中的语句组；否则，执行 otherwise 子句中的语句组 n.

2. 循环结构

（1）for 循环.

Matlab 与其他计算机语言一样有 for 循环语句，完成一个语句或一组语句在一定条件下反复使用的功能，其使用次数是预先设定的.

格式：

```
for 循环变量=初值 ： 步长 ： 终值
    循环体
end
```

功能：循环变量每循环一次，循环变量的值加 1，for 和 end 间的循环体（语句组）被执行一次.

例如：

```
M=4;N=3;
for i=1:M
    for j=1:N
      A(i,j)=1/(i+j-1);
    end
end
A
```

返回矩阵 $\boldsymbol{A}$ 的值为：

```
A=
     1.0000     0.5000     0.3333
     0.5000     0.3333     0.2500
     0.3333     0.2500     0.2000
```

```
    0.2500    0.2000    0.1667
```

从程序中可以看出，矩阵的元素 $a_{ij}=\dfrac{1}{i+j-1}$，程序通过一个二重循环给 $\boldsymbol{A}$ 的每一个元素赋了值.

（2）while 循环.

循环为一个语句或一组语句在一个逻辑条件的控制下重复未知次数的循环语句.

格式：

```
while  条件表达式
   循环体
end
```

功能：当条件表达式的所有运算为非零值时，循环体（语句组）将被执行. 如果判断条件是向量或矩阵的话，可能需要 all 或 any 函数作为判断条件.

例如，计算满足 1+2+⋯+ n<100 的最大正整数 n.

```
sum=0; n=0;              %赋初始值
while sum<100            %判断和式是否超出 100
   n=n+1;                %如果没有超出 100，加 1
   sum=sum+n;            %求和式
end;                     %循环终止语句
n=n-1                    %循环停止后，计算和式没有超出 100 的 n
```

结果为

```
n=
   13
```

3. 其他结构控制语句

系统还提供了几个用于控制结构的函数，如表 B-10 所示.

表 B-10　程序结构控制函数

函　数	功　能
pause	暂停执行，直到有击键动作
break	中断执行，用在循环体中则表示跳出循环
return	中断执行，返回上一层或命令窗口
error（字符串）	提示错误并显示“字符串”

4. M 文件与函数文件

M 文件是以后缀 m 结尾的文件. 它是普通的文本格式文件，可以使用任何文本编辑工具来进行编辑. Matlab 有两种 M 文件，一种是直接包含了一系列 Matlab 命令的文件，称为命令文件；第二种类型的 M 文件以 function 开始，提供了 Matlab 的外部函数，称为函数文件.

Matlab 有单独的 M 文件编辑环境，单击工具栏上的“新建”按钮或选择“File”→“New”→“M-File”命令即可打开编辑窗口，在此窗口可以进行 M 文件的编辑和调试. 保存 M 文件时要注意文件名应规范（命名规则与变量的命名规则基本相同），否则不能正常运行.

运行 M 文件的方法通常有三种：

① 在 M 文件编辑窗口单击工具栏上的“运行”按钮；

② 在 M 文件编辑窗口选择“Debug”→“Run”命令，或者按【F5】功能键；

③ 在命令窗口输入 M 文件的主名，回车.

（1）文本文件.

当一个文本文件被调用时，Matlab 将连续运行文件中出现命令，而不是交互地等待键盘输入，这对于分析解决问题及完成设计中所需要的一长串繁杂的命令是很有用的.

例如，在文件 fibo.m 中包含了如下文字：

```
% An M-file to calculate Fibonacci numbers
f=[1 1]; i=1:
while i<15
   f(i+2)=f(i)+f(i+1);
   i=i+1;
end
plot(f,' * k ' )
```

第一行以“%”符号开始的部分为注释部分，它的主要作用是描述本文件的目的和使用的算法等内容. 以后的内容计算了前 16 个 Fibonacci 数并绘图. 保存文件后，在命令窗口中键入 fibo 命令，运行结果如图 B-8 所示.

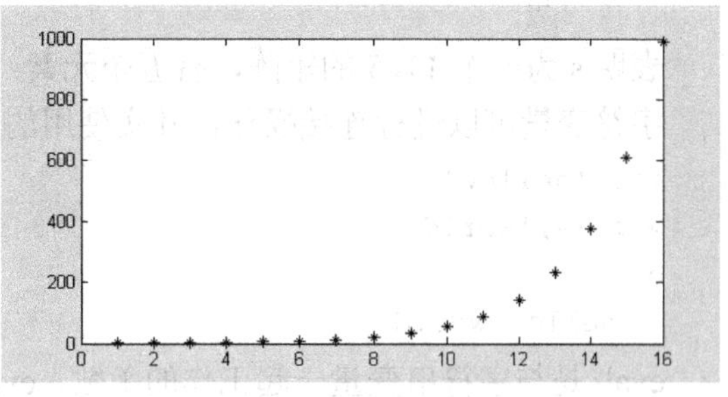

图 B-8　Fibonacci 数散点图

（2）函数文件.

如果 M 文件的第一行包含 function，这个文件就是函数文件，它与命令文件不同，它的变量可以定义，变量和运算都在文件内，而不在工作空间. 例如：

```
function y=mean(x)
%  MEAN Average or mean value, For vectors,
%  MEAN(x)returns the mean value
%  For matrix MEAN(x)is a row vector
%  containing the mean value of each column
[ m,n ]=size(x);
if m==1
   m=n;
end
y=sum(x)/m;
```

程序的第一行为函数名、输入变量、输出变量的说明，这是函数文件区别于文本文件的特征. 变量 m、n 和 y 是 mean()函数的局部变量，它们的值在执行完 mean()函数后将不会留下.

这个新定义的函数与 Matlab 的其他函数一样使用. 例如定义变量 z 为 1～99 的实函数向量：

```
z=1:99
```

使用 mean()函数计算均值为：

```
mean(z)
```

```
ans=
  50
```

5. 文字串处理

字符串用单个引号输入到 Matlab 中，例如：

```
>>s='Hello'
s=
  Hello
```

字符存在向量中，每个元素就是一个字符，例如：

```
>>size(s)
ans=
  1   5
```

表明 s 为一个 1×5 的矩阵，有五个元素．

字符变量可以进行连接操作，其实使用的是向量的扩充操作．例如：

```
>>s='hellow';
>>s=[s,'world']
s=
  hellow world
```

eval 是与字符串变量一起工作的函数．eval(t)执行包含在字符串 t 内的 Matlab 命令．例如：

```
t='eye(2)',eval(t)
ans=
  1   0
  0   1
```

Appendix C Answers to Selected Questions

Questions 1

1. $\ln x - \ln x^* \approx \varepsilon$
2. $0.01n$
3. 5 2 4 7
5. $\frac{1}{300}$
6. $\int_M^{M+1} \frac{1}{1+x^2} dx = \arctan \frac{1}{1+M(M+1)}$
7. $\frac{1}{2} \times 10^{-3}$
8. $x_1 \approx 26.04$ $x_2 \approx -0.038\,40$
9. The error of edge is less than 0.005
11. $|y_0 - y_0^*| \leqslant \frac{1}{2} \times 10^{-2} = \delta$ $|y_{10} - y_{10}^*| \leqslant 10^{10}\delta$ The computational process is not stable
12. (4) is the best result

Questions 2

1. $l_0(x) = \frac{1}{660}(x+6)(x+7)x$ $l_1(x) = \frac{-1}{84}(x-5)(x+6)x$

   $l_2(x) = \frac{1}{66}(x-5)(x+7)x$ $l_3(x) = \frac{-1}{210}(x-5)(x+6)(x+7)$

   $p_3(x) = l_0(x) - 23l_1(x) - 54l_2(x) - 954l_3(x) = 4x^3 + 35x^2 - 84x - 954$
2. $L_2(x) = \frac{5}{6}x^2 + \frac{3}{2}x - \frac{7}{3}$
3. $|\sin x - P(x)| \leqslant \frac{1}{10!} < 2.8 \times 10^{-7}$
6. $2x^3 - 10x^2 + 15x - 4$
7. $h \leqslant 0.006$
8. 2^n 2^{n-2}
12. $p(x) = 2 - (x+1) + x(x+1) - 2x(x+1)(x-1) + 2x(x+1)(x-1)(x-2)$
13. 1 0
14. $\begin{cases} 0.5x & 0 \leqslant x \leqslant 0.5 \\ 0.25 + 1.5(x-0.5) & 0.5 \leqslant x \leqslant 1 \\ 1 + 2.5(x-1) & 1 \leqslant x \leqslant 1.5 \\ 2.25 + 3.5(x-1.5) & 1.5 \leqslant x \leqslant 2 \end{cases}$

$|R_1(x)| \leqslant h^2/4 = 0.0625$

15. $P(x) = \frac{1}{4}x^2(x-3)^2$

16. $S(x) = x$

17. $a = c = d$ $\quad b$, e are arbitrary

18. $b=-1,\quad c=-3,\quad d=1$

19. $B=\frac{1}{4},\quad D=\frac{1}{4},\quad b=-\frac{1}{2},\quad d=\frac{1}{4}$

Questions 3

2. (1) $\|f\|_\infty = 1$， $\|f\|_1 = 1/4$， $\|f\|_2 = 1/\sqrt{7}$

   (2) $\|f\|_\infty = 1/2$， $\|f\|_1 = 1/4$， $\|f\|_2 = \frac{\sqrt{3}}{6}$

3. (1) $\|\cdot\|_\infty = \|\cdot\|_1 = 4$ $\qquad$ (2) $\|\cdot\|_\infty = 13$， $\|\cdot\|_1 = 14$

4. (1) $\lambda_1 = 0,\ x_1 = (1,-1)^{\mathrm{T}}$ $\qquad \lambda_2 = -1,\ x_1 = (1,-2)^{\mathrm{T}}$

   (2) $\lambda_1 = 3,\ x_1 = (-1,1,2)^{\mathrm{T}}$ $\qquad \lambda_2 = 4,\ x_2 = (0,1,2)^{\mathrm{T}}$

   $\lambda_3 = -2,\ x_2 = (-3,8,1)^{\mathrm{T}}$

5. $P_1(x) = \frac{2}{\pi}x + 0.105\,257$

6. $P_1(x) = (\mathrm{e}-1)x + \frac{1}{2}(\mathrm{e}-(\mathrm{e}-1)\ln(\mathrm{e}-1))$

7. $S(x) = 0.117\,187\,5 + 1.640\,625x^2 - 0.8203\,125x^4$

8. (1) $S_1^*(x) = -0.295\,8x + 1.141\,0$ $\qquad$ (2) $S_2^*(x) = 1.690\,3x + 0.873\,1$

9. $s = 22.253\,76t - 7.855\,047\,8$

10. $y = 0.972\,604\,6 + 0.050\,035\,1x^2$ $\qquad \delta = 0.123$

11. $y_2(x) = 0.408\,6 + 0.42x + 0.085\,7x^2$

    $y_3(x) = 0.408\,6 + 0.391\,67x + 0.085\,7x^2 + 0.008\,3x^3$

13. $x_1 \approx 15.266\,7,\quad x_2 \approx 5.866\,7$

Computer Questions 3

1. The least squares polynomials with their errors are, respectively

   $0.620\,895\,0 + 1.219\,621x,\quad \varphi = 2.719\times10^{-5}$

   $0.596\,580\,7 + 1.253\,293x - 0.010\,853\,43x^2,\quad \varphi = 1.801\times10^{-5}$

   $0.629\,019\,3 + 1.185\,010x + 0.035\,332\,52x^2 - 0.010\,047\,23x^3,\quad \varphi = 1.741\times10^{-5}$

2. (1) The least squares polynomials with their errors are, respectively

   $-194.138 + 72.084\,5x,\quad \varphi = 329$

   $1.235\,56 - 1.143\,52x + 6.618\,21x^2,\quad \varphi = 1.44\times10^{-3}$

   $3.429\,04 - 2.379\,191x + 6.845\,57x^2 - 0.013\,674\,2x^3,\quad \varphi = 5.27\times10^{-4}$

   (2) $24.258\,8\mathrm{e}^{0.372\,382x},\quad \varphi = 418$

Questions 4

1. (1) $a_{-1} = a_1 = h/3,\quad a_0 = 4h/3,$ order 3

(2) $x_1 = 0.689\,90$, $x_2 = -0.126\,60$ or $x_1 = -0.289\,90$, $x_2 = 0.526\,60$, order 2

(3) a=1/12, order 3

(4) a=1, $b = a$, c=1/3, d=−1/3, order 3

2. $\int_0^1 f(x)\mathrm{d}x \approx \frac{1}{2}\left(f\left(\frac{1}{4}\right)+f\left(\frac{3}{4}\right)\right) \qquad R(f)=\int_0^1 \frac{1}{2}f''(\xi)\left(x-\frac{1}{4}\right)\left(x-\frac{3}{4}\right)\mathrm{d}x$

3. $\int_0^3 f(x)\mathrm{d}x \approx \frac{3}{8}f(0)+\frac{9}{8}(1)+\frac{9}{8}f(2)+\frac{3}{8}f(3)$

4. (1) $T_8 = 0.111\,40$, $S_4 = 0.111\,57$

(2) $T_4 = 17.227\,74$, $S_2 = 17.322\,22$

5. $\int_1^2 \mathrm{e}^{\frac{1}{x}}\mathrm{d}x \approx T_4^{(0)} = 2.020\,058\,651$

7. (1) $b-a$ (2) 0

8. (1) 0.159 410 4 (2) 0.089 263 02 (3) −0.730 723 0 (4) −0.176 819 0

9. (1) $f'(0.0)\approx 3.707$, $f'(0.2)\approx 3.152\,0$, $f'(0.4)\approx 3.152\,0$

(2) $f'(0.5)\approx 0.852\,0$, $f'(0.6)\approx 0.852\,0$, $f'(0.7)\approx 0.796\,0$

10. (1) 0.311 57 (2) 0.255 25

Questions 5

2. (1) It needs 10 computations for using bisection method

(2) It needs 4 computations for using iteration method

4. x=2.625 n=1 error=0.125

x=2.618 06 n=2 error=0.006 944 44

x=2.618 03 n=3 error=0.000 021 566 6

x=2.618 03 n=4 error=$2.080\,07\times10^{-10}$

7. $\sqrt{115} \approx x_3$ =10.723 805 29

8. for $f(x)=x^n-a=0$, $\lim_{k\to\infty}(\sqrt[n]{a}-x_{k+1})/(\sqrt[n]{a}-x_k)^2=\frac{1-n}{2\sqrt[n]{a}}$

for $f(x)=1-\frac{a}{x^n}=0$, $\lim_{k\to\infty}(\sqrt[n]{a}-x_{k+1})/(\sqrt[n]{a}-x_k)^2=\frac{1+n}{2\sqrt[n]{a}}$

9. $\lim_{k\to\infty}\frac{\sqrt{a}-x_{k+1}}{(\sqrt{a}-x_k)^3}=\frac{1}{4a}\neq 0$

Questions 6

1. (1) $\boldsymbol{L}=\begin{pmatrix}1 & & \\ 1.5 & 1 & \\ 1.5 & 1 & 1\end{pmatrix}$, $\boldsymbol{U}=\begin{pmatrix}2 & -1 & 1\\ & 4.5 & 7.5\\ & & -4\end{pmatrix}$

(2) $\boldsymbol{L}=\begin{pmatrix}1 & & & \\ 0.5 & 1 & & \\ 0 & -2 & 1 & \\ 1 & -1.3333 & 2 & 1\end{pmatrix}$, $\boldsymbol{U}=\begin{pmatrix}2 & 0 & 0 & 0\\ & 1.5 & 0 & 0\\ & & 0.5 & 0\\ & & & 1\end{pmatrix}$

2. (1) $x_1=-13$, $x_2=8$, $x_3=2$ (2) $x_1=-7$, $x_2=3$, $x_3=2$, $x_4=2$
3. $x_1=0$, $x_2=-1$, $x_3=1$
4. (1) det $\boldsymbol{A}=-28$ (2) det $\boldsymbol{B}=-6$
5. (1) $x=(1,-1,2)^{\mathrm{T}}$ (2) $x=(2,1,-1)^{\mathrm{T}}$
6. (1) $x=(-1,2,1)^{\mathrm{T}}$ (2) $x=(1,-1,0)^{\mathrm{T}}$
7. $\boldsymbol{P}=\begin{pmatrix}0&1&0\\1&0&0\\0&0&1\end{pmatrix}$, $\boldsymbol{L}=\begin{pmatrix}1&0&0\\0&1&0\\0&2&1\end{pmatrix}$, $\boldsymbol{U}=\begin{pmatrix}2&-1&0\\0&1&2\\0&0&-3\end{pmatrix}$, $\det \boldsymbol{A}=6$
8. (1) $\alpha>8/7$ (2) $-2<\alpha<\dfrac{3}{2}$

Questions 7

1. (1) Jacobi method: $\boldsymbol{x}^{(1)}=(0.333\,333, 0., 0.571\,429)^{\mathrm{T}}$

$\boldsymbol{x}^{(2)}=(0.142\,857, -0.357\,143, 0.428\,571)^{\mathrm{T}}$

Seidel method: $\boldsymbol{x}^{(1)}=(0.333\,333, -0.166\,667, 0.5)^{\mathrm{T}}$

$\boldsymbol{x}^{(2)}=(0.111\,111, -0.222\,222, 0.619\,048)^{\mathrm{T}}$

(2) Jacobi method: $\boldsymbol{x}^{(1)}=(0.5, -0.25, 0., 0.333\,333)^{\mathrm{T}}$

$\boldsymbol{x}^{(2)}=(0.479\,167, -0.291\,667, -0.016\,666\,7, 0.083\,333\,3)^{\mathrm{T}}$

Seidel method: $\boldsymbol{x}^{(1)}=(0.5, -0.375, 0.025, 0.033\,333\,3)^{\mathrm{T}}$

$\boldsymbol{x}^{(2)}=(0.591\,667, -0.383\,333, 0.035, -0.003\,333\,33)^{\mathrm{T}}$

3. Both Jacobi and G-S methods are convergent

$$\text{Jacobi:}\ \begin{cases} x_1^{(k+1)}=\dfrac{1}{5}(-12-2x_2^{(k)}-x_3^{(k)})\\ x_2^{(k+1)}=\dfrac{1}{4}(20+x_1^{(k)}-2x_3^{(k)})\\ x_3^{(k+1)}=\dfrac{1}{10}(3-2x_1^{(k)}+3x_2^{(k)}) \end{cases}$$

$$\text{G-S:}\ \begin{cases} x_1^{(k+1)}=\dfrac{1}{5}(-12-2x_2^{(k)}-x_3^{(k)})\\ x_2^{(k+1)}=\dfrac{1}{4}(20+x_1^{(k+1)}-2x_3^{(k)})\\ x_3^{(k+1)}=\dfrac{1}{10}(3-2x_1^{(k+1)}+3x_2^{(k+1)}) \end{cases}$$

7. $\boldsymbol{x}^{(0)}=(0,0,0)^{\mathrm{T}}$, $\boldsymbol{x}^{(8)}=(-4.000\,027, 0.299\,998\,7, 0.200\,000\,3)^{\mathrm{T}}$

10. $\rho(\boldsymbol{J})=\sqrt{\dfrac{11}{12}}<1$ convergence

$\rho(\boldsymbol{G})=\dfrac{11}{12}<1$ convergence

Since $\rho(\boldsymbol{G})<\rho(\boldsymbol{J})$,therefore G-S method converges faster than Jacobi method

11. (1) $\rho(\boldsymbol{J})=0.5$

(2) $\rho(\boldsymbol{G})=0.25$

(3) Both Jacobi and G-S methods are convergent

Questions 8

1. $y_{n+1}=0.8y_n-0.2x_n y_n^2$

   $y(0.2)=0.8,\quad y(0.4)=0.614\,4,\quad y(0.6)=0.461\,321$

2. $y_{n+1}=\left(1-\dfrac{0.018}{1+2x_n}\right)y_n$

   $y(0.02)=0.982,\quad y(0.04)=0.965\,004$

   $y(0.06)=0.948\,92,\quad y(0.08)=0.933\,67,$

   $y(0.1)=0.919\,182$

3. $y_{j+1}=y_j+\dfrac{h}{2}(f(x_j,y_j)+f(x_{j+1},y_j+hf(x_j,y_j)))$

   $y(1.1)=0.523\,835,\quad y(1.2)=0.545\,5$

   $y(1.3)=0.565\,277,\quad y(1.4)=0.583\,404$

   $y(1.5)=0.600\,079$

4. $y(0.2)=1.242\,8,\quad y(0.4)=1.583\,64$

   $y(0.6)=2.044\,21,\quad y(0.8)=2.651\,04$

   $y(1.0)=3.436\,5$

6. $y(1.25)\approx y_1=1.687\,5,\quad y(1.50)\approx y_2=3.333\,298$

7. $y(0.2)=2.300\,4,\quad y(0.4)=2.465\,44$

   $y(0.6)=2.556\,11,\quad y(0.8)=2.605\,93$

   $y(1.0)=2.633\,3$

Reference

[1] RICHARD L B，J. DOUGLAS F. Numerical Analysis[M]. 7th ed. 北京：高等教育出版社，2001.

[2] ROBERT J S，SANDRA L H. Applied Numerical Methods for Engineers Using MATLAB and C[M]. 北京：机械工业出版社，2004.

[3] DAVID K，WARD C. Numerical Analysis[M]. 3th ed. 北京：机械工业出版社，2003.

[4] RAINER K. Numerical Analysis[M]. Berlin: Springer-Verlag，1998.

[5] STOER J，BULIRSCH R. Introduction to Numerical Analysis[M]. 7th ed. Berlin: Springer- Verlag，1993.

[6] SAUL A T，BRIAN P F. Numerical Recipes in C[M]. 7th ed. 北京：电子工业出版社，2004.

[7] 李庆扬，王能超，易大义. 数值分析[M]. 4 版. 北京：清华大学出版社，2001.

[8] 王兵团，桂文豪. 数学实验基础[M]. 北京：北方交通大学出版社，2003.

[9] 晨曦工作室，苟飞. Mathematica 4 实例教程[M]. 北京：中国电力出版社，2000.

[10] 杨珏，何旭洪，赵昊彤. Mathematica 应用指南[M]. 北京：人民邮电出版社，1999.

[11] 李丽，王振领. MATLAB 工程计算及应用[M]. 北京：人民邮电出版社，2001.

[12] 石博强，腾贵法，李海鹏. MATLAB 数学计算范例教程[M]. 北京：中国铁道出版社，2004.

[13] 陈渝，周璐，钱方，等. 数值方法：MATLAB 版[M]. 3 版. 北京：电子工业出版社，2002.

[14] 易大义，陈道琦. 数值分析引论[M]. 杭州：浙江大学出版社，1996.

[15] 马东升，雷永军. 数值计算方法[M]. 2 版. 北京：机械工业出版社，2006.